Complex Surveys

Parimal Mukhopadhyay

Complex Surveys

Analysis of Categorical Data

Springer

Parimal Mukhopadhyay
Indian Statistical Institute
Kolkata, West Bengal
India

ISBN 978-981-10-9272-5 ISBN 978-981-10-0871-9 (eBook)
DOI 10.1007/978-981-10-0871-9

Softcover reprint of the hardcover 1st edition 2016

Printed on acid-free paper

This Springer imprint is published by Springer Nature
The registered company is Springer Science+Business Media Singapore Pte Ltd.

To
The Loving Memory of My Wife
Manju

Preface

Most of the data a statistician uses is categorical in nature. In the realms of biomedicine and social sciences, ecology and demography, voting pattern and marketing, to name a few, categorical data dominate. They are the major raw materials for analysis for testing different hypotheses relating to the populations which generate such data. Such data are generally obtained from surveys carried under a complex survey design, generally a stratified multistage design. In analysis of data collected through sample surveys, standard statistical techniques are often routinely employed.

We recall the well-recited phrase which we chanted in our undergraduate days: Let $x_1,\ldots, x_n$ be a random sample of size n drawn from a population with probability density function $f(x)$ or probability mass function $p_M(x)$, etc. This means that the sampled units whose observations are x_1, $x_2,\ldots, x_n$, are drawn by simple random sampling with replacement (*srswr*). This also implies that observed variables $x_1,\ldots,$ x_n are independently and identically distributed (IID). In fact, most of the results in theoretical statistics, including those in usual analysis of categorical data, are based on these assumptions.

However, survey populations are often complex with different cell probabilities in different subgroups of the population, and this implies a situation different from the IID setup. Longitudinal surveys—where sample subjects are observed over two or more time points—typically lead to dependent observations over time. Moreover, longitudinal surveys often have complex survey designs that involve clustering which results in cross-sectional dependence among samples.

In view of these facts, it is, therefore, necessary to modify the usual tests of goodness of fit, like Pearsonian chi-square, likelihood ratio and Wald statistic to make them suitable for use in the context of data from complex surveys. A host of ardent researchers have developed a number of such tests for this purpose over more than last four decades.

There are already a myriad number of textbooks and research monographs on analysis of categorical data. Then why is another book in the area required? My humble answer is that all those treatise provide excellent description of the categorical data analysis under the classical setup (usual *srswr* or IID assumption), but none addresses the problem when the data are obtained through complex sample survey designs, which more often than not fail to satisfy the usual assumptions. The present endeavor tries to fill in the gap in the area.

The idea of writing this book is, therefore, to make a review of some of the ideas that have blown out in the field of analysis of categorical data from complex surveys. In doing so, I have tried to systematically arrange the results and provide relevant examples to illuminate the ideas. This research monograph is a review of the works already done in the area and does not offer any new investigation. As such I have unhesitatingly used a host of brilliant publications in this area. A brief outline of different chapters is indicated as under:

(1) Chapter 1: Basic ideas of sampling; finite population; sampling design; estimator; different sampling strategies; design-based method of making inference; superpopulation model; model-based inference
(2) Chapter 2: Effects of a true complex design on the variance of an estimator with reference to a *srswr* design or an IID-model setup; design effects; misspecification effects; multivariate design effect; nonparametric variance estimation
(3) Chapter 3: Review of classical models of categorial data; tests of hypotheses for goodness of fit; log-linear model; logistic regression model
(4) Analysis of categorical data from complex surveys under full or saturated models; different goodness-of-fit tests and their modifications
(5) Analysis of categorical data from complex surveys under log-linear model; different goodness-of-fit tests and their modifications
(6) Analysis of categorical data from complex surveys under binomial and polytomous logistic regression model; different goodness-of-fit tests and their modifications
(7) Analysis of categorical data from complex surveys when misclassification errors are present; different goodness-of-fit tests and their modifications
(8) Some procedures for obtaining approximate maximum likelihood estimators; pseudo-likelihood approach for estimation of finite population parameters; design-adjusted estimators; mixed model framework; principal component analysis
(9) Appendix: Asymptotic properties of multinomial distribution; asymptotic distribution of different goodness-of-fit tests; Neyman's (1949) and Wald's (1943) procedures for testing general hypotheses relating to population proportions

I gratefully acknowledge my indebtedness to the authorities of PHI Learning, New Delhi, India, for kindly allowing me to use a part of my book, *Theory and Methods of Survey Sampling*, 2nd ed., 2009, in Chap. 2 of the present book. I am thankful to Mr. Shamin Ahmad, Senior Editor for Mathematical Sciences at

Springer, New Delhi, for his kind encouragement. The book was prepared at the Indian Statistical Institute, Kolkata, to the authorities of which I acknowledge my thankfulness. And last but not the least, I must acknowledge my indebtedness to my family for their silent encouragement and support throughout this project.

January 2016 Parimal Mukhopadhyay

Contents

About the Author

Parimal Mukhopadhyay is a former professor of statistics at the Indian Statistical Institute, Kolkata, India. He received his Ph.D. degree in statistics from the University of Calcutta in 1977. He also served as a faculty member at the University of Ife, Nigeria, Moi University, Kenya, University of South Pacific, Fiji Islands and held visiting positions at the University of Montreal, University of Windsor, Stockholm University and the University of Western Australia. He has written more than 75 research papers in survey sampling, some co-authored and eleven books on statistics. He was a member of the Institute of Mathematical Statistics and elected member of the International Statistical Institute.

Chapter 1
Preliminaries

Abstract This chapter reviews some basic concepts in problems of estimating a finite population parameter through a sample survey, both from a design-based approach and a model-based approach. After introducing the concepts of finite population, sample, sampling design, estimator, and sampling strategy, this chapter makes a classification of usual sampling designs and takes a cursory view of some estimators. The concept of superpopulation model is introduced and model-based theory of inference on finite population parameters and model parameters is looked into. The role of superpopulation model vis-a-vis sampling design for making inference about a finite population has been outlined. Finally, a plan of the book has been sketched.

Keywords Finite population · Sample · Sampling frame · Sampling design · Inclusion probability · Sampling strategy · Horvitz–Thompson estimator · PPS sampling · Rao–Hartly–Cochran strategy · Generalized difference estimator · GREG · Multi-stage sampling · Two-phase sampling · Self-weighting design · Superpopulation model · Design-predictor pair · BLUP · Purposive sampling design

1.1 Introduction

The book has two foci: one is sample survey and the other is analysis of categorical data. The book is primarily meant for sample survey statisticians, both theoreticians and practitioners, but nevertheless is meant for data analysts also. As such, in this chapter we shall make a brief review of basic notions in sample survey techniques, while a cursory view of classical models for analysis of categorical data will be postponed till the third chapter.

Sample survey, finite population sampling, or survey sampling is a method of drawing inference about the characteristic of a finite population by observing only a part of the population. Different statistical techniques have been developed to achieve this end during the past few decades.

In what follows we review some basic results in problems of estimating a finite population parameter (like, its total or mean, variance) through a sample survey. We assume throughout most of this chapter that the finite population values are fixed

P. Mukhopadhyay, *Complex Surveys*, DOI 10.1007/978-981-10-0871-9_1

quantities and are not realizations of random variables. The concepts will be clear subsequently.

1.2 The Fixed Population Model

First, we consider a few definitions.

Definition 1.2.1 A finite (survey) population $\mathcal{P}$ is a collection of a known number N of identifiable units labeled $1, 2, \ldots, N$; $\mathcal{P} = \{1, \ldots, N\}$, where i denotes the physical unit labeled i. The integer N is called the size of the population.

The following types of populations, therefore, do not satisfy the requirements of the above definition: batches of industrial products of identical specifications (e.g., nails, screws) coming out from a production process, as one unit cannot be distinguished from the other, i.e., the identifiability of the units is lost; population of animals in a forest, population of fishes in a typical lake, as the population size is unknown. Collection of households in a given area, factories in an industrial complex, and agricultural fields in a village are examples of survey populations.

Let 'y' be a study variable having value y_i on i $(= 1, \ldots, N)$. As an example, in an industrial survey 'y_i' may be the value added by manufacture by a factory i. The quantity y_i is assumed to be fixed and not random. Associated with $\mathcal{P}$ we have, therefore, a vector of real numbers $\mathbf{y} = (y_1, \ldots, y_N)'$. The vector $\mathbf{y}$ therefore constitutes a parameter for the model of a survey population, $\mathbf{y} \in R^N$, the parameter space. In a sample survey one is often interested in estimating a parameter function $\theta(\mathbf{y})$, e.g., population total $T(\mathbf{y}) = T$ or $Y(= \sum_{i=1}^{N} y_i)$, population mean $\bar{Y}$ or $\bar{y}$ $(= T/N)$, population variance $S^2 = \sum_{i=1}^{N}(y_i - \bar{y})^2/(N-1)$. This is done by choosing a sample (a part of the population, defined below) from $\mathcal{P}$ in a suitable manner and observing the values of y only for those units which are included in the sample.

Definition 1.2.2 A sample is a part of the population, i.e., a collection of a suitable number of units selected from the assembly of N units which constitute the survey population $\mathcal{P}$.

A sample may be selected in a draw-by-draw fashion by replacing a unit selected at a draw to the original population before making the next draw. This is called sampling *with replacement* (*wr*).

Also, a sample may be selected in a draw-by-draw fashion without replacing a unit selected at a draw to the original population. This is called *sampling without replacement* (*wor*).

A sample when selected by a with replacement (*wr*)-sampling procedure may be written as a *sequence*:

$$S = i_1, \ldots, i_n, \ 1 \leq i_t \leq N \tag{1.2.1}$$

where i_t denotes the label of the unit selected at the tth draw and is not necessarily unequal to $i_{t'}$ for $t \neq t'(= 1, \ldots, n)$. For a without replacement (*wor*)-sampling procedure, a sample may also written as a sequence S, with i_t denoting the label of the unit selected at the tth draw. Thus, here,

$$S = \{i_1, \ldots, i_n\}, 1 \leq i_t \leq N, i_t \neq i_{t'} \text{ for } t \neq t'(= 1, \ldots, N) \quad (1.2.2)$$

since, here, repetition of unit in S is not possible.

Arranging the units in the (sequence) sample S in an increasing (decreasing) order of magnitudes of their labels and considering only the distinct units, a sample may also be written as a set s. For a *wr*-sampling by n draws, a sample written as a set is, therefore,

$$s = (j_1, \ldots, j_{\nu(S)}), 1 \leq j_1 < \cdots < j_{\nu(S)} \leq N \quad (1.2.3)$$

where $\nu(S)$ is the number of distinct units in S.

In a *wor*-sampling procedure, a sample of n-draws, written as a set, is

$$s = (j_1, \ldots, j_n), 1 \leq j_1 < \cdots < j_n \leq N. \quad (1.2.4)$$

As an example, if in a *wr*-sampling $S = \{2, 9, 4, 9\}$, the corresponding s is $s = (2, 4, 9)$ with $\nu(S) = 3$, since there are only three distinct units. Similarly, if for a *wor* sampling procedure, $S = \{4, 9, 1\}$, the corresponding s is $s = (1, 4, 9)$ with $\nu(S) = 3$. Clearly, information on the order of selection and repetition of units which is available in the (sequence) sample S is not available in the (set) sample s.

Definition 1.2.3 Number of distinct units in a sample is its *effective sample size*. Number of draws in a sample is its *nominal sample size*. In (1.2.3), $\nu(S)$ is the effective sample size, $1 \leq \nu(S) \leq n$. For a *wor*-sample of n-draws, $\nu(S) = \nu(s) = n$.

Note that a sample is a sequence or set of some units from the population and it does not include their y-values. Thus in an agricultural survey if the firms with labels 7, 3, and 11 are included in the sample and if the yields of these field are 231, 38, and 15 units, respectively, the sample comprised the firms labeled 7, 3, and 11 only; their yield figures have nothing to do with the sample in this case.

Definition 1.2.4 The sample space is the collection of all possible samples and is often denoted as $\mathcal{S}$. Thus $\mathcal{S} = \{S\}$ or $\{s\}$ accordingly as we are interested in S or s.

In a simple random sample with replacement (*srswr*) of n draws (defied in Sect. 1.3), $\mathcal{S}$ consists of N^n samples S. In a simple random sample without replacement (*srsowr*) (defined in Sect. 1.3), $\mathcal{S}$ consists of $(N)_n$ samples S and $\binom{N}{n}$ samples s where $(a)_b = a(a-1)\ldots(a-b+1)$, $a > b$. If the samples s of all possible sizes are considered in a *wor*-sampling procedure, there are 2^N samples in $\mathcal{S}$.

Definition 1.2.5 Let $\mathcal{A}$ be the minimal σ-field over $\mathcal{S}$ and p be a probability measure defined over $\mathcal{A}$ such that $p(s)$ [or $p(S)$] denotes the probability of selecting s [or S], satisfying

$$p(s)[p(S)] \geq 0$$

$$\sum_{\mathcal{S}} p(s)\left[\sum_{\mathcal{S}} p(S)\right] = 1. \tag{1.2.5}$$

In estimating a finite population parameter $\theta(\mathbf{y})$ through sample surveys, one of the main tasks of the survey statistician is to find a suitable $p(s)$ or $p(S)$. The collection $(\mathcal{S}, p)$ is called a *sampling design* (*s.d.*), often denoted as $D(S, p)$ or simply p. The triplet $(\mathcal{S}, \mathcal{A}, p)$ is the probability space for the model of the finite population.

The expected effective sample size of a s.d. p is, therefore,

$$E\{\nu(S)\} = \sum_{S\in\mathcal{S}} \nu(S)p(S) = \sum_{\mu=1}^{N} \mu P[\nu(S) = \mu] = \nu. \tag{1.2.6}$$

We shall denote by ρ_ν the class of all *fixed effective size* [FS(ν)] designs, i.e.,

$$\rho_\nu = \{p : p(s) > 0 \Longleftrightarrow \nu(S) = \nu\}.$$

A s.d. p is said to be *noninformative* if $p(s)[p(S)]$ does not depend on the y-values. In this treatise, unless otherwise stated, we shall consider noninformative designs only.

Basu (1958) and Basu and Ghosh (1967) proved that all the information relevant to making inference about the population characteristic is contained within the set sample s and the corresponding y-values. For this reason we shall mostly confine ourselves to the set sample s.

The quantities

$$\pi_i = \sum_{s\ni i} p(s),\ \pi_{ij} = \sum_{s\ni i,j} p(s)$$

$$\pi_{i_1,\ldots,i_k} = \sum_{s\ni i_1,\ldots,i_k} p(s) \tag{1.2.8}$$

are, respectively, the first order, second order, ..., kth order *inclusion probabilities* of units in a s.d. p. The following lemma states some relations among inclusion probabilities and expected effective sample size of a s.d.

Lemma 1.2.1 *For any s.d. p,*

(i) $\pi_i + \pi_j - 1 \le \pi_{ij} \le \min(\pi_i, \pi_j)$,
(ii) $\sum_{i=1}^{N} \pi_i = \nu$,
(iii) $\sum_{i\neq j=1}^{N} \pi_{ij} = \nu(\nu-1) + V(\nu(S))$.

If $p \in \rho_\nu$,

(iv) $\sum_{j(\neq i)=1}^{N} \pi_{ij} = (\nu-1)\pi_i$,
(v) $\sum_{i\neq j=1}^{N} \pi_{ij} = \nu(\nu-1)$.

Further, for any s.d. p,

$$\theta(1-\theta) \leq V\{\nu(S)\} \leq (N-\nu)(\nu-1) \tag{1.2.9}$$

where $\nu = [\nu] + \theta, 0 \leq \theta < 1, \theta$ *being the fractional part of* ν.

The lower bound in (1.2.9) is attained by a s.d. p for which

$$P[\nu(S) = [\nu]] = 1 - \theta \text{ and} P[\nu(S) = [\nu] + 1] = \theta.$$

Mukhopadhyay (1975) gave a s.d. with fixed nominal sample size $n[p(S) > 0 \Rightarrow n(S) = n \forall S]$ *such that* $V\{\nu(S)\} = \theta(1-\theta)/(n - [\nu])$, *which is very close to the lower bound in (1.2.9).*

It is seen, therefore, that a s.d. gives the probability $p(s)$ *[or* $p(S)$*] of selecting a sample s (or S), which, of course, belongs to the sample space. In general, it will be a formidable task to select a sample using only the contents of a s.d., because one has to enumerate all the possible samples in some order, calculate the cumulative probabilities of selection, draw a random number in [0, 1], and select the sample corresponding to the number so selected. It will be, however, of great advantages if one knows the conditional probabilities of selection of units at different draws.*

We shall denote by

$p_r(i)$ = *probability of selecting i at the rth draw,* $r = 1, \ldots, n$;

$p_r(i_r|i_1, \ldots, i_{r-1})$ = *conditional probability of drawing* i_r *at the rth draw given that* $i_1, \ldots, i_{r-1}$ *were drawn at the first,* $\ldots, (r-1)$*th draw, respectively;*

$p(i_1, \ldots, i_r)$ = *the joint probability that* $(i_1, \ldots, i_r)$ *are selected at the first,* $\ldots$, *rth draw, respectively.*

All these probabilities must be nonnegative and we must have

$$\sum_{i=1}^{N} p_r(i) = 1, r = 1, \ldots, n;$$

$$\sum_{i_r=1}^{N} p_r(i_r|i_1, \ldots, i_{r-1}) = 1.$$

Definition 1.2.6 A *sampling scheme* (s.s.) gives the conditional probability of drawing a unit at any particular draw given the results of the earlier draws.

A s.s., therefore, specifies the conditional probabilities $p_r(i_r|i_1, \ldots, i_{r-1})$, i.e., it specifies the values $p_1(i)(i = 1, \ldots, N)$, $p_r(i_r|i_1, \ldots i_{r-1}), i_r = 1, \ldots, N; r = 2, \ldots, n$.

The following theorem shows that any sampling design can be attained through a draw-by-draw mechanism.

Theorem 1.2.1 (Hanurav 1962; Mukhopadhyay 1972) *For any given sampling design, there exists at least one sampling scheme which realizes this design.*

Suppose now that the values $x_1, \ldots, x_N$ of a closely related (related to y) auxiliary variable x on units $1, 2, \ldots, N$, respectively, are available. The quantity $P_i = x_i/X, X = \sum_{i=1}^{N} x_k$ is called the size measure of unit $i (= 1, \ldots, N)$ and is often used in selection of samples. Thus in a survey of large-scale manufacturing industry, say, jute industry, the number of workers in a factory may be considered as a measure of size of the factory, on the assumption that a factory employing more manpower will have larger value of output.

Before proceeding to take a cursory view of different types of sampling designs we will now introduce some terms useful in this context.

Sampling frame: It is the list of all sampling units in the finite population from which a sample is selected. Thus in a survey of households in a rural area, the list of all the households in the area will constitute a frame for the survey. The frame also includes any auxiliary information like measures of size, which is used for special sampling techniques, such as stratification and probability proportional-to-size sample selections, or for special estimation techniques, such as ratio or regression estimates. All these techniques have been indicated subsequently.
However, a list of all the ultimate study units or ultimate sampling units may not be always available. Thus in a household survey in an urban area where each household is the ultimate sampling unit or ultimate study unit we do not generally have a list of all such households. But we may have a list of census block units within this area from which a sample of census blocks may be selected at the first stage. This list is the frame for sampling at the first stage. Each census block again may consist of several wards. For each selected census block one may prepare a list of such wards and select samples of wards. These lists are then sampling frames for sampling at the second stage. Multistage sampling has been discussed in the next section. Sarndal et al. (1992), among others, have investigated the relationship between the sampling frame and population.

Analytic and Descriptive Surveys: Descriptive uses of surveys are directed at the estimation of summary measures of the population such as means, totals, and frequencies. Such surveys are generally of prime importance to the Government departments which need an accurate picture of the population in terms of its location, personal characteristics, and associated circumstances. The analytic surveys are more concerned with identifying and understanding the causal mechanisms which underlie the picture which the descriptive statistics portray and are generally of interest to research organizations in the area. Naturally, the estimation of different superpopulation parameters, such as regression coefficients, is of prime interest in such surveys.
For descriptive uses the objective of the survey is essentially fixed. Target parameters, such as the total and ratio, are the objectives determined even before the data are collected or analyzed. For analytic uses, such as studying different parameters of the model used to describe the population, the parameters of interest are not generally fixed in advance and evolve through an adaptive process as the analysis progresses. Thus for analytic purposes the process is an evolutionary one where the final parameters to be estimated and the estimation procedures to be employed are

chosen in the light of the *superpopulation model* used to describe the population. Use of superpopulation model in sampling has been indicated in Sect. 1.5.

Strata: Sometimes, it may be necessary or desirable to divide the population into several subpopulations or strata to estimate population parameters like population mean and population total through a sample survey. The necessity of stratification is often dictated by administrative requirements or convenience. For a statewide survey, for instance, it is often convenient to draw samples independently from each county and carry out survey operations for each county separately. In practice, the population often consists of heterogeneous units (with respect to the character under study). It is known that by stratifying the population such that the units which are approximately homogeneous (with respect to 'y''), a better estimator of population total, mean, etc. can be achieved.
We shall often denote by y_{hi} the value of y on the ith unit in the hth stratum ($i = 1, \ldots, N_h; h = 1, \ldots, H$), $\sum_h N_h = N$, the population size; $\bar{Y}_h = \sum_{i=1}^{N_h} Y_{hi}/N_h$, $S_h^2 = \sum_{i=1}^{N_h}(Y_{hi} - \bar{Y}_h)^2/(N_h - 1)$, stratum population mean and variance, respectively; $W_h = N_h/N$, stratum proportion. The population mean is then $\bar{Y} = \sum_{h=1}^{H} W_h \bar{Y}_h$.

Cluster: Sometimes, it is not possible to have a list of all the units of study in the population so that drawing a sample of such study units is not possible. However, a list of some bigger units each consisting of several smaller units (study units) may be available from which a sample may be drawn. Thus, for instance, in a socioeconomic survey, our main interest often lies in the households (which are now study units or elementary units or units of our ultimate interest). However, a list of households is not generally available, whereas a list of residential houses each accommodating a number of households should be available with appropriate authorities. In such cases, samples of houses may be selected and all the households in the sampled houses may be studied. Here, a house is a 'cluster.' A cluster consists of a number of ultimate units or study units. Obviously, the clusters may be of varying sizes. Generally, all the study units in a cluster are of the same or similar character. In cluster sampling a sample of clusters is selected by some sampling procedure and data are collected from all the elementary units belonging to the selected clusters.

Domain: A domain is a part of the population. In a statewide survey, a district may be considered as a domain; in the survey of a city a group of blocks may form a domain, etc. After sampling has been done from the population as a whole and the field survey has been completed, one may be interested in estimating the mean or total relating to some part of the population. For instance, after a survey of industries has been completed, one may be interested in estimating the characteristic of the units manufacturing cycle tires and tubes. These units in the population will then form a domain. Clearly, sample size in a domain will be a random variable. Again, the domain size may or may not be known.

1.3 Different Types of Sampling Designs

The following types of sampling designs are generally used.

(a) Simple random sampling with replacement (*srswr*): Under this scheme units are selected one by one at random in n (a preassigned number) draws from the list of all available units such that a unit once selected is returned to the population before the next draw. As stated before, the sample space here consists of N^n sequences $S\{i_1, \ldots, i_n\}$ and the probability of selecting any such sequence (sample) is $1/N^n$.

(b) Simple random sampling without replacement (*srswor*): Here units are selected in n draws at random from the list of all available units such that a unit once selected is removed from the population before the next draw. Here again, as stated before the sample space consists of $(N)_n$ sequences S and $\binom{N}{n}$ sets s and the s.d. design allots to each of them equal probability of selection.

(c) Probability proportional to size with replacement (*ppswr*) sampling: a unit i is selected with probability p_i at the rth draw and a unit once selected is returned to the population before the next draw ($i = 1, \ldots, N; r = 1, 2, \ldots, n$). The quantity p_i is a measure of size of the unit i. This s.d. is a generalization of *srswr* s.d. where $p_i = 1/N \forall i$.

(d) Probability proportional to size without replacement (*ppswor*): a unit i is selected at the rth draw with probability proportional to its normed measure of size and a unit once selected is removed from the population before the next draw. Here,

$$p_1(i_1) = p_{i_1}$$

$$p_r(i_r|i_1, \ldots, i_{r-1}) = p_{i_r}/(1 - p_{i_1} - \cdots - p_{i_{r-1}}),\ r = 2, \ldots, n.$$

For $n = 2$, for this scheme,

$$\pi_i = p_i\left[1 + A - \frac{p_i}{1 - p_i}\right],$$

$$\pi_{ij} = p_i p_j\left(\frac{1}{1 - p_i} + \frac{1}{1 - p_j}\right),\ \text{where } A = \sum_{k=1}^{N}\frac{p_k}{1 - p_k}.$$

This sampling scheme is also known as '*successive sampling*.' The corresponding sampling design may also be attained by an inverse sampling procedure where units are drawn by *ppswr*, until for the first time n distinct units occur. The n distinct units each taken only once constitute the sample.

(e) Rejective sampling: n draws are made with *ppswr*; if all the units turn out to be distinct, the selected sequence constitutes the sample; otherwise, the whole selection is rejected and fresh draws made.

(f) Unequal probability without replacement (*upwor*) sampling: A unit i is selected at the rth draw with probability proportional to $p_i^{(r)}$ and a unit once selected is removed from the population. Here

$$p_1(i) = p_i^{(1)}$$

$$p_r(i_r|i_1,\ldots,i_{r-1}) = \frac{p_{i_r}^{(r)}}{1-p_{i_1}^{(r)}-p_{i_2}^{(r)}-\cdots p_{i_{r-1}}^{(r)}}, r = 2,\ldots,n. \quad (1.3.1)$$

The quantities $\{p_i^{(r)}\}$ are generally functions of p_i and p-values of the units already selected. In particular, *ppswor* sampling scheme described in item (d) above is a special case of this scheme, where $p_i^{(r)} = p_i, r = 1,\ldots,n$. The sampling design may also be attained by a inverse sampling procedure where units are drawn *wr*, with probability $p_i^{(r)}$ at the rth draw, until for the first time n distinct units occur. The n distinct units each taken only once constitute the sample.

(g) Generalized rejective sampling: Draws are made *wr* and with probability $\{p_i^{(r)}\}$ at the rth draw. If all the units turn out distinct, the solution is taken as a sample; otherwise, the whole sample is rejected and fresh draws are made. The scheme reduces to the scheme at (e) above, if $p_i^{(r)} = p_i \forall i$.

(h) Systematic sampling with varying probability (including unequal probability).

(k) Sampling from groups: The population is divided into L groups either at random or following some suitable procedure and a sample of size n_h is drawn from the hth group using any of the above-mentioned sampling designs such that the desired sample size $n = \sum_h n_h$ is attained. Examples are stratified sampling procedure and Rao–Hartley–Cochran (1962) (RHC) sampling procedure. Thus in stratified random sampling the population is divided into H strata of sizes $N_1,\ldots,N_H$ and a sample of size n_h is selected at random from the hth stratum $(h = 1,\ldots,H)$. The quantities n_h and $n = \sum_h n_h$ are suitably determined. RHC procedure has been discussed in the next section.

Based on the above methods, there are many unistage or multistage stratified sampling procedures. In a multistage procedure sampling is carried out in many stages. Units in a two-stage population consist of N first-stage units (fsu's) of sizes $M_1,\ldots,M_N$, with the bth second stage unit (ssu) in the ath fsu being denoted ab for $a = 1,\ldots,N; b = 1,\ldots,M_a$, with its associated y values being denoted y_{ab}. For a three-stage population the cth third stage unit (tsu) in the bth ssu in the ath fsu is labeled abc for $a = 1,\ldots,N; b = 1,\ldots,M_a; c = 1,\ldots,K_{ab}$. In a three-stage sampling a sample of n fsu's is selected out of N fsu's and from each of the ath selected fsu's, a sample of m_a ssu's is selected out of M_a fsu's in the selected fsu $(a = 1,\ldots,n)$. At the third stage from each of the selected ab ssu's, containing K_{ab} tsu's a sample of k_{ab} tsu's is selected $(a = 1,\ldots,n; b = 1,\ldots,m_a; c = 1,\ldots,k_{ab})$. The associated y-value is denoted as $y_{abc}, c = 1,\ldots,k_{ab}; b = 1,\ldots,m_a; a = 1,\ldots,n$.

The sampling procedure at each stage may be *srswr*, *srswor*, *ppswr*, *upwor*, systematic sampling, Rao–Hartley–Cochran sampling or any other suitable sampling procedure. The process may be continued to any number of stages. Moreover, the population may be initially divided into a number H of well-defined strata before undertaking the stage-wise sampling procedures. For stratified multistage population the label h is added to the above notation ($h = 1, \ldots, H$). Thus here the unit in the hth stratum, ath fsu, bth ssu, and cth tsu is labeled $habc$ and the associated y value as y_{habc}.

As is evident, samples for all the sampling designs may be selected by a whole sample procedure or mass-draw procedure in which a sample s is selected with probability $p(s)$.

A F.S.(n)-s.d. with π_i proportional to $p_i = x_i/X$, where x_i is the value of a closely related (to y) auxiliary variable on unit i and $X = \sum_{k=1}^{N} x_k$, is often used for estimating a population total. This is, because an important estimator, the Horvitz–Thompson estimator (HTE) has very small variance if y_i is nearly proportional to p_i. (This fact will be clear in the next section.) Such a design is called a πps design or $IPPS$ (inclusion probability proportional to size) design. Since $\pi_i \leq 1$, it is required that $x_i \leq X/n$ for such a design.

Many (exceeding seventy) unequal probabilities without replacement sampling designs have been suggested in the literature, mostly for use along with the HTE. Many of these designs attain the πps property exactly, some approximately. For some of these designs, such as the one arising out of Poisson sampling, sample size is a variable. Again, some of these sampling designs are sequential in nature (e.g., Chao 1982; Sunter 1977). Mukhopadhyay (1972), Sinha (1973), and Herzel (1986) considered the problem of realizing a sampling design with preassigned sets of inclusion probabilities of first two orders.

Again, in a sample survey, all the possible samples are not generally equally preferable from the point of view of practical advantages. In agricultural surveys, for example, the investigators tend to avoid grids which are located further away from the cell camps, which are located in marshy land, inaccessible places, etc. In such cases, the sampler would like to use only a fraction of the totality of all possible samples, allotting only a very mall probability to the non-preferred units. Such sampling designs are called Controlled Sampling Designs.

Chakraborty (1963) used a balanced incomplete block (BIB) design to obtain a controlled sampling design replacing a *srswor* design. For unequal probability sampling BIB designs and t designs have been considered by several authors (e.g., Srivastava and Saleh 1985; Rao and Nigam 1990; Mukhopadhyay and Vijayan 1996).

For a review of different unequal probability sampling designs the reader may refer to Brewer and Hanif (1983), Chaudhuri and Vos (1988), Mukhopadhyay (1996, 1998b), among others.

1.4 The Estimators

After the sample has been selected, the statistician collects data from the field. Here, again the data may be collected with respect to a *sequence* sample or *set* sample.

Definition 1.4.1 Data collected through a sequence sample S are

$$d' = \{(k, y_k), k \in S\}. \tag{1.4.1}$$

For the set sample data are

$$d = \{(k, y_k), k \in s\}. \tag{1.4.2}$$

It is known that data given in (1.4.2) are sufficient for making inference about θ, whether the sample is a sequence S or a set s (Basu and Ghosh 1967). Data are said to be unlabeled if after the collection of data its label part is ignored. Unlabeled data may be represented by a sequence or a set of the observed values without any reference to the labels.

It may not be possible, however, to collect the data from the sampled units correctly and completely. If the information is collected from a human population, the respondent may not be 'at home' during the time of collection of data or may refuse to answer or may give incorrect information, e.g., in stating income, age, etc. The investigators in the field may also fail to register correct information due to their own lapses.

We assume throughout that our data are free from such types of errors due to non-response and errors of measurement and it is possible to collect the information correctly and completely.

Definition 1.4.2 An estimator $e = e(s, \mathbf{y})$ or $e(S, \mathbf{y})$ is a function defined on $\mathcal{S} \times R^N$ such that for a given $(s, \mathbf{y})$ or $(S, \mathbf{y})$, its value depends on $\mathbf{y}$ only through those i for which $i \in s$ (or S).

Clearly, the value of e in a sample survey does not depend on the units not included in the sample.

An estimator e is unbiased for T with respect to a sampling design p if

$$E_p(e(s, \mathbf{y})) = T \ \forall \mathbf{y} \in R^N \tag{1.4.3}$$

i.e.,

$$\sum_{s \in \mathcal{S}} e(s, \mathbf{y}) p(s) = T \ \forall \mathbf{y} \in R^N$$

where E_p, V_p denote, respectively, expectation and variance with respect to the s.d. p. We shall often omit the suffix p when it is clear otherwise. This unbiasedness will sometimes be referred to as p-unbiasedness.

The mean square (MSE) of e around T with respect to a s.d. p is

$$\begin{aligned} M(e) &= E(e-T)^2 \\ &= V(e) + (B(e))^2 \end{aligned} \tag{1.4.4}$$

where $B(e)$ denotes the design bias, $E(e) - T$. If e is unbiased for T, $B(e)$ vanishes and (1.4.4) gives the variance of e, $V(e)$.

Definition 1.4.3 A combination (p, e) is called a *sampling strategy*, often denoted as $H(p, e)$. This is unbiased for T if (1.4.3) holds and then its variance is $V\{H(p, e)\} = E(e - T)^2$.

A unbiased sampling strategy $H(p, e)$ is said to be better than another unbiased sampling strategy $H'(p', e')$ in the sense of having smaller variance, if

$$V\{H(p, e)\} \leq V\{H'(p, e')\} \; \forall \mathbf{y} \in R^N \tag{1.4.5}$$

with strict inequality for at least one $\mathbf{y}$.

If the s.d. p is kept fixed, an unbiased estimator e is said to be better than another unbiased estimator e' in the sense of having smaller variance, if

$$V_p(e) \leq V_p(e') \; \forall \; \mathbf{y} \in R^N \tag{1.4.6}$$

with strict inequality holding for at least one $\mathbf{y}$.

We shall now consider different types of estimators for $\bar{y}$, when the s.d. is *srswor*, based on n draws.

(1) Mean per unit estimator: $\hat{\bar{Y}} = \bar{y}_s = \sum_{i \in s} y_i/n$
Variance: Var $(\bar{y}_s) = (1 - f)S^2/n$
$S^2 = \sum_{i=1}^N (y_i - \bar{Y})^2/(N-1)$
$\bar{Y} = \sum_{i=1}^N y_i/N, \; f = N/n$
(2) Ratio estimator: $\hat{\bar{y}}_R = (\bar{y}_s/\bar{x}_s)\bar{X}$
Mean square error: MSE $(\hat{\bar{y}}_R) \approx (\frac{1-f}{n})[S_y^2 + S_x^2 - 2RS_{yx}]$,
$R = Y/X$, $S_y^2 = S^2$, $S_x^2 = \sum_{i=1}^N (x_i - \bar{X})^2/(N-1)$, $X = \sum_{i=1}^N x_i$,
$\bar{X} = X/N \; S_{xy} = \sum_{i=1}^N (y_i - \bar{Y})(x_i - \bar{X})$.
(3) Difference estimator: $\hat{\bar{y}}_D = \bar{y}_s + d(\bar{X} - \bar{x}_s)$, where d is a known constant.
Variance : Var $(\hat{\bar{y}}_D) = (\frac{1-f}{n})(S_y^2 + d^2 S_x^2 - 2dS_{xy})$.
(4) Regression estimator: $\hat{\bar{y}}_{lr} = \bar{y}_s + b(\bar{X} - \bar{x}_s)$
$b = \sum_{i \in s} (y_i - \bar{y}_s)(x_i - \bar{x}_s)/\sum_{i \in s} (x_i - \bar{x}_s)^2$.
Mean square error: MSE $(\hat{\bar{y}}_{lr}) \approx (\frac{1-f}{n})S_y^2(1 - \rho^2)$
where ρ is the correlation coefficient between x and y.
(5) Mean of the ratios estimator: $\hat{\bar{y}}_{MR} = \bar{X}\bar{r}$
$\bar{r} = \sum_{i \in s} r_i/n$

Except for the mean per unit estimator and the difference estimator none of the above estimators is unbiased for $\bar{y}$. However, all these estimators are unbiased in large samples. Different modifications of the ratio estimator, regression estimator, product estimator, and the estimators obtained by taking convex combinations of these estimators have been proposed in the literature. Again, ratio estimator, difference estimator, and regression estimator, each of which depends on an auxiliary variable x, can be extended to $p(> 1)$ auxiliary variables $x_1, \ldots, x_p$.

In *ppswr*-sampling an unbiased estimator of population total is the Hansen–Hurwiz estimator,

$$\hat{T}_{pps} = \sum_{i \in S} \frac{y_i}{np_i}, \tag{1.4.9}$$

with

$$\begin{aligned} V(\hat{T}_{pps}) &= \tfrac{1}{n}\sum_{i=1}^{N}\left(\tfrac{y_i}{p_i} - T\right)^2 p_i \\ &= \tfrac{1}{2n}\sum\sum_{i \neq j=1}^{N}\left(\tfrac{y_i}{p_i} - \tfrac{y_j}{p_j}\right)^2 p_i p_j = V_{pps}. \end{aligned} \tag{1.4.10}$$

An unbiased estimator of V_{pps} is

$$v(\hat{T}_{pps}) = \frac{1}{n(n-1)} \sum_{i \in S}\left(\frac{y_i}{p_i} - \hat{T}_{pps}\right)^2 = v_{pps}.$$

We shall call the combination $(ppswr, \hat{T}_{pps})$ a *ppswr* strategy.

Clearly, different terms of an estimator will involve weights which arise out of the sampling designs used in estimation. It will therefore be of immense advantages if in the estimation formula all the units in the sample receive an identical weight. Before proceeding to further discussion on different types of estimators we therefore consider the situations when a sampling design can be made self-weighted.

Note 1.4.1 *Self-weighting Design*: A sample design which provides a single common weight to all sampled observations in estimating the population mean, total, etc. is called a self-weighting design and the corresponding estimator a self-weighted estimator. For example, consider two-stage sampling from a population consisting of N fsu's, the ath fsu containing M_a ssu's. A first-stage sample of n fsu's is selected by *srswor* and from the ath selected fsu m_a ssu's are selected also by *srswor*. It is known that for such a sampling design,

$$\hat{T} = \frac{N}{n} \sum_{a=1}^{n} \frac{M_a}{m_a} \sum_{b=1}^{m_a} y_{ab} = \frac{N}{n} \sum_{a=1}^{n} M_a \bar{y}_a \tag{1.4.11}$$

where $\bar{y}_a = \sum_{b=1}^{m_a} y_{ab}/m_a$ is the sample mean from the ath selected fsu, which is unbiased for population total T. This estimator is not generally self-weighted. If $m_a/M_a = \lambda$ (a constant), i.e., a constant proportion of ssu's is sampled from each selected fsu,

$$\hat{T} = \left(\frac{N}{n\lambda}\right)\left(\sum_{a=1}^{n}\sum_{b=1}^{m_a} y_{ab}\right)$$

becomes self-weighted. Again,

$$\lambda = \frac{\sum_{a=1}^{N} m_a}{\sum_{a=1}^{N} M_a} = \frac{N\bar{m}}{M_0} \tag{1.4.12}$$

where $\bar{m} = \sum_{a=1}^{N} m_a/N$, so that

$$\hat{T} = M_0 \sum_{a=1}^{n}\sum_{b=1}^{m_a} \frac{y_{ab}}{n\bar{m}}. \tag{1.4.13}$$

In particular, if $M_a = M \forall\, a$, a constant number m of ssu's must be sampled from each selected fsu in order to make the estimator (1.4.11) self-weighted.

A design can be made self-weighted at the field stage or at the estimation stage. If the selection of units is so done as to make all the weights in the estimator equal, the design is called self-weighted at the field stage. The case considered above is an example. Another example is the proportional allocation in stratified random sampling. If self-weighting is achieved using some technique at the estimation stage, the design is termed self-weighted at the estimation stage.

The procedures of designs self-weighted at the field stage have been considered by Hansen and Madow (1953) and Lahiri (1954). The technique of making designs self-weighted at the estimation stage has been considered by Murthy and Sethi (1959, 1961), among others.

1.4.1 A Class of Estimators

We now consider classes of *linear estimators* which are unbiased with respect to any s.d. For any s.d. p, consider a nonhomogeneous linear estimator of T,

$$e'_L(s, \mathbf{y}) = b_{0s} + \sum_{i \in s} b_{si} y_i \tag{1.4.14}$$

where the constants b_{0s} may depend only on s and b_{si} on $(s, i)(b_{si} = 0, i \notin s)$. The estimator e'_L is unbiased *iff*

$$\sum_{s \in \mathcal{S}} b_{0s}\, p(s) = 0 \tag{1.4.15a}$$

and

$$\sum y_i \sum_{s \ni i} b_{si}\, p(s) = T \ \forall \mathbf{y} \in R^N. \tag{1.4.15b}$$

Condition (1.4.15a) implies for all practical purposes

$$b_{0s} = 0 \forall\ s : p(s) > 0. \tag{1.4.16a}$$

Condition (1.4.15b) implies

$$\sum_{s \ni i} b_{si}\, p(s) = 1 \forall\ i = 1, \ldots, N \tag{1.4.16b}$$

Note that only the designs with $\pi_i > 0 \forall\ i$ admit an unbiased estimator of T.

Fattorini (2006) proposed a modification of the Horvitz–Thompson estimator for complex survey designs by estimating the inclusion probabilities by means of independent replication of the sampling scheme.

It is evident that the *Horvitz–Thompson estimator* (HTE) e_{HT} is the only unbiased estimator of T in the class of estimators $\{\sum_{i \in s} b_i y_i\}$,

$$e_{HT} = \sum_{i \in s} y_i / \pi_i. \tag{1.4.17}$$

Its variance is

$$\begin{aligned} V(e_{HT}) &= \sum_{i=1}^{N} y_i^2 \tfrac{1-\pi_i}{\pi_i} + \sum \sum_{i \neq j = 1}^{N} \tfrac{y_i y_j (\pi_{ij} - \pi_i \pi_j)}{\pi_i \pi_j} \\ &= V_{HT} \text{ (say)}. \end{aligned} \tag{1.4.18}$$

If $p \in \rho_n$, (1.4.18) can be written as

$$\begin{aligned} &\sum_{i<j=1}^{N} \left(\frac{y_i}{\pi_i} - \frac{y_j}{\pi_j} \right)^2 (\pi_i \pi_j - \pi_{ij}) \\ &= V_{YG} \text{ (say)}. \end{aligned} \tag{1.4.19}$$

The expression V_{HT} is due to Horvitz and Thompson (1952) and V_{YG} is due to Yates and Grundy (1953). An unbiased estimator of V_{HT} is

$$\sum_{i\in s}\frac{y_i^2(1-\pi_i)}{\pi_i^2}+\sum\sum_{i\neq j\in s}\frac{y_iy_j(\pi_{ij}-\pi_i\pi_j)}{\pi_i\pi_j\pi_{ij}}$$
$$= v_{HT}. \tag{1.4.20}$$

An unbiased estimator of V_{YG} is

$$\sum\sum_{i<j\in s}\left(\frac{y_i}{\pi_i}-\frac{y_j}{\pi_j}\right)^2\frac{\pi_i\pi_j-\pi_{ij}}{\pi_{ij}} \tag{1.4.21}$$
$$= v_{YG}.$$

The estimators v_{HT}, v_{YG} are valid provided $\pi_{ij} > 0 \forall i \neq j = 1,\ldots,N$. Both v_{HT} and v_{YG} can take negative values for some samples and this leads to the difficulty in interpreting the reliability of these estimators.

We consider some further estimators applicable to any sampling design.

(a) *Generalized Difference Estimator*: Basu (1971) considered an unbiased estimator of T,

$$e_{GD}(\mathbf{a}) = \sum_{i\in s}\frac{y_i-a_i}{\pi_i}+A,\ A=\sum_i a_i, \tag{1.4.22}$$

where $\mathbf{a} = (a_1,\ldots,a_N)'$ is a set of known quantities. The estimator is unbiased and has less variance than e_{HT} in the neighborhood of the point $\mathbf{a}$.

(b) *Generalized Regression Estimator* or GREG

$$e_{GR} = \sum_{i\in s}\frac{y_i}{\pi_i}+b\left(X-\sum_{i\in s}\frac{x_i}{\pi_i}\right) \tag{1.4.23}$$

where b is the sample regression coefficient of y on x. The estimator was first considered by Cassel et al. (1976) and is a generalization of linear regression estimator $N\hat{\bar{y}}_{lr}$ to any s.d. p.

(c) *Generalized Ratio Estimator*

$$e_{Ha} = X\frac{\sum_{i\in s} y_i/\pi_i}{\sum_{i\in s} x_i/\pi_i}. \tag{1.4.24}$$

The estimator was first considered by Ha'jek (1959) and is a generalization of $N\hat{\bar{y}}_R$ to any s.d. p.

The estimators e_{GR}, e_{Ha} are not unbiased for T. It is obvious that the estimators in (1.4.23) and (1.4.24) can be further generalized by considering $p(>1)$ auxiliary variables $x_1\ldots,x_p$ instead of just one auxiliary variable x. Besides all these, specific estimators have been suggested for specific procedures. An example is Rao–Hartley–Cochran (1962) estimator briefly discussed below.

(d) *Rao–Hartley–Cochran procedure* The population is divided into n groups $G_1 \ldots, G_n$ of size $N_1, \ldots, N_n$, respectively, at random. From the kth group a sample i is drawn with probability proportional to p_i, i.e., with probability p_i/Π_k where $\Pi_k = \sum_{i \in G_k} p_i$ if $i \in G_k$. An unbiased estimator of population total is

$$e_{RHC} = \sum_{i=1}^{n} \frac{y_i}{p_i} \Pi_i,$$

G_i denoting the group to which a sampled unit i belongs. It can be shown that

$$V(e_{RHC}) = \frac{n\left(\sum_{i=1}^{n} N_i^2 - N\right)}{N(N-1)} V(\hat{T}_{pps}) \tag{1.4.25}$$

and variance estimator

$$v(e_{RHC}) = \frac{\sum_{i=1}^{n} N_i^2 - N}{N^2 - \sum_{i=1}^{n} N_i^2} \sum_{i=1}^{n} \Pi_i \left(\frac{y_i}{p_i} - e_{RHC}\right)^2. \tag{1.4.26}$$

It has been found that the choice $N_1 = N_2 = \cdots = N_n = N/n$ minimizes $V(e_{RHC})$. In this case

$$V(e_{RHC}) = \left(1 - \frac{n-1}{N-1}\right) V(\hat{T}_{pps}). \tag{1.4.27}$$

We now briefly consider the concept of *double sampling*. We have seen that a number of sampling procedures require advanced knowledge about an auxiliary character. For example, ratio, difference, and regression estimator require a knowledge of the population total of x. When such information is lacking it is sometimes relatively cheaper to take a large preliminary sample in which the auxiliary variable x alone is measured and which is used for estimating the population characteristic like mean, total, or frequency distribution of x values. The main sample is often a subsample of the initial sample but may be chosen independently as well. The technique is known as *double sampling*.

All these sampling strategies have been discussed in details in Mukhopadhyay (1998b), among others.

1.5 Model-Based Approach to Sampling

So far our discussion has been under a fixed population approach. In the fixed population or design-based approach to sampling, the values $y_1, y_2, \ldots, y_N$ of the variable of interest in the population are considered as fixed but unknown constants. Randomness or probability enters the problem only through the deliberately imposed design by which the sample of units to observe is selected. In the design-based approach, with a design such as simple random sampling, the sample mean is a random variable only because it varies from sample to sample.

In the stochastic population or model-based approach to sampling, the values $\mathbf{y} = (y_1, y_2, \ldots, y_N)'$ are assumed to be a realization of a random vector $\mathbf{Y} = (Y_1, Y_2, \ldots, Y_N)'$, Y_i, being the random variable corresponding to y_i. The population model is then given by the joint probability distribution or density function $\xi_\theta = f(y_1, y_2, \ldots, y_N; \theta)$, indexed by a parameter vector $\theta \in \Theta$, the parameter space. Looked in this way population total $T = \sum_{i=1}^N Y_i$, population mean, $\bar{Y} = T/N$, etc. are random variables and not fixed quantities. One has, therefore, to predict $T, \bar{Y}$, etc. on the basis of the data and ξ, i.e., to estimate their model-expected values. Let $\hat{T}_s$ denote a predictor of T or $\bar{Y}$ based on s and $\mathcal{E}, \mathcal{V}, \mathcal{C}$ denote, respectively, expectation, variance, and covariance operators with respect to ξ_θ. Three examples of such superpopulations are

(i) $Y_1, \ldots, Y_N$ are independently and identically distributed (IID) normal random variables with mean μ and variance σ^2.
(ii) $\mathbf{Y}_1, \ldots, \mathbf{Y}_N$ are IID multinormal random vectors with mean vector $\boldsymbol{\mu}$ and covariance matrix $\boldsymbol{\Sigma}$. Here, instead of variable Y_i we have considered the p-variate vector $\mathbf{Y}_i = (Y_{i1}, \ldots, Y_{ip})'$.
(iii) Let $\mathbf{w} = (u, \mathbf{x}')'$, where u is a binary-dependent variable taking values 0 and 1 and $\mathbf{x}$ is a vector of explanatory variables. Writing $\mathbf{W}_i = (U_i, \mathbf{X}_i')'$, assume that $U_1, \ldots, U_N$ are IID with the logistic conditional distribution:

$$P[U_i = 1|\mathbf{W}_i = \mathbf{w}] = \frac{\exp(\mathbf{x}'\boldsymbol{\beta})}{[1 + \exp(\mathbf{x}'\boldsymbol{\beta})]}.$$

Superpopulation parameters are characteristics of ξ. In (i) parameters are μ and σ^2, in (ii) $\boldsymbol{\mu}$ and $\boldsymbol{\Sigma}$, and in (iii) $\boldsymbol{\beta}$.

As mentioned before, superpopulation parameters may often be preferred to finite population parameters as targets for inference in analytic surveys. However, if the population size is large there is hardly any difference between the two. For example, in (ii)

$$\bar{\mathbf{Y}}_{\mathcal{P}} = \boldsymbol{\mu} + O_p(N^{-1/2}),\ \mathbf{V}_{\mathcal{P}} = \boldsymbol{\Sigma} + O_p(N^{-1/2})$$

where $O_p(t)$ denotes terms of at most order t in probability and the suffix $\mathcal{P}$ stands for the finite population.

We shall briefly review here procedures of model-based inference in finite population sampling.

Definition 1.5.1 The predictor $\hat{T}_s$ is model-unbiased or ξ-unbiased or m-unbiased predictor of $\bar{Y}$ if

$$\mathcal{E}(\hat{T}_s) = \mathcal{E}(\bar{Y}) = \bar{\mu}(\text{say})\ \forall \theta \in \Theta \text{ and } \forall s : p(s) > 0 \tag{1.5.1}$$

where $\bar{\mu} = \sum_{k=1}^{N} \mu_k / N = \sum_{k=1}^{N} \mathcal{E}(Y_k)/N$.

Definition 1.5.2 The predictor $\hat{T}_s$ is design-model-unbiased (or $p\xi$-unbiased or pm unbiased) predictor of $\bar{Y}$ if

$$E\mathcal{E}(\hat{T}_s) = \bar{\mu}\ \forall \theta \in \Theta. \tag{1.5.2}$$

Clearly, a m-unbiased predictor is necessarily pm-unbiased.

For a non-informative design where $p(s)$ does not depend on the y-values, order of operators E and $\mathcal{E}$ can always be interchanged.

Two types of mean square errors (mse's) of a sampling strategy $(p, \hat{T}_s)$ for predicting T have been proposed:

(a)

$$\mathcal{E}MSE(p, \hat{T}) = \mathcal{E}E(\hat{T} - T)^2 = M(p, \hat{T}) \text{ (say);}$$

(b)

$$EMS\mathcal{E}(p, \hat{T}) = E\mathcal{E}(\hat{T} - \mu)^2 \text{ where } \mu = \textstyle\sum_k \mu_k = \mathcal{E}(T)$$

$$= M_1(p, \hat{T}) \text{ (say).}$$

If $\hat{T}$ is p-unbiased for T, M is model-expected p-variance of $\hat{T}$. If $\hat{T}$ is m-unbiased for T, M_1 is p-expected model variance of $\hat{T}$.

The following relation holds:

$$M(p, \hat{T}) = E\mathcal{V}(\hat{T}) + E\{\beta(\hat{T})\}^2 + \mathcal{V}(T) - 2\mathcal{E}\{(T - \mu)E(\hat{T} - \mu)\} \tag{1.5.3}$$

where $\beta(T) = \mathcal{E}(\hat{T} - T)$ is the model bias in $\hat{T}$.

Now, for the given data $d = \{(k, y_k), k \in s\}$, we have

$$T = \sum_s y_k + \sum_{\bar{s}} Y_i = \sum_s y_k + U_s \text{ (say)} \tag{1.5.4}$$

where $\bar{s} = \mathcal{P} - s$. Therefore, in predicting T one needs only to predict U_s, the part $\sum_s y_k$ being completely known.

A predictor

$$\hat{T}_s = \sum_s y_k + \hat{U}_s$$

will, therefore, be m-unbiased for T if

$$\mathcal{E}(\hat{U}_s) = \mathcal{E}\left(\sum_{\bar{s}} Y_k\right) = \sum_{\bar{s}} \mu_k = \mu_{\bar{s}} \text{ (say) } \forall \theta \in \Theta, \ \forall s : p(s) > 0. \tag{1.5.5}$$

In finding an optimal $\hat{T}$ for a given p, one has to minimize $M(p, \hat{T})$ (or $M_1(p, \hat{T})$) in a certain class of predictors. Now, for a m-unbiased $\hat{T}$,

$$\begin{aligned} M(p, \hat{T}) &= E\mathcal{E}(\hat{U}_s - \textstyle\sum_{\bar{s}} Y_k)^2 \\ &= E\left[\mathcal{V}(\hat{U}_s) + \mathcal{V}\left(\textstyle\sum_{\bar{s}} Y_k\right) - 2\mathcal{C}\left(\hat{U}_s, \textstyle\sum_{\bar{s}} Y_k\right)\right]. \end{aligned} \tag{1.5.6}$$

If Y_k's are independent, $\mathcal{C}(\hat{U}_s, \sum_{\bar{s}} Y_k) = 0(\hat{U}_s$ being a function of $Y_k, k \in s$ only). In this case, for a given s, an optimal m-unbiased predictor of T (in the minimum $\mathcal{E}(\hat{T}_s - T)^2$-sense) is (Royall 1970)

$$\hat{T}_s^+ = \sum_s y_k + \hat{U}_s^+ \tag{1.5.7}$$

where

$$\mathcal{E}(\hat{U}_s^+) = \mathcal{E}\left(\sum_{\bar{s}} Y_k\right) = \mu_{\bar{s}}, \tag{1.5.8a}$$

and

$$\mathcal{V}(\hat{U}_s^+) \le \mathcal{V}(\hat{U}_s') \tag{1.5.8b}$$

for any $\hat{U}_s$ satisfying (1.5.8a). It is clear that $\hat{T}_s$, when it exists, does not depend on the sampling design (unlike, the design-based estimator, e.g., e_{HT}.)

An optimal design-predictor pair $(p, \hat{T})$ in the class $(\rho, \hat{\tau})$ is, therefore, one for which

$$M(p^+, \hat{T}^+) \le M(p, \hat{T}) \tag{1.5.9}$$

for any $p \in \rho$, a class of sampling designs and any $\hat{T}$ which is an m-unbiased predictor belonging to $\hat{\tau}$. After $\hat{T}_s^+$ has been derived by means of (1.5.7)–(1.5.8b), an optimal sampling design is obtained through (1.5.9). The approach, therefore, is completely model-dependent, the emphasis being on the correct postulation of a superpopulation model that will efficiently describe the physical situation at hand and generating $\hat{T}_s$. After $\hat{T}_s$ has been specified, one makes a pre-sampling judgement of efficiency of $\hat{T}_s$ with respect to different sampling designs and obtain p^* (if it exists). The choice of a suitable sampling design is, therefore, relegated to secondary importance in this prediction-theoretic approach.

Note 1.5.1 We have attempted above to find the optimal strategies in the minimum $M(p, \hat{T})$ sense. The analogy may, similarly, be extended to finding optimality results in the minimum M_1 sense.

Example 1.5.1 (Thompson 2012) Suppose that our objective is to estimate the population mean, for example, the mean household expenditure for a given month in a geographical region. We may know from the economic theory that the amount a household may spend in a month follows a normal or lognormal distribution. In this case, the actual amount spent by a household in that given month is just one realization among many such possible realizations under the assumed distribution.

Considering a very simple population model, we assume that the population variables $Y_1, Y_2, \ldots, Y_N$ are independently and identically distributed (iid) random variables from a superpopulation distribution having mean θ, and a variance σ^2. Thus, for any unit i, Y_i is a random variable with expected value $\mathcal{E}(Y_i) = \theta$ and variance $\mathcal{V}(Y_i) = \sigma^2$, and for any two units i and j, the variables Y_i and Y_j are independent.

Suppose now that we have a sample s of n distinct units from this population and the object is to estimate the parameter θ of the distribution from which the finite population comes. For the given sample s, the sample mean

$$\bar{Y}_s = \frac{1}{n}\sum_{i \in s} Y_i$$

is a random variable whether or not the sample is selected at random, because for each unit i in the sample Y_i is a random variable. With the assumed model, the expected value of the sample mean is therefore $\mathcal{E}(\bar{Y}_s) = \theta$ and its variance is $\mathcal{V}(\bar{Y}_s) = \sigma^2/n$. Thus $\bar{Y}_s$ is a model-unbiased estimator of the parameter θ, since $\mathcal{E}(\bar{Y}_s) = \theta$. An approximate $(1-\alpha)$-point confidence interval for the parameter θ, based on the central limit theorem for the sample mean of independently and identically distributed random variables, is, therefore, given by

$$\bar{Y}_s \pm tS/\sqrt{n} \tag{1.5.10}$$

where S is the sample standard deviation and t is the upper $\alpha/2$ point of the t distribution with $n-1$ degrees of freedom. If further the Y_i's are assumed to have a normal distribution, then the confidence interval (1.5.10) is exact, even with a small sample size.

In the study of household expenditure the focus of interest may not be, however, on the superpopulation parameter θ of the model, but on the actual average amount spent by households in the community that month. That is, the object is to estimate (or predict) the value of the random variable

$$\bar{Y} = \frac{1}{N}\sum_{i=1}^{N} Y_i.$$

To estimate or predict the value of the random variable $\bar{Y}$ from the sample observations, an intuitively reasonable choice is again the sample mean $\hat{\bar{Y}} = \bar{Y}_s = \sum_{i=1}^{n} Y_i/n$. Both $\bar{Y}$ and $\bar{Y}_s$ have expected value θ, since the expected value of each of the Y_i is θ. Clearly, $\bar{Y}_s$ is model-unbiased for the population quantity $\bar{Y}$.

It can be shown that under the given conditions on the sample (i.e., the sample consists only of n distinct units), and under the given superpopulation model,

$$\mathcal{E}(\bar{Y}_s - \bar{Y})^2 = \frac{N-n}{nN}\sigma^2. \qquad (1.5.11)$$

An unbiased estimator or predictor of the mean square prediction error is, therefore,

$$\frac{N-n}{N}\frac{S^2}{n}$$

since $\mathcal{E}(S^2) = \sigma^2$.

Therefore, an approximate $(1-\alpha)$-point prediction interval for $\bar{Y}$ is given by

$$\bar{Y} \pm t\sqrt{\hat{\mathcal{E}}(\bar{Y}_s - \bar{Y})^2},$$

where t is the upper $\alpha/2$ point of the t-distribution with $n-1$ degrees of freedom. If, additionally, the distribution of the Y_i is assumed to be normal, the confidence level is exact.

We also note that for the given superpopulation model and for any FES(n) s.d. (including *srswor* s.d.), $M_1(p, \bar{Y}_s)$ is given by the right-hand side of (1.5.11). In fact even if the s.d. is such that it only selects a particular sample of n distinct units with probability unity, these results hold.

Example 1.5.2 Consider a superpopulation model ξ such that $Y_1, \ldots, Y_N$ are independently distributed with

$$\begin{aligned} \mathcal{E}(Y_i|x_i) &= \beta x_i \\ \mathcal{V}(x_i) &= \sigma^2 x_i, \end{aligned} \qquad (1.5.12)$$

where x_i is the (known) value of an auxiliary variable x on unit $i(=1,\ldots,N)$. An optimal m-unbiased predictor of population total T is

$$\hat{T}_s^* = \sum_s y_k + \hat{U}_s^*$$

where

$$\mathcal{E}(\hat{U}_s^*) = \mathcal{E}\left(\sum_{\bar{s}} Y_k\right) = \beta\sum_{\bar{s}} x_k \qquad (1.5.13)$$

and $\mathcal{V}(\hat{U}_s^*) \leq \mathcal{V}(\hat{U}_s')$ for all $\hat{U}_s'$ satisfying (1.5.13). Confining to the class of linear m-unbiased predictors, the BLUP (best linear (m)-unbiased predictor) of T is

$$\begin{aligned}\hat{T}^* &= \sum_s y_k + \hat{\beta}^* \sum_{\bar{s}} y_k \\ &= \sum_s y_k + \frac{\sum_s y_k}{\sum_s x_k} \sum_{\bar{s}} y_k = \frac{\bar{y}_s}{\bar{x}_s} X\end{aligned} \tag{1.5.14}$$

where $\bar{y}_s = \sum_k y_k/n, X = \sum_{k=1}^N x_k$ and where we have written y_k in place of $Y_k (k \in \bar{s})$. Again,

$$M(p, \hat{T}^*) = \sigma^2 E\left[\frac{\left(\sum_{\bar{s}} x_k\right)^2}{\sum_s x_k} + \sum_{\bar{s}} x_k\right]. \tag{1.5.15}$$

The model (1.5.12) roughly states that for fixed values of x, we have an array of values of the characteristic y such that both the conditional expectation and conditional variance of Y are each proportional to x. The regression equation of Y on x is therefore a straight line passing through the origin with conditional variance of Y proportional to x. In such cases, for any given sample, the ratio estimator would be the BLU estimator of the population total T. Again, (1.5.15) shows that a 'purposive' design which selects a combination of n units having the largest x-values with probability one will be the best s.d. to use the ratio statistic $\hat{T}$ in (1.5.14). We note that in contrast to the design-based approach where a probability sampling design has its pride of place, the model-based approach relegates the sampling design to a secondary consideration.

1.5.1 Uses of Design and Model in Sampling

Since, often very little are known about the nature of the population, design-based methods, especially *srs*-based methods, have been being used for a long time. In such a situation, most researchers find it reassuring to know that the estimation method used is unbiased no matter what the nature of the population may be. Such a method is called *design-unbiased*. The expected value of the estimator, taken over all samples which might have been selected (but is not all actually selected), is the correct population value. Here sampling design imposes a randomization which forms the basis of inference. Design-unbiased estimators of the variance, used for constructing confidence intervals, are also available for most sampling designs.

In many sampling situations involving auxiliary variables, it seems natural to postulate a theoretical model for the relationship between the auxiliary variables and the variable of interest. Thus in an agricultural context it is natural to postulate a linear regression model between yield of the crop and auxiliary variables like rainfall, fertilizer used, nature of firm land, etc. In such situations model-based methods are often (and should be) used in order to utilize the information contained in the population in sample selection and estimation. This is not to say that the design-based methods are not useful in such cases. Sampling designs often provide a protection against bias in case of model failures. Model-based approaches have, of late, been very popular in connection with ratio and regression estimation. A model can, of course, also be assumed for populations without auxiliary variables. For example, if the N variables $Y_1, \ldots, Y_N$ can be assumed to be independent and identically distributed, many standard statistical results apply without reference to how the sample is selected. Generally, the models become mathematically complex though often not suitably realistic. In particular, any model assuming that the Y-values are independent (or having an exchangeable distribution) ignores the tendency in many populations for nearby or related units to be correlated.

With some populations, however, it might have been found empirically and very convincingly that certain types of patterns are typical of the y-values of that type of population. For example, in spatially distributed geological and ecological populations, the y-values of nearby units may be positively correlated, with the strength of the relationship decreasing with the distance. If such tendencies are known to exist, they can be used in modeling the nature of the population, devising efficient sampling procedures and obtaining efficient predictors of unknown values of parameters. For detailed reviews of design-based, model-based, and design-model-based approaches to sampling interested readers may refer to Sarndal et al. (1992), Mukhopadhyay (1996, 2000, 2007), among others.

1.6 Plan of the Book

The book rests on two pillars: sample surveys and categorical data analysis. The first chapter makes a cursory review of the development in the arena of survey sampling. It introduces the notions of sampling designs and sampling schemes, estimators and their properties, various types of sampling designs and sampling strategies, and design-based methods of making inference about the population parameters in descriptive surveys. It then introduces the concept of superpopulation models and model-based methods of prediction of population parameters generally useful in analytic surveys.

It is known that the classical statistical models are based on the assumptions that the observations are obtained from samples drawn by simple random sampling with replacement (*srswr*) or equivalently the observations are independently and identically distributed (IID). In practice, in large-scale surveys samples are generally selected using a complex sampling design, such as a stratified multistage sampling design and this implies a situation different from a IID setup. Again, in large-scale sample surveys the finite population is often considered as a sample from a superpopulation. The sampling design may entail the situation that the sample observations are no longer subject to the same superpopulation model as the complete finite population. Thus, even if the IID assumption may hold for the complete population, the same generally breaks down for sample observations. After observing that the data obtained from a complex survey generally fail to satisfy IID assumption, Chap. 2 examines the effect of a true complex design on the variance of an estimator with respect to *srswr* design and/or IID-model assumption. It introduces the concepts of *design-effects* and *misspecification effect* of a parameter estimator and its variance estimator pair. The concepts have been extended to the multiparameter case.

Since estimation of variance of an estimator is one of the main problems encountered in various applications in this book, the chapter also makes a brief review of different nonparametric methods of variance estimation, like, linearization procedure, random group method, balanced repeated replication, Jack-knifing, and Bootstrap technique. Finally, we examine the impact of survey design on inference about a covariance matrix, and we consider the effect of a complex survey design on a classical test statistic for testing a hypothesis regarding a covariance matrix.

Chapter 3 makes a brief review of classical models of categorical data and their analysis. After a glimpse of general theory of fitting of statistical models and testing of parameters using goodness-of-fit tests, we return to the main distributions of categorical variables—multinomial distribution, Poisson distribution, and multinomial–Poisson distribution—and examine the associated test procedures. Subsequently, log-linear models and logistic regression models, both binomial and multinomial, are looked into and their roles in offering model parameters emphasized. Finally, some modifications of classical test procedures for analysis of data from complex surveys under logistic regression model have been introduced.

Chapter 4 proposes to investigate the effect of complex surveys on the asymptotic distributions of Pearson statistic, Wald statistic, log-likelihood ratio statistic for testing goodness-of-fit (simple hypothesis), independence in two-way contingency tables, and homogeneity of several populations in a saturated model. In particular, effects of stratification and clustering on these statistics have been examined. The model is called full or saturated, as we assume that the t population proportions or cell-probabilities $\pi_1, \ldots, \pi_t$ do not involve any other set of unknown parameters. In the next two chapters, the unknown population proportions are considered as generated out of some models through their dependence on $s(< t)$ of model parameters $\theta_1, \ldots, \theta_s$.

The core material of any categorical data analysis book is logistic regression and log-linear models. Chapter 5 considers analysis of categorical data from complex surveys using log-linear models for cell-probabilities in contingency tables. Noting that appropriate ML equations for the model parameter θ and hence of $\pi(\theta)$ are difficult to obtain for general survey designs, 'pseudo MLE's have been used to estimate the cell-probabilities. The asymptotic distributions of different goodness-of-fit (G-o-F) statistics and their modification are considered. Nested models have also been investigated.

Chapter 6 takes up the analysis of complex surveys categorical data under logistic regression models, both binomial and polytomous. Empirical logit models have been converted into general linear models which use generalized least square procedures for estimation. The model has been extended to accommodate cluster effects and procedures for testing of hypotheses under the extended model investigated.

So far we have assumed that there was no error in classifying the units according to their true categories. In practice, classification errors may be present and in these situations usual tests of goodness-of-fit, independence, and homogeneity become untenable. This chapter considers modifications of the usual test procedures under this context. Again, units in a cluster are likely to be related. Thus in a cluster sampling design where all the sampled clusters are completely enumerated, Pearson's usual statistic of goodness-of-fit seems unsuitable. This chapter considers modification of this statistic under these circumstances.

Noting that the estimation of model parameters of the distribution of categorical variables from data obtained through complex surveys is based on maximizing the pseudo-likelihood of the data, as exact likelihoods are rarely amenable to maximization, Chap. 8 considers some procedures and applications which are useful in obtaining approximate maximum likelihood estimates from survey data. Scott et al. (1990) proposed weighted least squares and quasi-likelihood estimators for categorical data. Maximum likelihood estimation (MLE) from complex surveys requires additional modeling due to information in the sample selection. This chapter reviews some of the approaches considered in the literature in this direction. After addressing the notion of ignorable sampling designs it considers exact MLE from survey data, the concept of weighted distributions, and its application in MLE of parameters from complex surveys. The notion of design-adjusted estimation due to Chambers (1986), the pseudo-likelihood approach to estimation of finite population parameters as developed by Binder (1983), Krieger and Pfeffermann (1991), among others, have also been revisited. Mixed model framework, which is a generalization of design-model framework, and the effect of sampling designs on the standard principal component analysis have also been revisited.

Since multinomial distribution is one of the main backbones of the analysis of categorical data collected from complex surveys, the Appendix makes a review of the asymptotic properties of the multinomial distribution and asymptotic distribution of Pearson chi-square statistic X_P^2 for goodness-of-fit based on this distribution. General theory of multinomial estimation and testing in case the population proportions $\pi_1, \ldots, \pi_{t-1}$ depend on several parameters $\theta_1, \ldots, \theta_s (s < t - 1)$, also unknown, is then introduced. Different minimum-distance methods of estimation, like, X_P^2, likelihood ratio statistic G^2, Freeman–Tukey (1950) statistic $(FT)^2$, and Neyman's (1949) statistic X_N^2 have been defined and their asymptotic distribution studied under the full model as well as nested models in the light of, among others, Birch's (1964) illuminating results. Finally, Neyman's (1949) and Wald's (1943) procedures for testing general hypotheses relating to population proportions have been revisited.

Chapter 2
The Design Effects and Misspecification Effects

Abstract It is known that the classical statistical models are based on the assumptions that the observations are obtained from samples drawn by simple random sampling with replacement (*srswr*) or equivalently the observations are independently and identically distributed (IID). As such the conventional formulae for standard statistical packages which implement these procedures are also based on IID assumptions. In practice, in large-scale surveys samples are generally selected using a complex sampling design, such as a stratified multistage sampling design and this implies a situation different from an IID setup. Again, in large-scale sample surveys the finite population is often considered as a sample from a superpopulation. Survey data are commonly used for analytic inference about model parameters such as mean, regression coefficients, cell probabilities, etc. The sampling design may entail the situation that the sample observations are no longer subject to the same superpopulation model as the complete finite population. Thus, even if the IID assumption may hold for the complete population, the same generally breaks down for sample observations. The inadequacy of IID assumption is well known in the sample survey literature. It has been known for a long time, for example, that the homogeneity which the population clusters generally exhibit tend to increase the variance of the sample estimator over that of the estimator under *srswr* assumption, and further estimates of this variance wrongly based on IID assumptions are generally biased downwards. In view of all these observations it is required to examine the effects of a true complex design on the variance of an estimator with reference to a *srswr* design or an IID model setup. Section 2.2 examines these effects, *design effect*, and *misspecification effect* of a complex design for estimation of a single parameter θ. The effect of a complex design on the confidence interval of θ is considered in the next section. Section 2.4 extends the concepts in Sect. 2.2 to multiparameter case and thus defines multivariate design effect. Since estimation of variance of estimator of θ, $\hat{\theta}$ (covariance matrix when θ is a vector of parameters) is of major interest in this chapter we consider different methods of estimation of variance of estimators, particularly nonlinear estimators in the subsequent section. The estimation procedures are very general; they do not depend on any distributional assumption and are therefore nonparametric in nature. Section 2.5.1 considers in detail a simple method of estimation of variance of a linear statistic. In Sects. 2.5.2–2.5.7 we consider Taylor series linearization procedure, random group (RG) method, balanced repeated replication (BRR), jackknife

P. Mukhopadhyay, *Complex Surveys*, DOI 10.1007/978-981-10-0871-9_2

(JK) procedure, JK repeated replication, and bootstrap (BS) techniques of variance estimation. Lastly, we consider the effect of a complex survey design on a classical test statistic for testing a hypothesis regarding a covariance matrix.

Keywords IID · Design effect · Misspecification effect · Design factor · Effective sample size · Multivariate design effect · Generalized design effect · Variance estimation · Linearization method · Random group · Balanced repeated replication · Jackknife (JK) procedure · JK repeated replication · Bootstrap · Wald statistic

2.1 Introduction

In analysis of data collected through sample surveys standard statistical techniques are generally routinely employed. However, the probabilistic assumptions underlying these techniques do not always reflect the complexity usually exhibited by the survey population. For example, in the classical setup, the log-linear models are usually based upon distributional assumptions, like Poisson, multinomial, or product-multinomial. The observations are also assumed to be independently and identically distributed (IID). On the other hand, survey populations are often complex with different cell probabilities in different subgroups of the population and this implies a situation different from the IID setup. A cross-tabulation of the unemployment data, for example, by age-group and level of education would not support the IID assumption of sample observations but would exhibit a situation far more complex in distributional terms. However, the conventional formulae for standard errors and test procedures, as implemented in standard statistical packages such as SPSS X or SAS are based on assumptions of IID observations or equivalently, that samples are selected by simple random sampling with replacement, and these assumptions are almost never valid for complex survey data.

Longitudinal surveys where sample subjects are observed over two or more time points typically lead to dependent observations over time. Moreover, longitudinal surveys often have complex survey designs that involve clustering which results in cross-sectional dependence among samples.

The inadequacy of IID assumption is well known in the sample survey literature. It has been known for a long time, for example, that the homogeneity which the population clusters generally exhibit tends to increase the variance of the sample estimator over that of the estimator under *srswr* assumption, and further estimates of this variance wrongly based on IID assumptions are generally biased downwards (Example 2.2.1). Hence consequences of wrong use of IID assumptions for cluster data are: estimated standard errors of the estimators would be too small and confidence intervals too narrow. For analytic purposes test statistic would be based on downwardly biased estimates of variance and the results would, therefore, appear to be more significant than was really the case. Hence such tests are therefore conservative in nature.

Again, in large-scale sample surveys the finite population is usually considered as a sample from a superpopulation. Survey data are commonly used for analytic inference about model parameters such as mean, regression coefficients, cell probabilities, etc. The sampling design may entail the situation that the sample observations are no longer subject to the same superpopulation model as the complete finite population. To illustrate the problem suppose that with each unit i of a finite population $\mathcal{P}$ is a vector $(Y_i, Z_i)'$ of measurements. Assume that $(Y_i, Z_i)'$ are independent draws from a bivariate normal distribution with mean $\boldsymbol{\mu}' = (\mu_Y, \mu_Z)$ and variance–covariance matrix $\boldsymbol{\Sigma}$. The values (y_i, z_i) are observed for a sample of n units selected by a probability sampling scheme. It is desirable to estimate mean μ_Y and variance σ_Y^2 of the marginal distribution of Y. We consider the following two cases.

(A) The sample is selected by *srswr* and only the values $\{(y_i, z_i), i \in s\}$ are known. This is the case of IID observations. Here, the maximum likelihood estimators (MLE's) of the parameters are

$$\hat{\mu}_Y = \bar{y}_s = \sum_{i \in s} y_i/n; \quad \hat{\sigma}_Y^2 = \sum_{i \in s} (y_i - \bar{y}_s)^2/n. \tag{2.1.1}$$

Clearly, $\mathcal{E}(\hat{\mu}_Y) = \mu_Y$ and $\mathcal{E}[n\hat{\sigma}_Y^2/(n-1)] = \sigma_Y^2$ where $\mathcal{E}(.)$ defines expectation with respect to the bivariate normal model. Thus, standard survey estimators are identical with the classical estimators in this case.

(B) The sample is selected with probability proportional to Z_i with replacement such that at each draw $i = 1, \ldots, n$, $P_i =$ Prob. $(i \in s) = Z_i / \sum_{i=1}^{N} Z_i$. The data known to the statistician are $\{(y_i, z_i), i \in s; z_j, j \notin s\}$. Suppose that the correlation coefficient $\rho_{Y,Z} > 0$. This implies that Prob.$(Y_i > \mu_Y | i \in s) > 1/2$ since the sampling scheme tends to select units with larger values of Z and hence large values of Y. Clearly, the distribution of the sample Y values, in this case, is different from the distribution in the population and the estimators defined in (2.1.1) are no longer MLE.

Recently, researchers in the social science and health sciences are increasingly showing interest in using data from complex surveys to conduct same sorts of analyses that they traditionally conduct with more straightforward data. Medical researchers are also increasingly aware of the advantages of well-designated subsamples when measuring novel, expensive variables on an existing cohort. Until recent times they would be analyzing the data using softwares based on the assumption that the data are IID.

In the very recent years, however, there have been some changes in the situation. All major statistical packages, like, STATA, SUDAAN, now include at least some survey analysis components and some of the mathematical techniques of survey analysis have been incorporated in widely used statistical methods for missing data and causal inference. The excellent book by Lumley (2010) provides a practical guide to analyzing complex surveys using R.

The above discussions strongly indicate that the standard procedures are required to be modified to be suitable for analysis of data obtained through sample surveys.

In Sects. 2.2–2.4 we consider the effects of survey designs on standard errors of estimators, confidence intervals of the parameters, tests of significance as well as the multivariate generalizations of these design effects.

Since the estimation of variance of an estimator under complex survey designs is one of the main subjects of interest in this chapter and in subsequent discussions we make a brief review of different nonparametric methods of estimation of variance in Sect. 2.5. Section 2.5.1 considers in detail a simple method of estimation of variance of a linear statistic. In Sects. 2.5.2–2.5.6 we consider Taylor series linearization procedure, random group method, balanced repeated replication, jackknife, and bootstrap techniques of variance estimation. All these procedures (except the bootstrap resampling) have been considered in detail in Wolter (1985). In this treatise we do not consider estimation of superpopulation-based variance of estimators. Interest readers may refer to Mukhopadhyay (1996) for a review in this area.

2.2 Effect of a Complex Design on the Variance of an Estimator

Let $\hat{\theta}$ be an estimator of a finite population parameter θ induced by a complex survey design of sample size n with $Var_{true}(\hat{\theta})$ as the actual design variance of $\hat{\theta}$. Let $Var_{SRS}(\hat{\theta})$ be the variance of $\hat{\theta}$ calculated under a hypothetical simple random sampling with replacement (*srswr*) (also, stated here as SRS) design of the same sample size (number of draws) n. The effect of the complex design on the variance of $\hat{\theta}$ (relative to the *srswr* design) is given by the design effect (*deff*) developed by Kish (1965),

$$\text{deff }(\hat{\theta})_{Kish} = \frac{Var_{true}(\hat{\theta})}{Var_{SRS}(\hat{\theta})}. \tag{2.2.1}$$

Clearly, if deff $(\hat{\theta})_{Kish} < 1$, the true complex design is a better design than a corresponding *srswr* design with respect to $\hat{\theta}$, the estimator of θ under the true design. Note that Kish's deff (2.2.1) is completely a design-based measure.

At the analysis stage one is, however, more interested in the effect of the design on the estimator of the variance. Let $v_0 = \hat{V}ar_{SRS}(\hat{\theta}) = \hat{V}ar_{IID}(\hat{\theta})$ be an estimator of $Var_{SRS}(\hat{\theta})$ which is derived under the SRS assumption or under the equivalent IID assumption, that is $E(v_0|SRS) = E(v_0|IID) = Var_{SRS}(\hat{\theta})$. Clearly, v_0 may be a design-based estimator or a model-based estimator. The effect of the true design on the estimator pair $(\hat{\theta}, v_0)$ is given by the bias of v_0,

$$E_{true}(v_0) - Var_{true}(\hat{\theta}), \tag{2.2.2}$$

where expectation in (2.2.2) is with respect to the actual complex design. However, for the sake of comparability with (2.2.1) we define the *misspecification effect* (*meff*) of $(\hat{\theta}, v_0)$ as

$$\text{meff}\ (\hat{\theta}, v_0) = \frac{Var_{true}(\hat{\theta})}{E_{true}(v_0)}. \tag{2.2.3}$$

This measure is given by Skinner (1989).

Note 2.2.1 Kish's design effect (2.2.1) is a design-based measure, while Skinner's misspecification effect may be defined either as a design-based measure or a as a model-based measure. When taken as a model-based measure, the quantities E_{true} and Var_{true} in (2.2.3) should be based on the true model distribution. Thus the measure (2.2.3) can also be used to study the effect of the assumed model on the variance of the estimator relative to the IID assumption. Clearly, under model-based approach meff$(\hat{\theta}, v_0)$ depends only on the model relationship between the units in the actual sample selected and not on how the sample was selected. □

It has been found that in large-scale sample surveys using stratified multistage sampling design with moderate sample sizes, $E_{true}(v_0) \approx Var_{SRS}(\hat{\theta})$. Hence, for such designs, the values of measures (2.2.1) and (2.2.3) are often very close. Also,

$$\hat{\text{deff}}\ (\hat{\theta}) \approx \hat{\text{meff}}\ (\hat{\theta}, v_0) = \frac{v}{v_0} \tag{2.2.4}$$

where v is a consistent estimator of $Var_{true}(\hat{\theta})$ under the true sampling design. Thus, even though the values of (2.2.1) and (2.2.3) may be unequal, the estimated values of $deff_{Kish}$ and *meff* are often equal.

We shall henceforth, unless stated otherwise, assume that all the effects on the variance are due to sampling designs only. The misspecification effect may now be called a *design effect*. Following Skinner (1989) we shall now define the design effect (deff) of $(\hat{\theta}, v_0)$ as

$$\text{deff}\ (\hat{\theta}, v_0) = \text{meff}\ (\hat{\theta}, v_0) = \frac{Var_{true}(\hat{\theta})}{E_{true}(v_0)}. \tag{2.2.5}$$

Note that measures in (2.2.5) may be based on both models and designs.

We can generalize (2.2.1) to define the *general design effect* (deff) of an estimator $\hat{\theta}$ as

$$\text{deff}\ (\hat{\theta}) = \frac{Var_{true}(\hat{\theta})}{Var_*(\hat{\theta})} \tag{2.2.6}$$

where $Var_*(\hat{\theta})$ is the variance of $\hat{\theta}$ under some benchmark design representing IID situations.

Example 2.2.1 In this example we shall clarify the distinction between design-based deff of Kish and model-based misspecification effect of Skinner.

Consider an infinite population of clusters of size 2 (elementary units) with mean θ (per elementary unit), variance σ^2 (per elementary unit), and intracluster correlation

(correlation between two units in a cluster) τ. Suppose that a sample of one cluster is selected for estimating θ and observations y_1, y_2 on the units in the cluster are noted. An estimator of θ under this cluster sampling design is $\hat{\theta} = (y_1 + y_2)/2$. The true variance of $\hat{\theta}$ is

$$Var_{true}(\hat{\theta}) = V\left[\frac{y_1 + y_2}{2}\right] = \frac{\sigma^2}{2}(1 + \tau).$$

Under the hypothetical assumption that the two elementary units have been drawn by *srswr* (or under the model assumption that y_1, y_2 are IID with mean and variance as above) from the population of elementary units in the hypothetical population,

$$Var_{SRS}(\hat{\theta}) = Var_{SRS}\left(\frac{y_1 + y_2}{2}\right) = \frac{\sigma^2}{2}.$$

Again, an estimator of $Var_{SRS}(\hat{\theta}) = Var_{IID}(\hat{\theta})$, also based on a srswr design, is

$$\begin{aligned} v_0 &= \tfrac{1}{2}[(y_1 - \hat{\theta})^2 + (y_2 - \hat{\theta})^2] \\ &= \frac{(y_1 - y_2)^2}{4}. \end{aligned}$$

Also,

$$E_{true}(v_0) = E_{true}\left[\frac{(y_1 - y_2)^2}{4}\right] = \sigma^2\frac{1 - \tau}{2}.$$

Therefore, by (2.2.1), Kish's design effect is

$$\text{deff }(\hat{\theta})_{Kish} = \frac{Var_{true}(\hat{\theta})}{Var_{SRS}(\hat{\theta})} = 1 + \tau.$$

Also, at the analysis stage, by (2.2.3),

$$\text{meff }(\hat{\theta}, v_0) = \frac{V_{true}(\hat{\theta})}{E_{true}(v_0)} = \frac{1 + \tau}{1 - \tau}.$$

Thus, if $\tau = 0.8$, deff $(\hat{\theta})_{Kish} = 1.8$, meff $_{Skinner}(\hat{\theta}, v_0) = 9$. This means that the true design variance is 80 % higher than the SRS-based variance under the design-based approach; but its true-model-based variance is 800 % higher than the average value of the IID-based variance estimator v_0. □

Example 2.2.2 Consider the problem of estimating the population mean θ by a simple random sample without replacement (*srswor*)-sample of size n. Here $\hat{\theta} = \bar{y}_s = \sum_{i \in s} y_i/n$, the sample mean with

$$Var_{true}(\hat{\theta}) = (N - n)\sigma^2/\{n(N - 1)\}, \; Var_{SRS}(\hat{\theta}) = \sigma^2/n,$$

$$v_0 = \hat{V}ar_{SRS}(\hat{\theta}) = \text{ estimator of } V_{SRS}(\bar{y}) \text{ under srswr } = s^2/n,$$

$$E_{true}(v_0) = N\sigma^2/\{(N-1)n\},$$

where $\sigma^2 = \sum_{i=1}^{N}(Y_i - \bar{Y})^2/N$, $s^2 = \sum_{i \in s}(y_i - \bar{y})^2/(n-1)$. Hence

$$\text{deff } (\hat{\theta}) = \frac{Var_{true}(\bar{y})}{Var_{SRS}(\bar{y})} = \frac{N-n}{N-1},$$

$$\text{deff } (\bar{y}, v_0) = \frac{Var_{true}(\bar{y})}{E_{true}(v_0)} = \frac{N-n}{N},$$

which is the finite population correction factor.

Example 2.2.3 Suppose we want to estimate $\theta = \bar{Y} = \sum_h W_h \bar{Y}_h$ by stratified random sampling of size n with proportional allocation, where $W_h = N_h/N$, etc. Here $\hat{\theta} = \bar{y}_{st} = \bar{y}$, the sample mean and the true variance,

$$Var_{true}(\bar{y}) = \frac{N-n}{nN}\sum_h W_h S_h^2,$$

where S_h^2 is the population variance of the hth stratum. If we assume that the sample has been drawn by *srswor*, an unbiased estimator of $Var_{SRS}(\bar{y})$, also calculated under *srswor*, is

$$v_{SRS} = \frac{N-n}{nN}s^2 = v_0.$$

Note that *srswor* is the benchmark design here. Its expectation under the true design is

$$E_{true}(v_0) \approx \frac{N-n}{nN}S^2 = \frac{N-n}{nN}\sum_h W_h\{S_h^2 + (\bar{Y}_h - \bar{Y})^2\}$$

where $S^2 = (N-1)^{-1}\sum_h \sum_i (y_{hi} - \bar{Y})^2$ is the finite population variance. Hence the design effect is

$$\text{deff } (\bar{y}, v_0) = \frac{\sum_h W_h S_h^2}{\sum_h W_h[S_h^2 + (\bar{Y}_h - \bar{Y})^2]}.$$

The deff is always less than or equal to one and can be further reduced by the use of an appropriate allocation rule.

Example 2.2.4 Consider the cluster sampling design in which n clusters are selected by *srswor* from a population of N clusters each of size M. An unbiased estimator of population mean (per element) $\theta = \sum_{c=1}^{N}\sum_{l=1}^{M} y_{cl}/(MN)$ is

$$\bar{y} = \sum_{c=1}^{n}\sum_{l=1}^{M} y_{cl}/(nM). \tag{2.2.7}$$

Hence,

$$V_{true}(\bar{y}) = \frac{N-n}{n(N-1)}\sigma_b^2 \text{ where } \sigma_b^2 = \frac{1}{N}\sum_{c=1}^{N}(\bar{y}_c - \theta)^2,\ \bar{y}_c = \frac{1}{N}\sum_{l=1}^{M} y_{cl}.$$

Now,

$$\sigma_b^2 = \frac{\sigma^2}{M}\{1 + (M-1)\tau\},\ \sigma^2 = \frac{1}{MN}\sum_{c=1}^{N}\sum_{l=1}^{M}(y_{cl} - \theta)^2$$

where τ is the intraclass correlation among units belonging to the same cluster (vide, Mukhopadhyay 2009). Hence,

$$\begin{aligned} Var_{true}(\bar{y}) &= \tfrac{N-n}{n(N-1)}\tfrac{\sigma^2}{M}\{1 + (M-1)\tau\} \\ &= \left(\tfrac{1}{n} - \tfrac{1}{N}\right)\tfrac{N\sigma^2}{(N-1)M}\{1 + (M-1)\tau\}. \end{aligned}$$

Also,

$$\begin{aligned} \tau &= \tfrac{E(y_{cl}-\theta)(y_{cm}-\theta)}{E(y_{cl}-\theta)^2} \\ &= \tfrac{1}{(M-1)MN\sigma^2}\sum_{c=1}^{n}\sum\sum_{l\neq m=1}^{M}(y_{cl} - \theta)(y_{cm} - \theta). \end{aligned} \tag{2.2.8}$$

In *srswor* of nM elements from the population of MN elements,

$$V_{wor}(\bar{y}) = \frac{N-n}{nNM}\sigma^2.$$

An estimator of $V_{wor}(\bar{y})$ based on without replacement sampling is

$$v_{wor}(\bar{y}) = \left(1 - \frac{n}{N}\right)\sum_{c=1}^{n}\sum_{l=1}^{M}(y_{cl} - \bar{y})^2/[nM(MN-1)], \tag{2.2.9}$$

(assuming $MN \approx MN - 1$). Again,

$$E_{true}[v_{wor}(\bar{y})] = \left(1 - \frac{n}{N}\right)\frac{\sigma^2}{(nM-1)}\left[1 - \frac{(N-n)\{1 + (M-1)\tau)\}}{Mm(N-1)}\right]. \tag{2.2.10}$$

Hence,

$$\text{deff }(\bar{y}, v_{wor}) = \frac{Var_{true}(\bar{y})}{E_{true}[v_{wor}(\bar{y})]}$$
$$= \frac{N(nM-1)\{1+(M-1)\tau\}}{nM(N-1)\{1-[(N-n)/(nN)(N-1)][1+(M-1)\tau]\}}. \tag{2.2.11}$$

If n is large, this gives approximately,

$$\text{deff }(\bar{y}, v_{wor}) = 1 + (M-1)\tau.$$

The above derivation of deff is based on the randomization due to sampling design. We now consider the corresponding result in a model-based setup. Consider the one-way random effect superpopulation model

$$y_{cl} = \theta + \alpha_c + \epsilon_{cl}, c = 1, \ldots, N; \quad l = 1, \ldots, M, \tag{2.2.12}$$

where θ is a constant overall effect, α_c is a random effect due to cluster, and ϵ_{cl} is a random error effect. We assume that α_c, ϵ_{cl} are mutually independent random variables with zero means and

$$V(\alpha_c) = \tau\sigma_0^2, \quad V(\epsilon_{cl}) = (1-\tau)\sigma_0^2.$$

The quantity τ can be interpreted as the intraclass correlation coefficient among the units belonging to the same cluster.

Here, $\hat{\theta} = \bar{y}$, the mean of the nM sampled elements. Under model (2.2.12),

$$Var_{true}(\bar{y}) = V\left(\frac{1}{nM}\sum_{c=1}^{n}\sum_{l=1}^{M} y_{cl}\right)$$
$$= \frac{1}{n^2M^2}[nM\sigma_0^2 + \sum_{c=1}^{n}\sum_{l\neq l'=1}^{M} \text{Cov }(y_{cl}, y_{cl'})]$$
$$= \frac{1}{n^2M^2}[Mn\sigma_0^2 + nM(M-1)\tau\sigma_0^2]$$
$$= \frac{\sigma_0^2[1+(M-1)\tau]}{nM}. \tag{2.2.13}$$

On the contrary, the IID model is

$$y_{cl} = \theta + e_{cl} \tag{2.2.14}$$

where e_{cl} are independently distributed random variables with mean 0 and variance σ_0^2. Hence

$$V_{IID}(\bar{y}) = \frac{\sigma_0^2}{nM}.$$

An unbiased estimator of $V_{IID}(\bar{y})$ under the IID assumption is

$$v_{IID}(\bar{y}) = \frac{1}{(nM-1)nM}\sum_{c=1}^{n}\sum_{l=1}^{M}(y_{cl}-\bar{y})^2. \tag{2.2.15}$$

$$\begin{aligned} E_{true}[v_{IID}(\bar{y})] &= \frac{1}{(nM-1)nM}\sum_{c=1}^{n}\sum_{l=1}^{M}E\{y_{cl}^2+\bar{y}^2-2\bar{y}(y_{cl})\}\\ &= \frac{1}{(nM-1)nM}\sum_{c=1}^{n}\sum_{l=1}^{M}\Bigg\{\theta^2+\sigma_0^2+\frac{\sigma_0^2}{nM}[1+(M-1)\tau]\\ &\quad +\theta^2-2\left(\frac{\sigma^2}{nM}+\frac{(M-1)\tau\sigma^2}{nM}+\theta^2\right)\Bigg\}\\ &= \frac{\sigma_0^2}{nM(nM-1)}\{nM-1-(M-1)\tau\}\\ &\approx \frac{\sigma_0^2}{nM} \end{aligned} \tag{2.2.16}$$

if n is large. Hence,

$$\text{deff}\,(\bar{y}, v_{IID}) \approx 1+(M-1)\tau$$

as in the case of design-based approach.

Example 2.2.5 Consider the linear regression model

$$E(Y|\mathbf{X}=\mathbf{x}) = \alpha+\mathbf{x}'\beta \tag{2.2.17}$$

where Y is the main variable of interest, $\mathbf{X}=(X_1,\ldots,X_k)'$, a set of k auxiliary variables. The ordinary least square (OLS) estimator of β which is best linear unbiased estimator (BLUE) when $V(Y|X)$ is a constant (model A) is

$$\hat{\beta}_{OLS} = \hat{\beta} = \mathbf{V}_{xx}^{-1}\mathbf{V}_{xy} \tag{2.2.18}$$

where

$$\mathbf{V}_{xx} = n^{-1}\sum_{i\in s}(\mathbf{x}_i-\bar{\mathbf{x}})(\mathbf{x}_i-\bar{\mathbf{x}})',\ \mathbf{V}_{xy} = n^{-1}\sum_{i\in s}(\mathbf{x}_i-\bar{\mathbf{x}})y_i,\ \bar{\mathbf{x}} = n^{-1}\sum_{i\in s}\mathbf{x}_i,$$

n being the size of the sample s, $\mathbf{x}_i=(x_{i1},\ldots,x_{ik})'$, y_i being observations on unit $i\in s$.

It is known that

$$v_{OLS}(\hat{\beta}) = \{n(n-k)\}^{-1}\left(\sum_{i\in s}e_i^2\right)\mathbf{V}_{xx}^{-1} \tag{2.2.19}$$

where $e_i = y_i - \bar{y} - (\mathbf{x}_i - \bar{\mathbf{x}})'\hat{\beta}$, $\bar{y} = \sum_{i \in s} y_i/n$, has expectation

$$E\{v_{OLS}(\hat{\beta})|A\} = V(\hat{\beta}|A). \tag{2.2.20}$$

Now, if heteroscedasticity is present, i.e., if $V(Y_i|\mathbf{X} = \mathbf{x}_i) = \sigma^2(\mathbf{x}_i)$ (model B), then $v_{OLS}(\hat{\beta})$ may be inconsistent for $V(\hat{\beta}|B)$ even under simple random sampling and

$$E[v_{OLS}(\hat{\beta})|B] \approx \{n(n-k)\}^{-1}\left\{\sum_{i \in s} \sigma^2(\mathbf{x}_i)\right\}\mathbf{V}_{xx}^{-1}. \tag{2.2.21}$$

Hence, in the multivariate case,

$$meff(\hat{\beta}, v_{OLS}(\hat{\beta})) = (E\{v_{OLS}(\hat{\beta})|B\})^{-1}V(\hat{\beta}|B). \tag{2.2.22}$$

For $k = 1$,

$$meff(\hat{\beta}, v_{OLS}(\hat{\beta})) = \frac{V(\hat{\beta}|B)}{E(v_{OLS}(\hat{\beta})|B)} \approx 1 + \rho C_\sigma C_x \tag{2.2.23}$$

where C_σ is the coefficient of variation (cv) of $\sigma^2(x_i)$, C_x is the cv of $(x_i - \bar{x})^2$, and ρ is the mutual correlation between x and y. This misspecification effect is due to the inconsistency of $v_{OLS}(\hat{\beta})$ under heteroscedastic model B. This inconsistency occurs even under simple random sampling and hence it is not proper to call Eq. (2.2.23) a design effect.

Now, under simple random sampling with replacement, a linearization estimator which is unbiased for $V(\hat{\beta}|B)$ in large samples is

$$v_B(\hat{\beta}) = n^{-2}\mathbf{V}_{xx}^{-1}\sum_{i \in s}(\mathbf{x}_i - \bar{\mathbf{x}})e_i^2(\mathbf{x}_i - \bar{\mathbf{x}})'\mathbf{V}_{xx}^{-1}. \tag{2.2.24}$$

Therefore, in large samples,

$$meff(\hat{\beta}, v_B(\hat{\beta})) = \{E\{v_B(\hat{\beta})|B)\}^{-1}V(\hat{\beta}|B) = \{V(\hat{\beta}|B)\}^{-1}V(\hat{\beta}|B) = \mathbf{I}_k. \tag{2.2.25}$$

Hence, there is no inconsistency under simple random sampling in this case.

2.3 Effect of a Complex Design on Confidence Interval for θ

Let $\tilde{\theta}$ be an unbiased estimator of θ under the hypothetical SRS design (IID model assumption) and v_0 an estimate of $Var_{SRS}(\tilde{\theta})$. Then

$$t_0 = \frac{\tilde{\theta} - \theta}{\sqrt{v_0}} \tag{2.3.1}$$

is approximately distributed as a $N(0, 1)$ variable and 95 % confidence interval for θ under the IID assumption is

$$C_0 = \{\theta : |\tilde{\theta} - \theta| \leq 1.96\sqrt{v_0}\}. \tag{2.3.2}$$

Our aim is to study the properties of C_0 under the effect of true complex design.

Under the true design $\tilde{\theta}$ may be assumed to be normal with mean θ and variance $Var_{true}(\tilde{\theta})$. Again, in large samples, $v_0 \approx E_{true}(v_0)$ so that, from (2.3.1),

$$t_0 \approx \frac{\tilde{\theta} - \theta}{\sqrt{E_{true}(v_0)}} = \frac{\tilde{\theta} - \theta}{\sqrt{Var_{true}(\tilde{\theta})}} \sqrt{\frac{Var_{true}(\tilde{\theta})}{E_{true}(v_0)}}.$$

Hence the distribution of t_0 under the true design would be approximately

$$t_0 \sim_{true} N\left(0, \frac{Var_{true}(\tilde{\theta})}{E_{true}(v_0)} = \text{deff}\ (\tilde{\theta}, v_0)\right). \tag{2.3.3}$$

Therefore, under the complex design, true 95 %-confidence interval for θ is

$$\tilde{\theta} \pm 1.96\sqrt{\text{deff}\ (\tilde{\theta}, v_0) \cdot E_{true}(v_0)}. \tag{2.3.4}$$

Hence, the actual coverage probability of a confidence interval obtained from the IID assumption would be different from its nominal value depending on the deff $(\hat{\theta}, v_0)$. If an estimated $\widehat{\text{deff}}\ (\hat{\theta}, v_0)$ is available then an adjusted confidence interval for θ with approximately 95 % coverage is

$$\tilde{\theta} \pm 1.96\sqrt{v_0 \cdot \widehat{\text{deff}}}. \tag{2.3.5}$$

Thus the deff$(\hat{\theta}, v_0)$ measures the inflation or deflation of IID-based pivotal statistic due to the use of true design. Table 2.1 adopted from Skinner et al. (1989) shows some such values.

We note that if deff $= 1$, C_0 has the same coverage probability as its nominal value. If deff $> (<) 1$, C_0 has coverage less (more) than its nominal value and hence its significance level is more (less) than the nominal significance level.

Suppose we want to test the null hypothesis $H_0 : \theta = \theta_0$ using data collected through a sampling design whose design effect is 1.5 and we shall use tests with nominal level 95 %. If we assume *srswr* or IID assumption, ignoring the true complex design we will use the confidence interval C_0 whose true coverage probability is 89 %, much below the nominal 95 % value. Therefore, in many cases H_0 will be

Table 2.1 Coverage of IID-based confidence intervals C_0

Design effect	Nominal level 95 %	Nominal level 99 %
0.9	96	99.3
1.0	95	99
1.5	89	96
2.0	83	93
2.5	78	90
3.0	74	86

rejected though we should have accepted the same in those cases. Test based on IID assumption is therefore conservative.

In practice, it is generally considered more desirable to have a conservative test (actual coverage probability less than the nominal coverage probability), than to use a liberal test. Therefore when using data from a complex survey, one should be careful of the large design effect. Even a design effect of 1.5 can make the actual significance level more than double its nominal value.

We now consider two definitions.

Definition 2.3.1 The *design factor* (deft) of a survey design is defined as

$$\text{deft} = \sqrt{\text{deff}}. \tag{2.3.6}$$

This is the appropriate inflation factor for standard errors and confidence intervals.

Definition 2.3.2 The *IID-effective sample size* or simply, *effective sample size* is defined as

$$n_e = \frac{n}{\text{deff}}, \tag{2.3.7}$$

and has the property that the SRS formula given by (2.3.2) becomes correct for the true design if n is replaced by n_e. (This definition is not to be confused with the Definition 1.2.3 which is concerned with the with-replacement sampling.)

Say

$$v_0 = \frac{A}{n}.$$

Then, if we replace n by n_e, v_0 becomes

$$v_0' = \frac{A}{n_e} = \left(\frac{A}{n}\right)(\text{deff}).$$

Therefore, if we use v_0' in place of v_0 in (2.3.1), and use the modified statistic

$$t_0' = (\hat{\theta} - \theta)/v_0'^{1/2}$$

the adjusted confidence interval $(\tilde{\theta} - 1.96\sqrt{v_0'}, \tilde{\theta} + 1.96\sqrt{v_0'})$ obtained from (2.3.2) has approximately the correct coverage probability.

2.4 Multivariate Design Effects

Suppose now that $\boldsymbol{\theta}$ is a $p \times 1$ vector, $\hat{\boldsymbol{\theta}}$ an estimator of $\boldsymbol{\theta}$ under the true design, and $\mathbf{V}_0$ a $p \times p$ matrix of estimators of covariance matrix of $\hat{\boldsymbol{\theta}}$ derived under the IID assumption or equivalently under the simple random sampling with replacement (SRS) assumption. The estimator $\mathbf{V}_0$ is also derived under the IID assumption. We may define the *multivariate design effects matrix* (in Skinner's sense) of the estimator-pair $\hat{\boldsymbol{\theta}}$ and $\mathbf{V}_0$ as

$$\text{deff}\ (\hat{\boldsymbol{\theta}}, \mathbf{V}_0) = (E_{true}(\mathbf{V}_0))^{-1} Cov_{true}(\hat{\boldsymbol{\theta}}). \tag{2.4.1}$$

The eigenvalues of this matrix $\delta_1 \geq \delta_2 \geq \cdots \geq \delta_p$ are called the *generalized design effects* (in Skinner's sense) and has the property that δ_1, δ_p denote the bounds for the univariate design effects of any linear combination $\mathbf{c}'\hat{\boldsymbol{\theta}}$ of elements of $\hat{\boldsymbol{\theta}}$,

$$\begin{aligned} \delta_1 &= \max\ \text{deff}_{\mathbf{c}}(\mathbf{c}'\hat{\boldsymbol{\theta}}, \mathbf{c}'\mathbf{V}_0\mathbf{c}), \\ \delta_p &= \min\ \text{deff}_{\mathbf{c}}(\mathbf{c}'\hat{\boldsymbol{\theta}}, \mathbf{c}'\mathbf{V}_0\mathbf{c}). \end{aligned} \tag{2.4.2}$$

In the special case when the deff $(\hat{\boldsymbol{\theta}}, \mathbf{V}_0)$ is a $p \times p$ identity matrix, $\delta_1 = \cdots = \delta_p = 1$ so that the univariate design effects of all linear combinations of elements of $\hat{\boldsymbol{\theta}}$ are unity.

Note 2.4.1 The calculation of the design effect involves variance estimation and hence requires second-order inclusion probabilities. It also depends on how auxiliary information is used, and needs to be estimated one at a time for different scenarios. Wu et al. (2010) present bootstrap procedures (discussed in Sect. 2.5.7) for constructing pseudo empirical likelihood ratio confidence intervals for finite population parameters. The proposed method bypasses the need for design effects and is valid under general single-stage unequal probability sampling designs with small sampling fractions. Different scenarios in using auxiliary information are handled by simply including the same type of benchmark constraints with the bootstrap procedures.

Since estimation of variance of $\hat{\theta}$ (covariance matrix of $\hat{\theta}$, when θ is a vector parameter) is of major interest in this context we shall in the next section consider different methods of estimation of variance of estimators, particularly for nonlinear estimators. The estimation procedures are very general, they do not depend on any distributional assumption and are, therefore, nonparametric in nature.

2.5 Nonparametric Methods of Variance Estimation

Modern complex surveys often involve estimation of nonlinear functions, like population ratio, difference of ratios, regression coefficient, correlation coefficient, etc. Therefore, the usual formulae for unbiased estimation of sampling variance of simple (linear) estimators of, say, totals and means are inadequate for such surveys. There are two approaches to the estimation of variance of a nonlinear estimator. One is linearization, in which the nonlinear estimator is approximated by a linear one for the purpose of variance estimation. The second is replication in which several estimators of the population parameter are derived from different comparable parts of the original sample. The variability of these estimators is then used to estimate the variance of the parameter estimator.

We review these results in this chapter. Section 2.5.1 considers in detail a simple method of estimation of variance of a linear statistic. In Sects. 2.5.2–2.5.7 we consider Taylor series linearization procedure, random group (RG) method, balanced repeated replication (BRR), jackknife (JK) procedure, JK repeated replication, and bootstrap (BS) techniques of variance estimation. All these procedures (except the bootstrap resampling) have been considered in detail in Wolter (1985). Sarndal et al. (1992) have also considered in detail the problem of variance estimation in their wonderful book. We do not consider estimation of superpopulation-based variance, the topic being outside the scope of this book. Interested readers may refer to Mukhopadhyay (1996) for a review in this area. We review these results in this section.

2.5.1 A Simple Method of Estimation of Variance of a Linear Statistic

In a stratified three-stage sampling consider a linear statistic of the form

$$\hat{\theta} = \sum_{h=1}^{H} \sum_{a=1}^{n_h} \sum_{b=1}^{m_{ha}} \sum_{c=1}^{k_{hab}} u_{habc} \tag{2.5.1}$$

where u_{habc} is the value associated with the cth unit (ultimate-stage unit) belonging to the bth sampled ssu (second-stage unit) in the ath sampled fsu (first-stage unit) belonging to the hth stratum. For example, $\hat{\theta}$ may be the estimator of a population mean of a variable 'y', when

$$u_{habc} = \frac{y_{habc}}{N\pi_{habc}} \tag{2.5.2}$$

where N is the number of ultimate units and π_{habc} is the inclusion probability of the unit ($habc$).

A simple unbiased estimator of the design variance of $\hat{\theta}$ can be obtained under the following assumptions:

(1) Samples are selected independently from one stratum to the other.
(2) The n_h sampled psu's within stratum h are selected with replacement (wr). (At each of the n_h draws there is a finite probability p_{ha} of selecting the ath psu, $\sum_{a=1}^{N_h} p_{ha} = 1$, where N_h is the total number of psu's in the hth stratum.)
(3) $n_h \geq 2$.

We may rewrite Eq. (2.5.1) as

$$\hat{\theta} = \sum_{h=1}^{H} \sum_{a=1}^{n_h} u_{ha} \tag{2.5.3}$$

where

$$u_{ha} = \sum_{b=1}^{m_{ha}} \sum_{c=1}^{k_{hab}} u_{habc}. \tag{2.5.4}$$

By assumption (2), the variables $u_{h1}, \ldots, u_{hn_h}$ are identically and independently distributed (IID) random variables within stratum h and therefore, by virtue of assumption (1),

$$\text{Var}\,(\hat{\theta}) = \sum_{h=1}^{H} n_h \text{ Var}\,(u_{ha}). \tag{2.5.5}$$

Therefore, by (1) and (3), an unbiased estimator of Var $(\hat{\theta})$ is

$$v(\hat{\theta}) = \sum_{h=1}^{H} n_h \frac{1}{n_h - 1} \sum_{a=1}^{n_h} (u_{ha} - \bar{u}_h)^2, \tag{2.5.6}$$

where $\bar{u}_h = \sum_{a=1}^{n_h} u_{ha}/n_h$.

The estimator $v(\hat{\theta})$ can be readily computed from the aggregate quantities u_{ha} formed from the ultimate units. If the psu's are selected with replacement, one need not care about in how many subsequent stages sampling is carried out and/or if the sampling at the ultimate stage is by systematic sampling or any other procedure.

For the special case where $n_h = 2 \forall\, h$,

$$v(\hat{\theta}) = \sum_{h=1}^{H} (u_{h1} - u_{h2})^2. \tag{2.5.7}$$

Even in surveys with $n_h > 2$, the ultimate sampled units can often be grouped in two groups (on the basis of some criteria), the assumptions (1) and (2) made and the formula (2.5.7) applied. The groups are often called *Keyfitz groups* after Keyfitz (1957). The grouping of clusters, however, lead to some loss of efficiency.

The assumption (1) is often valid. In case $n_h = 1$ for some strata, such strata are often collapsed to form the new strata for which $n_h \geq 2$. Defining $v(\hat{\theta})$ with respect to the new strata then gives a conservative variance estimator.

Assumption (2) is almost always violated since the n_h psu's are generally selected by some without replacement procedure. In this case an unbiased variance estimator of $\hat{\theta}$ often involves complex formula with components for each stage of sampling. Some simplified procedures for the case $n_h = 2$ have been proposed by Durbin (1967) and Rao and Lanke (1984). One approximation is based on the assumption that the n_h psu's within stratum h are selected by srswor ($h = 1, \dots, H$). In this case an estimator of $\text{Var}(\hat{\theta})$ is

$$v_{wor}(\hat{\theta}) = \sum_{h=1}^{H} \left(1 - \frac{n_h}{N_h}\right) \frac{n_h}{n_h - 1} \sum_{a=1}^{n_h} (u_{ha} - \bar{u}_h)^2. \tag{2.5.8}$$

obtained by inserting a finite population correction factor in (2.5.6). Often the sampling fraction n_h/N_h is small and the difference between $v(\hat{\theta})$ and $v_{wor}(\hat{\theta})$ is negligible. In any case, $v(\hat{\theta})$ is often a conservative estimator.

In analytic surveys, where the parameter of interest is often a superpopulation model parameter, the finite population correction n_h/N_h is inappropriate and $v(\hat{\theta})$ is to be used.

We shall now show that under a measurement error model, the estimator $v(\hat{\theta})$ is a better estimator of the total variance rather than $v_{wor}(\hat{\theta})$. Consider the model

$$u_{ha} = U_{ha} + \epsilon_{ha} \tag{2.5.9}$$

where U_{ha} are the true values and ϵ_{ha} are random variables distributed independently with mean 0 and variance σ_h^2. The errors ϵ_{ha} arise, for example, from interviewers' errors and other non-sampling errors. Now, by (2.5.3),

$$\mathcal{V}_p(\hat{\theta}) = \mathcal{V}_p\left(\sum_{h=1}^{H}\sum_{a=1}^{n_h} U_{ha}\right) + \mathcal{V}_p\left(\sum_{h=1}^{H}\sum_{a=1}^{n_h} \epsilon_{ha}\right) \tag{2.5.10}$$

where $\mathcal{V}_p$ means variance due to joint randomization of sampling design and measurement error distribution. Now, if the psu's are selected by *srswosr*,

$$\begin{aligned}\mathcal{V}_p\left(\sum_{h=1}^{H}\sum_{a=1}^{n_h} U_{ha}\right) &= \sum_{h=1}^{H} n_h^2 \left(\frac{N_h - n_h}{N_h n_h}\right) S_h^2 \\ &= \sum_{h=1}^{H} n_h \left(1 - \frac{n_h}{N_h}\right) S_h^2 \end{aligned} \tag{2.5.11}$$

where

$$S_h^2 = (N_h - 1)^{-1} \sum_{a=1}^{N_h} (U_{ha} - \bar{U}_h)^2, \ \bar{U}_h = \sum_{a=1}^{Nh} U_{ha}/N_h.$$

Also

$$\begin{aligned} \mathcal{V}_p\left(\sum_{h=1}^{H}\sum_{a=1}^{n_h}\epsilon_{ha}\right) &= \mathcal{V}\left(\sum_{h=1}^{H}\sum_{a=1}^{n_h}\epsilon_{ha}\right) \\ &= \sum_{h=1}^{H} n_h\sigma_h^2, \end{aligned} \tag{2.5.12}$$

where $\mathcal{V}(.)$ denotes variance wrt error distribution and σ_h^2 denotes the fixed variance of ϵ_{ha}. Hence,

$$\mathcal{V}_p(\hat{\theta}) = \sum_{h=1}^{H} n_h \left[\left(1 - \frac{n_h}{N_h}\right) S_h^2 + \sigma_h^2\right]. \tag{2.5.13}$$

Here, from (2.5.6)

$$Ev(\hat{\theta}) = \sum_{h=1}^{H} n_h \left(S_h^2 + \sigma_h^2\right). \tag{2.5.14}$$

From (2.5.8)

$$Ev_{wor}(\hat{\theta}) = \sum_{h=1}^{H} n_h \left(1 - \frac{n_h}{N_h}\right)\left(S_h^2 + \sigma_h^2\right). \tag{2.5.15}$$

Therefore,

$$E[v_{wor}(\hat{\theta})] \le \mathcal{V}_p(\hat{\theta}) \le E(v(\hat{\theta})). \tag{2.5.16}$$

The estimator $v(\hat{\theta})$ is preferred, since it is a conservative estimator.

In the multivariate case, where $\hat{\boldsymbol{\theta}} = (\hat{\theta}_1, \ldots, \hat{\theta}_p)'$, we can write $\hat{\boldsymbol{\theta}}$ as

$$\hat{\boldsymbol{\theta}} = \sum_h \sum_a \sum_b \sum_c \mathbf{u}_{habc} \tag{2.5.17}$$

where $\mathbf{u}_{habc}$ is a vector of values associated with the unit '*habc*'. Corresponding to $v(\hat{\theta})$ in (2.5.6) in the univariate case, we have the covariance matrix estimator

$$v(\hat{\boldsymbol{\theta}}) = \sum_{h=1}^{H} \frac{n_h}{n_h - 1} \sum_{a=1}^{n_h} (\mathbf{u}_{ha} - \bar{\mathbf{u}}_h)(\mathbf{u}_{ha} - \bar{\mathbf{u}}_h)' \tag{2.5.18}$$

where

$$\mathbf{u}_{ha} = \sum_b \sum_c \mathbf{u}_{habc}, \quad \bar{\mathbf{u}}_h = \sum_{a=1}^{n_h} \mathbf{u}_{ha}/n_h.$$

Clearly, assumptions (1) and (2) above are of vital importance and the procedure can be applied to any sampling design based on sampling at any arbitrary number of stages. The above results are derived following Wolter (1985) and Skinner et al (1989).

2.5.2 *Linearization Method for Variance Estimation of a Nonlinear Estimator*

We now consider the problem of estimation of variance of a nonlinear estimator, like ratio estimator, regression estimator. In the estimation of variance of a nonlinear estimator we adopt the method based on Taylor series expansion. The method is also known as *linearization method*.

Let $\mathbf{Y} = (Y_1, \ldots, Y_p)'$ where Y_j is a population total (or mean) of the jth variable and let $\hat{\mathbf{Y}} = (\hat{Y}_1, \hat{Y}_2, \ldots, \hat{Y}_p)'$ where $\hat{Y}_j$ is a linear estimator of Y_j. We consider a finite population parameter $\theta = f(\mathbf{Y})$ with a consistent estimator $f(\hat{\mathbf{Y}})$. A simple example is a population subgroup ratio, $\theta = Y_1/Y_2$ with $\hat{\theta} = \hat{Y}_1/\hat{Y}_2$, Y_1, Y_2 are population totals for groups 1 and 2.

Suppose that continuous second-order derivatives exist for the function $f(\mathbf{Y})$. Now,

$$f(\hat{\mathbf{Y}}) = f(\mathbf{Y}) + \sum_{j=1}^{p} (\hat{Y}_j - Y_j)\frac{\partial f}{\partial Y_j}$$
$$+ \sum \sum_{j,k=1}^{p} (\hat{Y}_j - Y_j)(\hat{Y}_k - Y_k)\frac{\partial f}{\partial Y_j}\frac{\partial f}{\partial Y_k} + \ldots. \tag{2.5.19}$$

Thus using only the linear terms of the Taylor series expansion, we have an approximate expression

$$\hat{\theta} - \theta = \sum_{j=1}^{p} (\hat{Y}_j - Y_j)\frac{\partial f}{\partial Y_j}. \tag{2.5.20}$$

Using the linearized equation (2.5.20), an approximate expression for variance of $\hat{\theta}$ is

$$E(\hat{\theta} - \theta)^2 = V(\hat{\theta}) \approx \sum_{j=1}^{p} \left(\frac{\partial f}{\partial Y_j}\right)^2 V(\hat{Y}_j) + \sum \sum_{j \neq k=1}^{p} \left(\frac{\partial f}{\partial Y_j}\right)\left(\frac{\partial f}{\partial Y_k}\right) \text{Cov}\,(\hat{Y}_j, \hat{Y}_k). \tag{2.5.21}$$

We have thus reduced the variance of a nonlinear estimator to the function of the variance and covariance of p linear estimators $\hat{Y}_j$. A variance estimator $v(\hat{\theta})$ is obtained from (2.5.21) by substituting the variance and covariance estimators $v(\hat{Y}_j, \hat{Y}_k)$ for the corresponding parameters $V(\hat{Y}_j, \hat{Y}_k)$. The resulting variance estimator is a first-order Taylor series approximation. The justification for ignoring the remaining higher order terms has to be sought from practical experience derived from various complex surveys in which sample sizes are sufficiently large. Krewski and Rao (1981) have shown that the linearization estimators are consistent.

Basic principles of the linearization method for the variance estimation of a nonlinear estimator under complex sampling designs are due to Keyfitz (1957) and other. A criticism against the method is about the convergence of the Taylor series used to develop (2.5.20). For ratio estimator Koop (1972) gave a simple example where the convergence condition is violated. Again, for complex estimators, the analytic partial differentiation needed to derive the linear substitute has been found to be intractable. Woodruff and Causey (1976) describes a solution to this problem that uses a numerical procedure to obtain the necessary partial derivative. Binder (1983) provides a general approach to the analytic derivation of variance estimators for linear Taylor series approximations for a wide class of estimators. Empirical evidences have shown, however, that the linearization variance estimators are generally of adequate accuracy, particularly, when the sample size is large. The approximation may be unreliable in the case of highly skewed population.

Example 2.5.1 Ratio Estimator: Let

$$\mathbf{Y} = (Y_1, Y_2)', \theta = f(\mathbf{Y}) = \frac{Y_1}{Y_2}, \hat{\theta} = f(\hat{\mathbf{Y}}) = \frac{\hat{Y}_1}{\hat{Y}_2},$$

$$\frac{\partial f(\mathbf{Y})}{\partial Y_1} = \frac{1}{Y_2}, \frac{\partial f(\mathbf{Y})}{\partial Y_2} = -\frac{Y_1}{Y_2^2}, \frac{\partial f(\mathbf{Y})}{\partial Y_1}\frac{\partial f(\mathbf{Y})}{\partial Y_2} = -\frac{Y_1}{Y_2^3}.$$

Hence,

$$\begin{aligned} V(\hat{\theta}) &= \frac{V(\hat{Y}_1)}{Y_2^2} + \frac{Y_1^2 V(\hat{Y}_2)}{Y_2^4} - \frac{2Y_1}{Y_2^3}\text{Cov}(\hat{Y}_1, \hat{Y}_2) \\ &= \frac{Y_1^2}{Y_2^2}\left[\frac{V(\hat{Y}_1)}{Y_1^2} + \frac{V(\hat{Y}_2)}{Y_2^2} - \frac{2\,\text{Cov}(\hat{Y}_1, \hat{Y}_2)}{Y_1 Y_2}\right]. \end{aligned} \tag{2.5.22}$$

Example 2.5.2 Combined and Separate Ratio Estimator in Stratified Two-Stage Sampling: The population consists of H strata, the hth stratum containing N_h clusters (which consist of M_h elements, $\sum_{h=1}^{H} M_h = M$). A first-stage sample of $n_h (\geq 2)$ clusters is drawn from the hth stratum and a second-stage sample of m_h elements is drawn from the n_h sampled clusters, $m = \sum_h m_h$. The quantity m_h is a random variable.

We assume that the sampling design is self-weighing, i.e., inclusion probability of each of M elements in the population is a constant over the strata and adjustments for nonresponse is not necessary. Let

m_{ha} = number of elements in the ath cluster in the hth stratum in the sample;
$y_{ha} = \sum_{b=1}^{m_{ha}} y_{hab}$ = sum of the response variable y over the m_{ha} elements in the ath cluster belonging to the hth stratum in the sample ($a = 1, \ldots, n_h$; $h = 1, \ldots, H$).

Let Y_{ha}, M_{ha} denote the respective population totals. A combined ratio estimator of population ratio (mean per element)

$$r = \frac{\sum_{h=1}^{H} \sum_{a=1}^{N_h} Y_{ha}}{\sum_{h=1}^{H} \sum_{a=1}^{N_h} M_{ha}} = \frac{T}{M} \tag{2.5.23}$$

is

$$\hat{r}_{com} = \frac{\sum_{h=1}^{H} \sum_{a=1}^{n_h} y_{ha}}{\sum_{h=1}^{H} \sum_{a=1}^{n_h} m_{ha}} = \frac{\sum_{h=1}^{H} y_h}{\sum_{h=1}^{H} m_h} = \frac{y}{m} \tag{2.5.24}$$

where $y = \sum_{h=1}^{H} y_h$, $y_h = \sum_{a=1}^{n_h} y_{ha}$ is the sample sum of the response variable for the hth stratum and $m = \sum_{h=1}^{H} m_h$ and m_h the number of sample elements in the hth stratum. For a binary 0–1 variable, $r = P = M_1/M$, the population proportion where M_1 is the count of elements each having value 1. In estimator (2.5.24) not only the quantities y_{ha} vary, but also the quantities m_{ha} and m in the denominator. Hence, $\hat{r}_{com} = \hat{r}$ is a nonlinear estimator.

A separate ratio estimator is a weighted sum of stratum sample ratios, $\hat{r}_h = y_h/m_h$ which themselves are ratio estimators of the population stratum ratios, $r_h = \sum_{a=1}^{N_h} Y_{ha} / \sum_{a=1}^{N_h} M_{ha}$. Thus,

$$\hat{r}_{sep} = \sum_{h=1}^{H} W_h r_h \tag{2.5.25}$$

with $W_h = M_h/M$. A linearized variance for the combined ratio estimator $\hat{r}_{com} = \hat{r} = y/x$ in (2.5.24) is, according to (2.5.22),

$$V(\hat{r}) = r^2 \left[\frac{V(y)}{y^2} + \frac{V(m)}{m^2} - \frac{2\,\mathrm{Cov}\,(y, m)}{ym} \right]. \tag{2.5.26}$$

Hence, an estimator of $V(\hat{r})$ can be written as

$$v_{des} = \hat{r}^2 [y^{-2} \hat{V}(y) + m^{-2} \hat{V}(m) - 2m^{-1} y^{-1} \widehat{Cov}(y, m)] \tag{2.5.27}$$

as the design-based variance estimator of $\hat{r}$ is based on the linearization method. The estimators $\hat{V}$'s depend on the sampling design.

The variance estimator (2.5.27) is a large-sample approximation in that good performance can be expected if not only the number of sampled elements is large, but also the number of sampled clusters is so. In case of a small number of sampled clusters the variance estimator can be unstable.

The variance estimator v_{des} is consistent if $\hat{V}(y)$, $\hat{V}(m)$, $\hat{V}(m, y)$ are consistent estimators. The cluster sample sizes should not vary too much for the reliable performance of the approximate variance estimator (2.5.27). The method can be safely used if the coefficient of variation of m_{ha} is less than 0.2. If the cluster sample sizes m_{ha} are all equal, $\hat{V}(m) = 0$, $\hat{V}(y, m) = 0$, and $\hat{V}(\hat{r}) = \hat{V}(y)/m^2$. For a 0–1 binary response variable and for sampling under IID conditions, $\hat{r} = p = m_1/m$, sample proportion, where m_1 is the number of elements in the sample each having value 1 and m is the number of elements in the sample. Assuming m is a fixed quantity, the variance estimator (2.5.27) reduces to the binomial variance estimator $v_{des}(p) = v_{bin}(\hat{p}) = p(1 - p)/n$.

Assuming that $n_h(\geq 2)$ clusters are selected by srswr from each stratum we obtain relatively simple variance and covariance estimators in (2.5.27). We have

$$\hat{V}(y) = \sum_h n_h s^2_{yh}, \quad \hat{V}(m) = \sum_h n_h s^2_{mh},$$

$$\hat{V}(y, m) = \sum_h n_h s^2_{y,mh}$$

where

$$\begin{aligned} s^2_{yh} &= (n_h - 1)^{-1} \textstyle\sum_{a=1}^{n_h} (y_{ha} - \bar{y}_h)^2, \\ s^2_{y,mh} &= (n_h - 1)^{-1} \textstyle\sum_{a=1}^{n_h} (y_{ha} - \bar{y}_h)(m_{ha} - \bar{m}_h)^2, \end{aligned} \tag{2.5.28}$$

$\bar{y}_h = \sum_a y_{ha}/m_h$ and s^2_{mh}, $\bar{m}_h$ have similar meanings. Note that by assuming srswr of clusters we only estimate the between-cluster components of variances and do not account for the within-cluster variances. As such the variance estimator (2.5.27) obtained using (2.5.28) will be an underestimate of the true variance. This bias is negligible if the first-stage sampling fraction n_h/N_h in each stratum is small. This happens if N_h is large in each stratum.

2.5.3 *Random Group Method*

The random group (RG) method was first developed at the US Bureau of Census. Here, an original sample and other $k(\geq 2)$ samples, also called random groups, are drawn from the population, usually using the same sampling design. The task of these last k random samples or random groups is to provide an estimate for the variance of an estimator of population parameter of interest based on the original sample. We shall distinguish two cases:

(a) *Samples or Random Groups are mutually independent*: Let $\hat{\theta}, \hat{\theta}_1, \ldots, \hat{\theta}_k$ be the estimators obtained from the original sample and k random groups respectively, all the estimators using the same estimating procedure. Here $\hat{\theta}_1, \ldots, \hat{\theta}_k$ are mutually independent. We want to estimate Var($\hat{\theta}$), variance of the estimator $\hat{\theta}$ based on the original sample.

The RG estimate of θ is $\bar{\hat{\theta}} = \sum_i \hat{\theta}_i/k$. If $\hat{\theta}$ is linear, $\bar{\hat{\theta}} = \hat{\theta}$. Now an estimate of Var ($\bar{\hat{\theta}}$) is

$$v(\bar{\hat{\theta}}) = \frac{1}{k(k-1)} \sum_{i=1}^{k} (\hat{\theta}_i - \bar{\hat{\theta}})^2. \tag{2.5.29}$$

Note that for the above formula to hold it is neither required to assume that all $\hat{\theta}_i$'s have the same variance nor to assume that $\hat{\theta}_i$ are independent. It is sufficient to assume that all $\hat{\theta}_i$'s have finite variances and that they are pairwise uncorrelated.

Now, by Cauchy–Schwarz inequality

$$0 \le \left[\sqrt{Var(\bar{\hat{\theta}})} - \sqrt{Var(\hat{\theta})}\right]^2 \le Var[\bar{\hat{\theta}} - \hat{\theta}] \tag{2.5.30}$$

and Var($\bar{\hat{\theta}} - \hat{\theta}$) is generally small relative to both Var($\bar{\hat{\theta}}$) and Var($\hat{\theta}$). Thus, the two variances are usually of similar magnitude.

To estimate Var ($\hat{\theta}$) one may use either $v_1(\hat{\theta}) = v(\bar{\hat{\theta}})$ or

$$v_2(\hat{\theta}) = \frac{1}{k(k-1)} \sum_{i=1}^{k} (\hat{\theta}_i - \hat{\theta})^2. \tag{2.5.31}$$

Note that $v_2(\hat{\theta})$ does not depend on $\bar{\hat{\theta}}$. When the estimator of θ is linear $v_1(\hat{\theta})$ and $v_2(\hat{\theta})$ are identical. For nonlinear estimators we have,

$$\sum_{i=1}^{k} (\hat{\theta}_i - \hat{\theta})^2 = \sum_{i=1}^{k} (\hat{\theta}_i - \bar{\hat{\theta}})^2 + k(\bar{\hat{\theta}} - \hat{\theta})^2. \tag{2.5.32}$$

Thus,

$$v_1(\hat{\theta}) \le v_2(\hat{\theta}). \tag{2.5.33}$$

If a conservative estimator of Var ($\hat{\theta}$) is desired, $v_2(\hat{\theta})$ is, therefore, preferable to $v_1(\hat{\theta})$. However, as noted above, Var($\bar{\hat{\theta}} - \hat{\theta}$) $= E(\bar{\hat{\theta}} - \hat{\theta})^2$ will be unimportant in many complex surveys and there should be little difference between v_1 and v_2. It has been shown that the bias of v_1 as an estimator of Var($\hat{\theta}$) is less than or equal to the bias of v_2.

Inferences about parameter θ are usually based on normal theory or Student's t distribution. The results are stated in the following theorem.

Theorem 2.5.1 *Let $\hat{\theta}_1, \ldots, \hat{\theta}_k$ be independently and identically distributed (iid) $N(\theta, \sigma^2)$ variables. Then*

(i) $\frac{\sqrt{k}(\bar{\hat{\theta}}-\theta)}{\sigma}$ *is distributed as a N(0,1) variable. (Obvious modification will follow if $\hat{\theta}_i$'s have different but known variances.)*

(ii) $\frac{\sqrt{k}(\bar{\hat{\theta}}-\theta)}{\sqrt{v_1(\hat{\theta})}}$ *is distributed as a $t_{(k-1)}$ variable.*

If Var $(\bar{\hat{\theta}}) = \sigma^2/k$ is known, or k is large, $100(1-\alpha)$ % confidence interval for θ is

$$\bar{\hat{\theta}} \pm \tau_{\alpha/2}\sigma/\sqrt{k}$$

where $\tau_{\alpha/2}$ is the upper $100(\alpha/2)$ percentage point of the $N(0, 1)$ distribution. When Var $\hat{\theta}_i$ is not known or k is not large $100(1-\alpha)$ % confidence interval for θ is

$$\bar{\hat{\theta}} \pm t_{k-1;\alpha/2}\sqrt{v(\bar{\hat{\theta}})}$$

where $t_{k-1;\alpha/2}$ is the upper $100(\alpha/2)$ percentage point of the $t_{(k-1)}$ distribution.

(b) *Random groups are not independent*: In practical sample surveys, samples are often selected as a whole using some form of without replacement sampling instead of in the form of a series of independent random groups. Random groups are now formed by randomly dividing the parent sample into k groups. The random group estimators $\hat{\theta}_i$'s are no longer uncorrelated because sampling is performed without replacement. Theorem 2.5.1 is no longer valid. Here also $\bar{\hat{\theta}}$, $v_1(\hat{\theta})$, $v_2(\hat{\theta})$ as defined above are used for the respective purposes. However, because the random group estimators are not independent, $v_1(\hat{\theta}) = v(\bar{\hat{\theta}})$ is not an unbiased estimator of Var $(\bar{\hat{\theta}})$. The following theorem describes some properties of $v(\bar{\hat{\theta}})$.

Theorem 2.5.2 *If $E(\hat{\theta}_i) = \mu_i (i = 1, \ldots, k)$,*

$$E\{v(\bar{\hat{\theta}})\} = Var\ (\bar{\hat{\theta}}) + \frac{1}{k(k-1)}\left[\sum_{i=1}^{k}(\mu_i - \bar{\mu})^2 - 2\sum\sum_{i<j=1}^{k} \text{Cov}(\hat{\theta}_i, \hat{\theta}_j)\right]. \tag{2.5.34}$$

Proof It is obvious that

$$E(\bar{\hat{\theta}}) = \bar{\mu} = \sum_{i=1}^{k} \mu_i/k.$$

Again,

$$\begin{aligned} v(\bar{\hat{\theta}}) &= \frac{1}{k(k-1)}\left[\sum_{i=1}^{k} \hat{\theta}_i^2 - k\bar{\hat{\theta}}^2\right] \\ &= \bar{\hat{\theta}}^2 - \frac{2}{k(k-1)}\sum\sum_{i<j=1}^{k} \hat{\theta}_i\hat{\theta}_j. \end{aligned}$$

Now,

$$E[\bar{\hat{\theta}}^2] = Var(\bar{\hat{\theta}}) + \bar{\mu}^2$$

and

$$E[\hat{\theta}_i\hat{\theta}_j] = \text{ Cov }(\hat{\theta}_i, \hat{\theta}_j) + \mu_i\mu_j.$$

Therefore, the result follows. □

Theorem 2.5.2 gives the bias of $v(\bar{\hat{\theta}})$ as an estimator of Var $(\bar{\hat{\theta}})$. For large populations and small sampling fractions, the term $2\sum\sum_{i<j}$ Cov $(\hat{\theta}_i, \hat{\theta}_j)$ will tend to be a relatively small negative quantity. The quantity

$$\frac{1}{k(k-1)}\sum_{i=1}^{k}(\mu_i - \bar{\mu})^2$$

will also be relatively small if $\mu_i \approx \bar{\mu}(i = 1, \ldots, k)$. Thus the bias of $v(\bar{\hat{\theta}})$ will be unimportant in many large-scale sample surveys and will tend to be slightly positive. Work by Frankel (1971) suggests that the bias of $v(\bar{\hat{\theta}})$ is often small and decreases as the size of the groups increase (or equivalently as the number of groups decreases).

The RG procedure was initiated by Mahalanobis (1946) and Deming (1956). Mahalanobis called the various samples as *Interpenetrating samples*, Deming proposed the term *replicated samples*. They selected k independent samples using the same sampling design and used the estimator of the type (2.5.29) to estimate the variance of the overall estimator. In RG method, the major difference is that the replicates are not necessarily formed independently.

It has been found that if $\hat{\theta}_1, \ldots, \hat{\theta}_k$ are independently and identically distributed random variables, then coefficient of variation (cv) of the RG estimator $v(\bar{\hat{\theta}})$, which measures its stability is

$$\begin{aligned} cv[v(\bar{\hat{\theta}})] &= [Var\{v(\bar{\hat{\theta}})\}]^{1/2}/Var(\bar{\hat{\theta}}) \\ &= \left\{\frac{\beta_4(\hat{\theta}_1)-(k-3)/(k-1)}{k}\right\}^{1/2}. \end{aligned}$$

The cv is thus an increasing function of kurtosis $\beta_4(\hat{\theta}_1)$ of the distribution of $\hat{\theta}_1$ and a decreasing function of k for a wide range of complex surveys (when N, n are large and $n/N \approx 0$, the result holds even for nonindependent RG's). As a result, the larger the number (k) of groups, the higher the precision, though computational cost will increase at the same time. The optimum value of k is a trade-off between cost and precision. The RG method is suitable for surveys using a large number of primary-stage units (psu's) where many psu's are selected per stratum.

2.5.4 Balanced Repeated Replications

The method of balanced half-sample repeated replications (BRR) has proved very useful for surveys in which two primary-stage units (psu's) are selected per stratum. Following Plackett and Burman (1946), McCarthy (1966, 1969a, b) introduced the concept of BRR, also known as balanced half-samples, balanced fractional samples, and pseudoreplication.

Suppose that two units are selected by *srswr* from each of H strata for estimating $\bar{Y} = \sum_h W_h \bar{Y}_h$ where $W_h = N_h/N$, $N_h(n_h)$ is the stratum population (sample) size, $\bar{Y}_h = \sum_{i=1}^{N_h} Y_{hi}/N_h$, Y_{hi}, being the value of 'y' on the ith unit in stratum h.

By selecting one unit from the sampled units in each stratum at random we can form 2^H sets of two half-samples (HS's) each such that each set forms a complete replicate.

In a set α denote the two HS's as S_α, $S_{\alpha'}$ with the corresponding estimates $\bar{y}_{st,\alpha} = \sum_h W_h y_{h1,\alpha}$ and $\bar{y}_{st,\alpha'} = \sum_h W_h y_{h2,\alpha}$. (A more complicated notation is given below). The customary estimator is

$$\bar{y}_{st(\alpha)} = \frac{\bar{y}_{st,\alpha} + \bar{y}_{st,\alpha'}}{2}.$$

The αth replicate estimate of $V(\bar{y}_{st})$ is

$$\begin{aligned} v_\alpha(\bar{y}_{st}) &= \tfrac{1}{2}[(\bar{y}_{st,\alpha} - \bar{y}_{st(\alpha)})^2 + (\bar{y}_{st,\alpha'} - \bar{y}_{st(\alpha)})^2] \\ &= \tfrac{1}{4}(\bar{y}_{st,\alpha} - \bar{y}_{st,\alpha'})^2. \end{aligned} \tag{2.5.35}$$

The estimator v_α is unbiased for $V(\bar{y}_{st})$ (Exercise 2.2).

Now,

$$\bar{y}_{st,\alpha} = \sum_h W_h y_{h1} = \sum_h W_h \{Y_{h1}\delta_{h1\alpha} + Y_{h2}\delta_{h2\alpha}\} \tag{2.5.36}$$

wherein we denote the values of y on the two units selected from the hth stratum as Y_{h1}, Y_{h2}, respectively, in some well-defined manner. The term Y_{h1} becomes y_{h1} if the corresponding unit goes to S_α. Also,

$$\begin{aligned} \delta_{h1\alpha} &= 1(0) \text{ if the unit } (h, 1) \in S_\alpha \text{ (otherwise)} \\ \delta_{h2\alpha} &= 1 - \delta_{h1\alpha}. \end{aligned} \tag{2.5.37}$$

Now,

$$\bar{y}_{st,\alpha} - \bar{y}_{st} = \frac{1}{2} \sum_h W_h \delta_h^\alpha d_h \tag{2.5.38}$$

where

$$\begin{aligned} \delta_h^\alpha &= 2\delta_{h1\alpha} - 1 \\ d_h &= y_{h1} - y_{h2}. \end{aligned} \tag{2.5.39}$$

Hence,

$$\begin{aligned} v_\alpha &= \frac{1}{4}\left(\sum_h W_h \delta_h^\alpha d_h\right)^2 \\ &= \frac{1}{4}\left[\sum_h W_h^2 d_h^2 + 2\sum_{h<h'}\sum W_h W_{h'} \delta_h^\alpha \delta_{h'}^\alpha d_h d_{h'}\right]. \end{aligned} \tag{2.5.40}$$

It follows that

$$\frac{1}{2^H}\sum_{\alpha=1}^{2^H} \bar{y}_{st,\alpha} = \bar{y}_{st}, \quad \frac{1}{2^H}\sum_{\alpha=1}^{2^H} v_\alpha = v(\bar{y}_{st}). \tag{2.5.41}$$

When H is large, computation of $v(\bar{y}_{st})$ as the average of v_α over 2^H HS's becomes formidable. However, if we choose a set η of K HS's such that

$$\sum_{\alpha\in\eta} \delta_h^\alpha \delta_{h'}^\alpha = 0, h < h' = 1, \ldots, H, \tag{2.5.42}$$

then

$$\bar{v}_{(K)} = \sum_{\alpha\in\eta} v_\alpha / K = v(\bar{y}_{st}). \tag{2.5.43}$$

Plackett and Burman (1946) developed a method for constructing $m \times m$ orthogonal matrices with entries $+1, -1$ where m is a multiple of 4. These can be used directly to obtain values of δ_h^α satisfying (2.5.42). The orthogonal matrix of size K where K is a multiple of 4, between H and $H+3$, can be used dropping the last $K-H$ columns. The entries in the matrix can be substituted as δ_h^α, each column standing for a stratum. McCarthy referred to the set η as balanced. If, further, the condition

$$\sum_{\alpha\in\eta} \delta_h^\alpha = 0, h = 1, \ldots, H \tag{2.5.44}$$

is satisfied, then $\bar{y}_{st,\alpha}/K = \bar{y}_{st}$. The set of replicates satisfying (2.5.42) and (2.5.44) is set to be in full orthogonal balance.

For designs with *wor* sampling of psu's, v_α is positively biased. In this case, a separate adjustment is necessary to account for this bias, though the bias is generally negligible.

In the nonlinear case, in which the BRR is most useful, let $\hat{\theta}$, $\hat{\theta}_\alpha$, $\hat{\theta}_{\alpha'}$ be the estimates of θ based on the whole sample, S_α, and $S_{\alpha'}$, respectively. Let $\bar{\hat{\theta}}_\alpha = (\hat{\theta}_\alpha + \hat{\theta}_{\alpha'})/2$. We note that even for a balanced set of HS's, $\sum_{\alpha \in \eta} \hat{\theta}_\alpha / K = \hat{\hat{\theta}}_\alpha \neq \hat{\theta}$ in general. Empirical studies by Kish and Frankel (1970), among others, however, show that $\bar{\hat{\hat{\theta}}}_\alpha$ is very close to $\hat{\theta}$ in general. Writing

$$
\begin{aligned}
\bar{v}_{(K)}(\hat{\theta}) &= \sum_{\alpha \in \eta} (\hat{\theta}_\alpha - \hat{\theta})^2 / K \\
\bar{v}'_{(K)}(\hat{\theta}) &= \sum_{\alpha \in \eta} (\hat{\theta}_{\alpha'} - \hat{\theta})^2 / K,
\end{aligned}
\tag{2.5.45}
$$

we have the following alternative variance estimators:

$$
\begin{aligned}
&\text{(i)} \;\; \bar{v}_{(K)}(\hat{\theta}) \\
&\text{(ii)} \;\; \bar{v}'_{(K)}(\hat{\theta}) \\
&\text{(iii)} \;\; [\bar{v}_{(K)}(\hat{\theta}) + \bar{v}'_{(K)}(\hat{\theta})]/2 = \bar{\bar{v}}_{(K)}(\hat{\theta}) \\
&\text{(iv)} \;\; \sum_{\alpha \in \eta} (\hat{\theta}_\alpha - \hat{\theta}_{\alpha'})^2 / (4K) = \bar{v}^{+}_{(K)}(\hat{\theta}).
\end{aligned}
\tag{2.5.46}
$$

The estimators (i), (ii), (iii) are sometimes regarded as estimators of mse$(\hat{\theta})$, while (iv) is regarded as estimator of Var $(\hat{\theta})$.

Since (iii) is the average of (i) and (ii), it is at least as precise as the others and equally biased. However, (iii) is comparatively costlier than (i) (and (ii)) and perhaps, significantly so, when many estimators are produced.

Another set of variance estimators can be attained by replacing $\hat{\theta}$ by $\hat{\hat{\theta}} = \sum_\alpha \hat{\theta}_\alpha / K$ or $\sum_\alpha \hat{\theta}_{\alpha'} / K$ in (i), (ii), and (iii) of (2.5.46). Such estimators are unbiased for linear $\hat{\theta}$ only if the number of HS's is $T > H$. If H is a multiple of 4, $T (= H + 4)$ HS's must be used to maintain the unbiasedness (Lemeshow and Epp 1977). The estimators using $\hat{\hat{\theta}}$ are generally not preferred to those using $\hat{\theta}$, since they give smaller and less conservative estimates of mse, as they do not include the components for bias of $\hat{\theta}$. Empirical works of McCarthy (1969a, b), Kish and Frankel (1970), Levy (1971), Frankel (1971) and others show that BRR provides satisfactory estimates of the true variance.

All the above-mentioned BRR estimators become identical in the linear case.

Two modifications of BRR has been proposed, that require fewer replicates but the corresponding estimates are less precise and equally biased as the full BRR estimate. In one modification strata are combined into groups, not necessarily of the same size. For each replicate all strata into a group g are assigned the same value δ_h^α. The constraints (2.5.42) are imposed for pairs h, h' of strata which are not in the same group g. Thus if G groups are formed, the number of replicates required is the multiple of 4, which lies in the range G to $G + 3$. The Plackett and Burman matrices of size K may then be used to derive the values of δ_h^α.

The second procedure for reducing the number of replicates, discussed by McCarthy (1966) and developed by Lee (1972, 1973) is the method of partially balanced repeated replications (PBRR). Here the strata are divided into groups and full balancing are applied to the strata within each group. If T replicates are required for H strata, $G = H/T$ groups are formed with T strata in each. A $T \times T$ orthogonal matrix is then used to ensure a full balance within each group. Lee (1972, 1973), Rust (1984) suggested methods of implementing PBIB that would minimize the loss in precision over fully balanced BRR.

Rust (1984, 1986) shows that the method of PBRR and combined strata are equivalent. However, the combined strata method has a greater flexibility in the sense that the number of strata per group varies.

For general designs in which strata sample sizes vary, BRR can be implemented by dividing the psu's in each stratum into two groups of equal sizes (assuming $n_h = 2m_h$, m_h an integer), and then using these groups as units (Kish and Frankel 1970). In this case the BRR variance estimator is somewhat less precise than the customary variance estimator. Valliant (1987) considers the large-sample prediction properties of the BRR separate ratio and regression estimator under a superpopulation model when n_h is large and compares these with jackknife and linearization procedure.

Example 2.5.3 Let us consider BRR for estimation of population ratio $R = Y/X$. A ratio estimator of R based on the set S_α is

$$\hat{r}_\alpha = \frac{\sum_h y_{h1}}{\sum_h x_{h1}} = \frac{\sum_h (Y_{h1}\delta_{h1\alpha} + Y_{h2}\delta_{h2\alpha})}{\sum_h (X_{h1\alpha}\delta_{h1\alpha} + X_{h2\alpha}\delta_{h2\alpha})}, \quad \alpha = 1, \ldots, 2^H.$$

Consider variance estimator for the mean of α-HS estimators

$$\bar{\hat{r}}_\alpha = \sum_{\alpha=1}^{2^H} \hat{r}_\alpha / 2^H.$$

The parent estimator of population ratio R is

$$\hat{r} = \frac{\sum_h (y_{h1} + y_{h2})}{\sum_h (x_{h1} + x_{h2})}.$$

Estimator of $V(\hat{r})$ is

$$\begin{aligned}
(i)\ \ \bar{v}(\hat{r}) &= \sum_{\alpha=1}^{2^H} (\hat{r}_\alpha - \hat{r})^2/2^H, \\
(ii)\ \ \bar{v}'(\hat{r}) &= \sum_{\alpha'=1}^{2^H} (\hat{r}_{\alpha'} - \hat{r})^2/2^H, \\
(iii)\ \ \bar{\bar{v}}(\hat{r}) &= [\bar{v}(\hat{r}) + \bar{v}'(\hat{r})]/2, \\
(iv)\ \ \bar{v}^{+}(\hat{r}) &= \sum_{\alpha=1}^{2^H} (\hat{r}_\alpha - \hat{r}_{\alpha'})^2/[4(2^H)].
\end{aligned} \qquad (2.5.47)$$

Three other estimators are obtained by replacing in (i), (ii), and (iii) of (2.5.47) $\hat{r}$ by $\hat{\bar{r}}(=\sum_{\alpha}\hat{r}_{\alpha}/2^H$ or $\sum_{\alpha'}\hat{r}_{\alpha'}/2^H)$. Since these estimators are nonlinear, they are not identical. For example,

$$\bar{\bar{v}}(\hat{r}) = \bar{v}^{+}(\hat{r}) + \sum_{\alpha=1}^{2^H}(\bar{\hat{r}}_{\alpha} - \hat{r})^2/(2^H)$$

where $\bar{\hat{r}}_{\alpha} = (\hat{r}_{\alpha} + \hat{r}_{\alpha'})/2$ and hence

$$\bar{\bar{v}}(\hat{r}) \geq \bar{v}^{+}(\hat{r}).$$

One problem that occasionally arises in BRR is that one or more replicate estimates will remain undefined due to division by zero. This happens particularly often when ratio estimator has been used with very small cell sizes. Fay suggested a solution to this problem: Instead of increasing the weight of one HS by 100 % and decreasing the weight of the other HS to zero, he recommended perturbing the weights by +/− 50 %. Judkins (1990) evaluated Fay's techniques through simulation and also discusses further modification to the techniques that are used for variance estimation when only one psu is selected per stratum.

2.5.5 *The Jackknife Procedures*

Quenouille (1949, 1956) originally introduced jackknife (JK) as a method of reducing the bias of an estimator. Tukey (1958) suggested the use of this technique for variance estimation. Durbin (1953) first considered its use in finite population. Extensive discussion of JK method is given in Miller (1964, 1974), Gray and Schucany (1972) and in a monograph by Efron (1982).

Let θ be the parameter to be estimated. An estimator $\hat{\theta}$ is obtained from the full sample. Assuming $n = mk(m, k$ integers), we partition the sample into k groups of m original observations each. Let $\hat{\theta}_{(\alpha)}$ be the estimator of θ computed from the whole sample except the αth group. Define pseudo-values $\hat{\theta}_{\alpha}$ as

$$\hat{\theta}_{\alpha} = k\hat{\theta} - (k-1)\hat{\theta}_{(\alpha)}. \tag{2.5.48}$$

Quenouille's estimator is

$$\hat{\bar{\theta}} = \frac{1}{k}\sum_{i=1}^{k}\hat{\theta}_{\alpha}. \tag{2.5.49}$$

Tukey suggested that $\hat{\theta}_\alpha$ are approximately independently and identically distributed. The JK estimator of variance is

$$\begin{aligned} v_1(\hat{\theta}) &= \frac{1}{k(k-1)}\sum_{\alpha=1}^{k}(\hat{\theta}_\alpha - \bar{\hat{\theta}})^2 \\ &= \frac{k(k-1)}{k}\sum_{\alpha=1}^{k}(\hat{\theta}_{(\alpha)} - \hat{\theta}_{(.)})^2 \end{aligned} \tag{2.5.50}$$

where $\hat{\theta}_{(.)} = \sum_{\alpha=1}^{k} \hat{\theta}_{(\alpha)}/k$. In practice, $v_1(\bar{\hat{\theta}})$ is used not only to estimate the variance of $\bar{\hat{\theta}}$, but also of $\hat{\theta}$. Alternatively, one may use

$$v_2(\hat{\theta}) = \frac{1}{k(k-1)}(\hat{\theta}_\alpha - \hat{\theta})^2 \tag{2.5.51}$$

which is always at least as large as $v_1(\hat{\theta})$.

The number of groups k is determined from the point of view of computational cost and the precision of the resulting estimator. The precision is maximized when each dropout group is of size one and each unit is dropped only once. Rao (1965), Rao and Webster (1966), Chakraborty and Rao (1968), Rao and Rao (1971) in their studies on ratio estimator based on superpopulation models showed that both bias and variance of $\bar{\hat{\theta}}$ are maximized for the choice $k = n$.

Brillinger (1966) showed that both v_1 and v_2 give plausible estimates of the asymptotic variance. Shao and Wu (1989), Shao (1989) considered the efficiency and consistency of JK variance estimators. For nonlinear statistic $\hat{\theta}$ that can be expressed as functions of estimated means of p variables, such as ratio, regression, correlation coefficient, Krewski and Rao (1981) established the asymptotic consistency of variance estimators from JK, the linearization, and BRR methods. In the case of two samples psu's per stratum, Rao and Wu (1985) showed that the linearization and JK variance estimators are asymptotically efficient. In case of item nonresponse in sample surveys, Rao and Shao (1999) considered jackknife variance estimation for stratified multistage surveys which is obtained by first adjusting the hot deck imputed values for each pseudoreplicate and then applying the standard jackknife formulae. Rao and Tausi (2004) considered variance estimation for the *generalized regression estimator* (GREG) of a total based on p auxiliary variables under stratified multistage sampling. Customary resampling procedures, like jackknife, balanced repeated replication, and bootstrap (Sect. 2.5.7) for estimating the variance of a GREG estimator requires the inversion of a $p \times p$ matrix for each subsample. This may result in illconditioned matrices for some subsamples. The authors applied the estimating function resampling methods to obtain variance estimators using jackknife resampling.

2.5.6 The Jackknife Repeated Replication (JRR)

This is a combination of JK and BRR techniques. We assume as in the case of BRR, that two clusters are selected with replacement from each of H strata. We construct the pseudo-samples following the method suggested by Frankel (1971).

For the first pseudo-sample, we exclude the cluster (1, 1) (i.e., the cluster 1 in stratum 1), weigh the second cluster (1, 2) by 2 and leave the sampled clusters in the remaining $H-1$ strata unchanged. By repeating the procedure for all the strata we get a total of H pseudo-samples. These are

$$\begin{aligned}
\text{First Pseudo-sample} &: \{(2y_{12});\ (y_{21}, y_{22}),\ (y_{31}, y_{32}), \ldots (y_{H1}, y_{H2})\},\\
\text{Second Pseudo-Sample} &: \{(y_{11}, y_{12}),\ (2y_{22}),\ (y_{31}, y_{32}), \ldots, (y_{H1}, y_{H2})\},\\
&\ldots\\
H\text{th Pseudo-Sample} &: \{(y_{11}, y_{12}),\ (y_{21}, y_{22}),\ (y_{31}, y_{32}), \ldots, (2y_{H2})\}.
\end{aligned}$$

Changing the order of excluded clusters we get another set of H pseudo-samples.

The JRR variance estimators are derived using these two sets of pseudo-samples. We illustrate this by the example of finding the variance estimator of combined ratio estimator $\hat{r}$.

For this, we first construct ratio estimator for each pseudo-sample. The estimator of population ratio r based on the hth pseudo-sample in the first set is

$$\hat{r}_h = \frac{2y_{h2} + \sum_{h'(\neq h)=1}^{H} \sum_{\alpha=1}^{2} y_{h'\alpha}}{2x_{h2} + \sum_{h'(\neq h)=1}^{H} \sum_{\alpha=1}^{2} x_{h'\alpha}}, \quad h = 1, \ldots, H. \tag{2.5.52}$$

Similarly, the estimator of the population ratio based on the hth pseudo-sample of the second set is

$$\hat{r}_h^c = \frac{2y_{h1} + \sum_{h'(\neq h)=1}^{H} \sum_{\alpha=1}^{2} y_{h'\alpha}}{2x_{h1} + \sum_{h'(\neq h)=1}^{H} \sum_{\alpha=1}^{2} x_{h'\alpha}}, \quad h = 1, \ldots, H. \tag{2.5.53}$$

Using the pseudo-sample estimators $\hat{r}^h$, $\hat{r}_h^c$, $h = 1, \ldots, H$ we get different JRR estimators of variance of $\hat{r}$. These are

$$v_{1,jrr}(\hat{r}) = \frac{1}{H} \sum_{h=1}^{H} (\hat{r}_h - \hat{r})^2, \tag{2.5.54}$$

$$v_{2,jrr}(\hat{r}) = \frac{1}{H} \sum_{h=1}^{H} (\hat{r}_h^c - \hat{r})^2, \tag{2.5.55}$$

$$v_{3,jrr}(\hat{r}) = \frac{1}{2}(v_{1,jrr} + v_{2,jrr}). \tag{2.5.56}$$

Another set of variance estimators can be obtained by using the estimator $\hat{r}$ first corrected for its bias using pseudo-values as in the JK procedure. The pseudo-value of $\hat{r}_h$ is, following (2.5.52),

$$\hat{r}_h^p = 2\hat{r} - \hat{r}_h, \quad h = 1, \ldots, H. \tag{2.5.57}$$

A bias-corrected estimator of r is, therefore,

$$\bar{\hat{r}}^p = \frac{1}{H}\sum_{h=1}^{H} \hat{r}_h^p. \tag{2.5.58}$$

Similarly, the pseudo-value of $\hat{r}_h^c$ is

$$\hat{r}_h^{pc} = 2\hat{r} - \hat{r}_h^c, \quad h = 1, \ldots, H. \tag{2.5.59}$$

A bias-corrected estimator based on $\hat{r}_h^c (h = 1, \ldots, H)$ is, therefore,

$$\bar{\hat{r}}^{pc} = \frac{1}{H}\sum_{h=1}^{H} \hat{r}_h^{pc}. \tag{2.5.60}$$

Following (2.5.54)–(2.5.56) we, therefore, get the following variance estimators of $\hat{r}$:

$$v_{4,jrr}(\hat{r}) = \sum_{h=1}^{H} (\hat{r}_h^p - \bar{\hat{r}}^p)^2/\{H(H-1)\}, \tag{2.5.61}$$

$$v_{5,jrr}(\hat{r}) = \sum_{h=1}^{H} (\hat{r}_h^{pc} - \bar{\hat{r}}^{pc})^2/\{H(H-1)\}, \tag{2.5.62}$$

$$v_{6,jrr}(\hat{r}) = (v_{4,jrr} + v_{5,jrr})/2. \tag{2.5.63}$$

Finally, from all the $2H$ pseudo-samples we obtain

$$v_{7,jrr}(\hat{r}) = \sum_{h=1}^{H} (\hat{r}_h - \hat{r}_h^c)^2/4. \tag{2.5.64}$$

For a nonlinear estimator, the bias-corrected JRR estimators and the parent estimator coincide. In practice all the JRR variance estimators should give closely related results.

The method can be extended to a more general case where more than two clusters are selected from each stratum without replacement (see Wolter 1985, Sect. 4.6).

2.5.7 The Bootstrap

The bootstrap (BS) method is the most recent technique of variance estimation for complex sample surveys. The technique uses a highly computer-intensive resampling procedure to mimic the theoretical distribution from which the sample is derived. The method does not need any prior assumption about the distribution of observations or the estimators. It provides estimates of bias and standards errors and other distributional properties of the estimators, however complex it may be.

The naive BS technique was suggested by Efron (1979) who indicated that the method may be better than its competitors. The BS method for finite population sampling was introduced and discussed by Gross (1980), Bickel and Freedman (1984), Chao and Lo (1985), McCarthy and Snowdon (1985), Booth et al. (1991), among others. Rao and Wu (1988) showed the application of the BS in design-based survey sampling under different sampling designs including stratified cluster sampling with replacement, stratified simple random sampling without replacement, unequal probability random sampling without replacement, and two-stage cluster sampling with equal probabilities and without replacement. Rao (2006) showed that BS method provides an alternative option to the analysis of complex surveys for taking account of the design effects and weight adjustments. We consider here the elements of variance estimation by BS.

Suppose we have p variables $y_1, \ldots, y_p$ with $Y_{hij}(y_{hij}), \bar{Y}_j, \bar{Y}_{hj}, \bar{y}_{hj}$ as the value of y_j on the ith unit in the hth stratum in the population (sample), population mean, stratum population mean, stratum sample mean of y_j, respectively ($h = 1, \ldots, H; j = 1, \ldots, p$; $\bar{Y}_j = \sum_h W_h \bar{Y}_{hj}$, $\bar{Y}_{hj} = \sum_{i=1}^{N_h} Y_{hij}/N_h$, $\bar{y}_{hj} = \sum_{i \in s_h} y_{hij}/n_h$, N_h, n_h being, respectively, the size of sample s_h and population in stratum h). Suppose we want to estimate $\theta = g(\bar{Y}_1, \ldots, \bar{Y}_p) = g(\bar{\mathbf{Y}})$, a nonlinear function of $\bar{\mathbf{Y}} = (\bar{Y}_1, \ldots, \bar{Y}_p)'$. This includes population ratio, regression, correlation coefficient, etc. A natural estimator of θ, whenever $n_h (\geq 2)$ psu's are selected with replacement from each stratum is $g(\hat{\bar{\mathbf{Y}}}) = g(\bar{\mathbf{y}})$ where $\bar{\mathbf{y}} = (\bar{y}_1, \ldots, \bar{y}_p)', \bar{y}_j = \sum_h W_h \bar{y}_{hj}$. We denote by $\bar{\mathbf{y}}_h = (\bar{y}_{h1}, \ldots, \bar{y}_{hp})'$.

In this case, the random vector $\mathbf{y}_{hi} = (y_{hi1}, \ldots, y_{hip})', i = 1, \ldots, n_h$ are iid with $E(\mathbf{y}_{hi}) = \bar{\mathbf{Y}}_h = (\bar{Y}_{h1}, \ldots, \bar{Y}_{hp})'$. The vectors $\mathbf{y}_{hi}, \mathbf{y}_{h'k} (h \neq h')$ are independently but not necessarily identically distributed. The BS sampling procedure is as follows:

(a) Draw a random sample wr $\{\mathbf{y}^*_{hi}, i = 1, \ldots, n_h\}$ of size n_h from the given sample $\{\mathbf{y}_{hi}, i = 1, \ldots, n_h\}$ independently from each stratum. Calculate $\bar{\mathbf{y}}^*_h = \sum_i \mathbf{y}^*_{hi}/n_h, \bar{\mathbf{y}}^* = \sum_h W_h \bar{\mathbf{y}}^*_h$ and $\hat{\theta}^* = g(\bar{\mathbf{y}}^*)$.
(b) Repeat step (a) a large number of times, say B times and calculate the corresponding estimates $\hat{\theta}^{*1}, \ldots, \hat{\theta}^{*B}$ of θ.

(c) Calculate the Monte Carlo estimate of $V(\hat{\theta})$,

$$v_b(a) = \sum_{b=1}^{B} (\hat{\theta}^{*b} - \hat{\theta}^{*\cdot})^2/(B-1) \tag{2.5.65}$$

where $\hat{\theta}^{*\cdot} = \sum_{b=1}^{B} \hat{\theta}^{*b}/B$.

The estimator $v_b(a)$ is a fair approximation to the BS variance estimator of $\hat{\theta}$,

$$v_b = var_*(\hat{\theta}^*) = E_*(\hat{\theta}^* - E_*(\hat{\theta}^*))^2 \tag{2.5.66}$$

where E_* denotes expectation with respect to BS sampling. The BS estimator $E_*(\hat{\theta}^*)$ of θ is approximated by $\hat{\theta}^{*\cdot}$.

In the linear case with $p = 1, \theta = \bar{Y}, \hat{\theta}^* = \sum_h W_h \bar{y}_h^* = \bar{y}^*$ and v_b reduces to

$$var_*(\bar{y}^*) = \sum_h W_h^2 \sigma_h^2/n_h \tag{2.5.67}$$

where $\sigma_h^2 = \sum_{i=1}^{n_h}(y_{hi} - \bar{y}_h)^2/n_h$. Comparing (2.5.67) with the customary estimator $v(\bar{y}_{st}) = \sum_h W_h^2 s_h^2/n_h$ where $s_h^2 = \sum_i (y_{hi} - \bar{y}_h)^2/(n_h - 1)$, it follows that $var_*(\bar{y}^*)/v(\bar{y}_{st})$ does not converge in probability to 1 when n_h is bounded. Hence, $var_*(\bar{y}^*)$ is not a consistent estimator of $V(\bar{y}_{st})$ unless n_h and $f_h = n_h/N_h$ are constants for all h. Moreover, v_b in (2.5.66) is not a consistent estimator of the variance of a general nonlinear estimator.

Recognizing this problem Efron (1982) suggested to draw BS sample of size $n_h - 1$ with *srswr* sampling design instead of n_h independently from each stratum. The rest of the procedure is the same as before.

To get rid of this difficulty, Rao and Wu (1988) suggested the following resampling procedure.

(i) Draw a random sample $\{y_{hi}^*, i = 1, \ldots, m_h\}$ with replacement ($m_h \geq 1$) from the original sample $\{y_{hi}, i = 1, \ldots, n_h\}$. Calculate

$$\begin{aligned} \tilde{y}_{hi} &= \bar{y}_h + \frac{\sqrt{m_h}}{\sqrt{n_h - 1}}(y_{hi}^* - \bar{y}_h) \\ \tilde{y}_h &= \sum_i \tilde{y}_{hi}/m_h = \bar{y}_h + \frac{\sqrt{m_h}}{\sqrt{n_h - 1}}(\bar{y}_h^* - \bar{y}_h) \\ \tilde{y} &= \sum_h W_h \tilde{y}_h, \quad \tilde{\theta} = g(\tilde{y}). \end{aligned} \tag{2.5.68}$$

(ii) Repeat the step (i) B times independently and calculate the corresponding estimates $\tilde{\theta}^1, \ldots, \tilde{\theta}^B$. The BS estimator $E_*(\tilde{\theta})$ of θ is approximated by $\tilde{\theta}^{\cdot} = \sum_b \tilde{\theta}^b/B$.

(iii) The BS variance estimator of $\hat{\theta}$,

$$\tilde{v}_b = E_*(\tilde{\theta} - E_*(\tilde{\theta}))^2$$

is approximated by the Monte Carlo estimator

$$\tilde{v}_b(a) = \sum_{b=1}^{B} (\tilde{\theta}^b - \tilde{\theta}^{\cdot})^2/(B-1).$$

In the linear case of $\theta = \bar{Y}$ with $p = 1$, $\tilde{v}_b$ reduces to the customary unbiased variance estimator $v(\bar{y}_{st})$ for any choice of m_h, because,

$$\tilde{v}_b = E_*(\tilde{y} - \bar{y})^2 = \sum_h W_h^2 \cdot \frac{m_h}{n_h-1} E_*(\bar{y}_h^* - \bar{y}_h)^2$$

$$= \sum_h W_h^2 \cdot \frac{m_h}{n_h-1} \cdot \frac{(n_h-1)s_h^2}{m_h n_h} = \sum_h W_h^2 \frac{s_h^2}{n_h} = v(\bar{y}_{st}).$$

Thus, Rao and Wu (1988) applied the previously stated algorithm of a naive BS procedure with a general sample size m_h not necessarily equal to n_h, but rescaled the resampled values appropriately so that the resulting variance estimator is the same as the usual unbiased variance estimator in the linear case.

In the nonlinear case it has been shown that under certain conditions

$$\tilde{v}_b = v_L + 0(n^{-2})$$

where v_L is the customary linearization variance estimator,

$$v_L = \sum_{j,k=1}^{p} g_j(\bar{\mathbf{y}}) g_k(\bar{\mathbf{y}}) \sum_{h=1}^{H} \frac{W_h^2}{n_h} s_{hjk},$$

where for $\mathbf{t} = (t_1, \ldots, t_p)' g_j(\mathbf{t}) = \frac{\partial g(\mathbf{t})}{\partial t_j}$, $s_{hjk} = \sum_{i \in s_h} (y_{hij} - \bar{y}_{hj})(y_{hik} - \bar{y}_{hk})$. Since v_L is a consistent estimator of the variance of $\hat{\theta}$, $\tilde{v}_b$ is consistent for Var $(\hat{\theta})$.

It has also been found that the estimate of bias of $\hat{\theta}$, $B(\hat{\theta}) = E(\hat{\theta}) - \theta$, based on the suggested BS procedure, which is $\tilde{B}(\hat{\theta}) = E_*(\tilde{\theta}) - \hat{\theta} = \tilde{\theta}^{\cdot} - \hat{\theta}$, is consistent, while the same based on the naive BS procedure is not consistent.

The choice $m_h = n_h - 1$ gives $\tilde{y}_{hi} = y_{hi}^* (i = 1, \ldots, m_h)$ and the method reduces to the naive BS. For $n_h = 2$ and $m_h = 1$, the method reduces to the well-known random half-sample replication. For $n_h \geq 5$, $m_h \approx n_h - 3$ Rao and Wu made some empirical studies on the choice of m_h.

The method can be easily extended to simple random sampling within stratum by changing $\tilde{y}_{hi}$ in (2.5.68) to

$$\tilde{y}_{hi} = \bar{y}_h + \frac{\sqrt{m_h}}{\sqrt{n_h - 1}} \cdot \sqrt{1 - f_h}(y^*_{hi} - \bar{y}_h) \tag{2.5.69}$$

where $f_h = n_h/N_h$, $h = 1, \ldots, H$. Here, even by choosing $m_h = n_h - 1$, we do not get $\tilde{y}_{hi} = y^*_{hi}$. Hence, the naive BS using y^*_{hi} will still give a wrong scale.

The method has been extended to any unbiased sampling strategy including Rao–Hartley–Cochran sampling procedure.

Apart from these with replacement procedures, a without replacement BS technique was proposed by Gross (1980) in the case of a single stratum. His method assumes that $N = Rn$ for some integer R and creates a pseudopopulation of size N by replicating the data R times. However, the method does not yield the usual unbiased estimate of variance in the linear case. The difficulty was corrected by Bickel and Freedman (1984) who proposed a randomization between two pseudopopulations and also allowed an extension of the method for $H > 1$. Sitter (1992) developed a BS procedure which retains the desirable properties of both with replacement BS and without replacement BS techniques but extends to more complex without replacement sampling designs.

Hall (1989) considered three efficient bootstrap algorithms: these are balanced bootstrap and the linear approximation method proposed by Davison et al. (1986) and a centering method proposed by Efron (1982) in the context of bias estimation. He compares the asymptotic performance of these methods and show that they are asymptotically equivalent. Hall prove that the variances and mean square errors of all these three algorithms are asymptotic to the same constant multiple of $(Bn^2)^{-1}$, where B denotes the number of bootstrap resamples and n is the size of the original sample. The convergence rate $(Bn^2)^{-1}$ represents a significant improvement on that for the more usual, unbalanced bootstrap algorithm, which has mean square error of only $(Bn)^{-1}$. These results apply to smooth functions of means.

Ahmad (1997) suggested a new bootstrap variance estimation technique, *rescaling bootstrap without replacement* technique and also proposed an optimum choice of bootstrap sample size for his proposed procedure. Canty and Davison (1999) considered labor force surveys to demonstrate the advantages of resampling methods in estimation of variance. Labor force surveys are conducted to estimate, among others, quantities such as unemployment rate and the number of people at work. Interest focusses typically both in estimates at a given time and in changes between two successive time points. Calibration of the sample to ensure agreement with the known population margins results in random weights being assigned to each response, but the usual method of variance estimation do not account for this. The authors describe how resampling methods, such as jackknife, jackknife linearisation, balanced repeated replication and bootstrap can be used to do so. Robert et al. (2004) suggested a design-based bootstrapping method for the estimation of variance of an estimator in longitudinal surveys.

Multiple imputation is a method of estimating the variance of estimators that are constructed with some imputed data. Kim et al. (2006) give an expression for the bias of the multiple imputation variance estimator for data that are collected with a complex survey design. A bias-adjusted variance estimator is also suggested.

2.6 Effect of Survey Design on Inference About Covariance Matrix

In this section we shall look into the effect of sampling design on a classical test statistic for testing a hypothesis regarding a covariance matrix.

Let $\mathbf{V} = ((v_{ij}))$ be a consistent estimator of a $p \times p$ covariance matrix $\mathbf{\Sigma}$ under the IID assumption. For example, for a self-weighing design, $\mathbf{V}$ may be the usual sample covariance matrix. Let

$$\tilde{\omega} = Vech(\mathbf{V}) = (v_{11}, v_{21}, v_{22}, v_{31}, v_{32}, v_{33}, \ldots, v_{pp})' \tag{2.6.1}$$

be the $u \times 1$ vector of distinct elements of $\mathbf{V}$ where $u = p(p+1)/2$. Suppose we may write $\tilde{\omega}$ as

$$\tilde{\omega} = \frac{1}{n}\sum_{i \in s} \omega_i \tag{2.6.2}$$

where ω_i is a vector of sample square and cross-product terms, each term centered around the corresponding sample mean, and also possibly weighted, say, for unequal selection probabilities and n is the sample size. Thus, for equal selection probability sampling,

$$\omega_k(k \in s) = ((y_{1k} - \bar{y}_1)^2, \ldots, (y_{pk} - \bar{y}_p)^2, (y_{1k} - \bar{y}_1)(y_{2k} - \bar{y}_2),$$
$$\ldots, (y_{p-1,k} - \bar{y}_{p-1})(y_{p,k} - \bar{y}_p))'$$

where y_{jk} denotes the value of y_j on the unit k in the sample. Then $\tilde{\omega}$ is consistent for $Vech\mathbf{\Sigma} = \mu$ (say).

Under IID assumptions, $\tilde{\omega}$ will generally be asymptotically normally distributed with mean μ and the linearization asymptotic covariance matrix estimator of $\tilde{\omega}$, $\text{Var}(\tilde{\omega})$ may be expressed as the $u \times u$ matrix

$$\mathbf{V}^* = \frac{1}{n(n-1)}\sum_{i \in s}(\omega_i - \tilde{\omega})(\omega_i - \tilde{\omega})'. \tag{2.6.3}$$

Consider a linear hypothesis about $\mathbf{\Sigma}$ which may be expressed as, say, $\mathbf{A}\boldsymbol{\mu} = \mathbf{0}$ where $\mathbf{A}$ is a given $q \times u$ matrix of rank q. An IID procedure for testing H_0 is the Wald statistic

$$X_W^2 = (\mathbf{A}\tilde{\omega})'[\mathbf{A}\mathbf{V}^*\mathbf{A}']^{-1}(\mathbf{A}\tilde{\omega}), \tag{2.6.4}$$

which follows the central chi-square distribution, $\chi^2_{(q)}$ in large sample. (Wald statistic has been introduced in Chap. 4).

Under a complex design this procedure may be modified by replacing $\mathbf{V}^*$ by the linearization estimator $\mathbf{V}_L$ of $Var(\tilde{\omega})$ which accommodates the sampling design. This gives a modified Wald statistic X^2_{W0} (Pervaiz 1986). This approach which also assumes near normality of $\tilde{\omega}$ is constrained by the fact that the d.f. ν of $V_L(\tilde{\omega})$ may be low compared to u, particularly if p is moderate to large, in which case $\mathbf{V}_L(\tilde{\omega})$ may become very unstable and even singular. As a result X^2_W may deviate considerably from $\chi^2_{(q)}$. It is therefore suggested to correct X^2_W for its first moment, that is, to refer

$$X^2_{W1} = \frac{(q)X^2_W}{tr[(\mathbf{AV^*A'})^{-1}\mathbf{AV}_L(\tilde{\omega})\mathbf{A'}]} \tag{2.6.5}$$

as $\chi^2_{(q)}$ (Layard 1972).

Note 2.6.1 Graubard and Korn (1993) considered the problem of testing the null hypotheses $H_0 : \theta = \mathbf{0}$, where the p-dimensional parameter $\theta = \mathbf{g}(\boldsymbol{\lambda})$ and $\boldsymbol{\lambda}$ is a r-dimensional vector of means. The authors used replicated estimates of the variances that take into account the complex survey design. The Wald statistic can be used to test H_0, but inference for θ may have very poor power. They used an alternative procedure based on classical quadratic test statistic. Another reference in the area is due to Korn and Graubard (1990).

2.7 Exercises and Complements

2.1 Suppose that y_{abc} is the value of the cth sampled unit belonging to the bth second-stage unit sampled from the ath sampled first-stage unit in a three-stage sample $(a = 1, \ldots, n;\ b = 1, \ldots, m;\ c = 1, \ldots, k.)$ Consider the superpopulation model

$$y_{abc} = \theta + \alpha_a + \beta_b + \epsilon_{abc}$$

where $\alpha_a, \beta_b, \epsilon_{abc}$ are independent random variables with mean zero and

$$\text{Cov}\,(y_{abc}, y_{a'b'c'}) = \begin{cases} \sigma_0^2 & \text{if } (a,b,c) = (a',b',c') \\ \tau_2\sigma_0^2 & \text{if } (a,b) = (a'\cdot b'), c \neq c' \\ \tau_1\sigma_0^2 & \text{if } a = a', b \neq b', c \neq c' \\ 0 & \text{if } a \neq a', b \neq b', c \neq c' \end{cases}$$

Here τ_2 is the inter-second-stage-unit correlation, τ_1 is the inter-first-stage-unit correlation. Then show that

$$\begin{aligned} Var_{true}(\bar{y}) &= Var_{true}\left[\sum_{a=1}^{n}\sum_{b=1}^{m}\sum_{c=1}^{k} y_{abc}/nmk\right] \\ &= \frac{\sigma_0^2}{nmk}[1 + (m-1)k\tau_1 + (k-1)\tau_2]. \end{aligned}$$

For the IID model

$$y_{abc} = \theta + e_{abc}$$

with e_{abc} independently distributed with zero mean and variance σ_0^2, show that (using the notations of Example 2.2.4)

$$E_{true}[v_{IID}(\bar{y})] \approx \frac{\sigma_0^2}{nmk},$$

if n is large. Hence, deduce that

$$\text{deff}\ (\bar{y}, v_{IID}) \approx 1 + (m-1)k\tau_1 + (k-1)\tau_2.$$

(Skinner 1989)

2.2: Show that the estimator v_α given in (2.5.35) is unbiased for $V(\bar{y}_{st})$.

Chapter 3
Some Classical Models in Categorical Data Analysis

Abstract This chapter makes a brief review of classical models of categorical data and their analysis. After a glimpse of general theory of fitting of statistical models and testing of parameters using goodness-of-fit tests, Wald's maximum likelihood statistic, Rao's statistic, likelihood ratio statistic, we return to the main distributions of categorical variables—multinomial distribution, Poisson distribution, and multinomial-Poisson distribution and examine the associated test procedures. Subsequently, log-linear models and logistic regression models, both binomial and multinomial, are looked into and their roles in offering model parameters emphasized. Lastly, some modifications of classical test procedures for analysis of data from complex surveys under logistic regression model have been introduced.

Keywords Categorical random variable · Full model · Nested model · Information matrix · Goodness-of-fit statistics · Wald's maximum likelihood statistic · Rao's statistic · Likelihood ratio statistic · Multinomial model · Poisson model · Log-linear models · Binomial logistic regression models · Polytomous logistic regression models

3.1 Introduction

In this chapter, we will make a brief review of the classical theory of categorical data analysis which is based on the assumption that the variables are independently and identically (IID) distributed. Equivalently, the samples are assumed to be drawn by simple random sampling with replacement.

The purpose of most investigations is to assess relationships among a set of variables. The choice of an appropriate technique for that purpose depends on the type of variables under investigation. Suppose we have a set of numerical values for a variable.

(i) If each element of this set may lie only at a few isolated points, we have a discrete or categorical data set. In other words, a categorical variable is one for which measurement scale consists of a set of categories. Examples are: race, sex, age-group, etc.

P. Mukhopadhyay, *Complex Surveys*, DOI 10.1007/978-981-10-0871-9_3

(ii) If each element of this set may theoretically lie anywhere in this numerical scale, we have a continuous data set. Examples are blood pressure, blood sugar, cholesterol level, etc.

Here we shall consider analysis of categorical data under classical setup.

A *categorical random variable* X is a random variable that takes values in one of k categories. The entire probabilistic behavior of X is summarized by its probability distribution

$$P(X = i) = \pi_i, \ i = 1, \ldots, k$$

where the π_i are (usually unknown) parameters that add up to unity. There are thus really only $k - 1$ of these fundamental parameters. For two categorical variables X_1, X_2, we introduce the notation

$$P(X_1 = i, X_2 = j) = \pi_{ij}, \ i = 1, \ldots r; \ j = 1, \ldots, l$$

so that the marginal probability $P(X_1 = i) = \pi_{i0} = \sum_{j=1}^{l} \pi_{ij}$, etc. A two-way table of categorical data is called a contingency table.

3.2 Statistical Models

We are concerned here with categorical data and so we shall take the data to be a set of counts $Y_1, \ldots, Y_k$ of k different categories of events. A *statistical model* is a set of assumptions about the joint distribution of the data. This set of assumptions usually has two components.

The first component of a statistical model is an assumption that the distribution of the data, commonly called the *error distribution* comes from a specific set, often a *parametric family* of distributions. The error distribution describes the random variation of the data about any systematic features or patterns. For categorical data, the most common error distributions are Poisson and Multinomial. The error distribution describes the random variation of the data Y_i about their mean values $E(Y_i) = e_i$.

The second component of a statistical model, called the *systemic component*, is a statement about the underlying pattern of the data. Commonly, this is a statement about the mean value of the Y_i, called a *regression function*. For instance we might assume $E(Y_i)$ is a linear function of x_i, the value of a covariate x on units in the ith category. One goal of statistical analysis is to discover such simple regression functions which summarize the main pattern of the data reasonably well. (We note that the expectation used above is in the superpopulation-sense (Chap. 1) and we shall use the same notations E, V, etc., when there is no ambiguity.)

A statistical model with as many parameters as the data values are called a *full* or *saturated* model. Though this model is rarely of intrinsic interest in itself, we compare more refined models with it in goodness-of-fit tests. A model is *intermediate* if it neither specifies all the fundamental parameters nor leaves them entirely unrestricted, i.e., it is neither a full model nor a simple model.

We say that one model is a *sub-model* of another if it is a special case of the other model. A sequence of models $M_1, M_2, M_3, \ldots$ where each model is a sub-model of the previous one is said to be *nested*. Nested models have also been defined in the Appendix (Sect. A.6).

3.2.1 Fitting Statistical Models

The general problem is that we have a set of data $Y_1, \ldots, Y_k$ and a model which specifies the distributions of these random variables in terms of a set of p unknown parameters $\theta = (\theta_1, \ldots, \theta_p)$. The aim is to estimate θ.

The mean value of Y_i depends on θ and once θ is estimated we can estimate this mean value. We use the notation

$$e_i(\theta) = E(Y_i; \theta), \qquad \hat{e}_i = e_i(\hat{\theta}),$$

where the $\hat{e}_i$'s are called *fitted* values. If one imagines the data as being made up of a systematic component or trend with random error added, then the best-fitted values estimate the trend as closely as possible. They are a *smoothed* version of the data. Throughout this section, we shall only consider the maximum likelihood (ML) method of estimation of θ.

3.2.2 Large Sample Estimation Theory

Let $X = (X_1, \ldots, X_n)$ be a set of independent random variables with distribution depending on a vector of parameters $\theta = (\theta_1, \ldots, \theta_p) \in \Theta$. We emphasize that in this case the sample is drawn by simple random sampling with replacement so that the above assumption is satisfied. The variables $X_1, \ldots, X_n$ are therefore IID and all the results in this chapter therefore hold under these assumptions only. These results do not necessarily hold if the sample is drawn by some complex design, say stratified multistage cluster sampling design.

It is convenient to work with the logarithm of the likelihood function rather than the likelihood function itself. The log-likelihood function

$$\mathcal{L}(\theta) = \log L(\theta) = \sum_{i=1}^{n} \log p_i(X_i; \theta), \quad \theta \in \Theta \tag{3.2.1}$$

where p_i is the probability function of the ith random variable X_i. The derivative of $\mathcal{L}$ with respect to θ_j is denoted by

$$U_j = \sum_{i=1}^{n} \frac{\partial \log p_i(X_i; \theta)}{\partial \theta_j}, \quad j = 1, \ldots, p \tag{3.2.2}$$

and its mean is zero almost in all cases. The p-dimensional derivative vector $U = (U_1, \ldots, U_p)'$ is called the *score function*. The likelihood is maximized by simultaneously equating all elements of the score function to their zero expectation. This system of equations is known as the *likelihood equations*.

We now introduce the concept of *information* about a parameter θ. The observed information matrix is a symmetric matrix $J(\theta)$ with the (j, k)th element

$$J_{jk} = -\frac{\partial U_j}{\partial \theta_k} = -\frac{\partial^2 \mathcal{L}}{\partial \theta_j \partial \theta_k}(X; \theta). \tag{3.2.3}$$

Since $\hat{\theta}$ is a maximum of the log-likelihood function, $J(\hat{\theta})$ is also nonnegative definite when evaluated at $\hat{\theta}$. The *expected information matrix* is given by I whose (j, k)th element is

$$I_{jk} = -E\left(\frac{\partial U_j}{\partial \theta_k}\right). \tag{3.2.4}$$

It can be shown that

$$\text{Cov}\,(U_j(\theta), U_k(\theta)) = -E\left(\frac{\partial U_j(\theta)}{\partial \theta_k}\right)$$

so that I may be viewed as the variance-covariance matrix of U.

3.2.3 Asymptotic Properties of ML Estimates

It follows that U is asymptotically distributed as

$$N_p(0, I(\theta)). \tag{3.2.5}$$

Now, by Taylor expansion

$$U_j(\hat{\theta}) = U_j(\theta) + \sum_{k=1}^{p} \frac{\partial U_j(\theta)}{\partial \theta_k}(\hat{\theta}_k - \theta_k) + \cdots. \tag{3.2.6}$$

Since $U_j(\hat{\theta}) = 0\ \forall j$ where $\hat{\theta}$ is the mle of θ, we have

$$U(\theta) \approx J(\theta)(\hat{\theta} - \theta) \approx I(\theta)(\hat{\theta} - \theta).$$

Isolating $\hat{\theta}$ and noting that $U(\theta)$ is asymptotically normal with variance $I(\theta)$, we get two variations of a central limit theorem for $\hat{\theta}$. These are

$$\hat{\theta} \rightarrow^d N_p(\theta, I^{-1}(\theta)),$$

and

$$\hat{\theta} \rightarrow^d N_p(\theta, J^{-1}(\theta)), \tag{3.2.7}$$

where the inverse of either the expected or observed information may be used as the asymptotic variance of $\hat{\theta}$. One may also estimate θ by $\hat{\theta}$ in these information matrices.

3.2.4 Testing of Parameters

Suppose we want to test an hypothesis H_0 which specifies exactly $r \leq p$ restrictions on the parameters $\theta_1, \ldots, \theta_p$. Consider a function $g(\theta) = \omega$ from $\mathcal{R}^p$ to $\mathcal{R}^r$ such the for the hypothesized value of θ, $g(\theta) = \omega_0$. Therefore, such an hypothesis can be written as $H_0 : g(\theta) = \omega = \omega_0$. The components of ω, namely, $\omega_1, \ldots, \omega_r$ are now the parameters of interest and the null hypothesis specifies the values of these parameters. We are interested in testing the H_0 against the bothsided alternative hypothesis $H_1 : \omega \neq \omega_0$. Let $\hat{\omega} = g(\hat{\theta})$ be the *ml* estimate of ω with jth component $\hat{\omega}_j$. We can certainly test each ω_j separately using $\hat{\omega}_j$, but we want to test them simultaneously.

Goodness-of-Fit Statistics: Goodness-of-fit statistics assess the distance between the observed distribution and the distribution that a model proposes. Under the null hypothesis, this distance is only randomly higher than zero. Using a significance test, researchers estimate the probability that the observed frequencies or frequencies with even larger distances from those estimated based on the model under the null hypothesis occurs. This probability or P-value is termed the size of the test. The P-value can indeed be used to make a formal test of size α by simply rejecting H_0 if and only if it is smaller than α.

Many tests have been proposed for the evaluation of model parameters. Some of these tests are exact, others are asymptotic. In the context of categorical data, most software packages offer exact tests only for small contingency tables, such as 2×2 tables and asymptotic tests for log-linear modeling, logistic regression modeling. We shall therefore consider here some asymptotic tests.

Wald'd Maximum Likelihood Statistic

We shall write the parameter vector θ as $\theta = (\omega, \psi)$ where ψ represents all the other $p - r$ parameters besides the interest parameter ω. Let $\hat{\psi}_0$ denote the restricted *ml* estimator of ψ assuming $\omega = \omega_0$ and let $\hat{\theta}_0 = (\omega_0, \hat{\psi}_0)$.

As we have seen, the *ml* estimator $\hat{\theta}$ has an approximately multivariate normal distribution $N_p(0, I^{-1}(\theta))$. Let $V(\omega, \psi)$ denote the $r \times r$ sub-matrix in the upper left-hand corner. Then the *ml* estimator of ω, $\hat{\Omega}$ is approximately r-variate normal with mean ω and variance-covariance matrix $V(\omega, \psi)$. Under H_0 we may estimate this variance by $\hat{V}_0 = V(\hat{\theta}_0)$. The quadratic form

$$W = (\hat{\Omega} - \omega_0)'\hat{V}_0^{-1}(\hat{\Omega} - \omega_0) \tag{3.2.8}$$

is called the Wald statistic and was proposed by Wald (1941). Under the null hypothesis, W follows approximately a $\chi^2_{(r)}$ distribution. In the simple case, where ω is a scalar, testing H_0 is a test of a single parameter, and W is the square of the usual z-statistic where $z = (\hat{\Omega} - \omega_0)/s.e(\hat{\Omega})$.

Rao's Score Statistic

The score unction $U(\theta)$ is a vector of length p. We have seen that under mild restrictions, $U(\theta)$ has mean zero and variance equal to the expected information matrix and that further the distribution is approximately multivariate normal. The *score statistic*, proposed by Rao (1947) is the standardized quadratic form

$$S = U(\hat{\theta}_0)'\hat{I}_0^{-1}U(\hat{\theta}_0)$$

where $\hat{I}_0 = I(\hat{\theta}_0)$ is the information matrix with $\hat{\theta}_0$ substituted for θ.

The score statistic judges the hypothesis H_0 by assessing how far $U(\hat{\theta}_0)$ is away from its full mean zero. The null distribution of S is approximately $\chi^2_{(r)}$ and it tends to be larger when H_0 is false. A computational advantage of S is that only the model H_0 needs to be fitted.

The Likelihood Ratio Statistic

The likelihood function $L(\theta)$ assesses the likelihood of a particular value of θ in the light of the given data. Under the hypothesis H_0, the likelihood function takes its maximum value $L(\hat{\theta}_0)$. Under the more general hypothesis H_1, the maximum value of the likelihood is $L(\hat{\theta})$ where $\hat{\theta}$ is the *mle* of θ under H_1. Clearly, $L(\hat{\theta}) > L(\hat{\theta}_0)$, because H_1 imposes less severe restrictions on θ than H_0. The ratio $L(\hat{\theta})/L(\hat{\theta}_0)$ measures how much more probable are the data under H_1 than under H_0. A sufficiently large ratio casts doubt on the tenability of H_0 itself. The statistic

$$LR = -2(\mathcal{L}(\hat{\theta}_0) - \mathcal{L}(\hat{\theta})) \tag{3.2.9}$$

is often called the *generalized* likelihood ratio statistic or simply likelihood ratio statistic as it generalizes the simple likelihood ratio for testing two simple hypotheses as proposed by Neyman and Pearson. The null distribution of LR in large samples is again $\chi^2_{(r)}$ and it tends to be larger when H_0 is false. A computational disadvantage of the LR statistic is that both the null and alternative models must be fitted to compute the LR statistic.

Pearson's Goodness-of-Fit Statistic holds only for categorical data and we shall introduce it after a glimpse of a multinomial or Pearson formulated categorical data model.

3.2.5 Transformation of the Central Limit Theorem

It can be shown that if $\hat{\theta}$ has an approximate normal distribution, then any reasonable function $\hat{\Omega} = \mathbf{g}(\hat{\theta})$ of $\hat{\theta}$ has also an approximate normal distribution (vide, e.g., Lloyd 1999).

Suppose that $g(\theta)$ is differentiable in θ and let $\mathbf{G}(\theta)$ be the Jacobian, $\mathbf{G} = \frac{\partial g(\theta)}{\partial \theta}$. Then,

$$Var(\hat{\Omega}) = \mathbf{GVG}'$$

where $\mathbf{V}$ is the covariance matrix of $\hat{\theta}$. It follows that

$$\hat{\Omega} \rightarrow^{d} N_r(\omega, \mathbf{G}\hat{\mathbf{V}}\mathbf{G}'). \qquad (3.2.10)$$

3.3 Distribution Theory for Count Data

We shall now outline the basic distribution theory of multinomial and Poisson models for a single categorical data set, and subsequently give the specific form of the statistical data set associated with one of these models.

3.3.1 Multinomial Models

If X is multinomially distributed and if the probability parameters in $\boldsymbol{\pi} = (\pi_1, \ldots, \pi_k)'$ are fully known, we can make precise statements about the future behavior of X, which are subject only to the unavoidable natural random effects.

In practice, the probability parameters in $\boldsymbol{\pi}$ are at least partially unknown and to infer their values we need data. A random sample is a single sample of n independent observations on X. These n observations form simply a list of n responses on the k categories. Denoting by Y_i the number of responses in category i, the statistics $Y_1, \ldots, Y_k$, $(\sum_k Y_k = n)$ are jointly sufficient for $\boldsymbol{\pi}$. Hence, by sufficiency principle, the data may be summarized by these counts. The joint distribution of the sufficient statistics $Y_1, \ldots, Y_k$ is given by the probability mass function

$$P_M(y_1, \ldots, y_n; n, k, \boldsymbol{\pi}) = \frac{n!}{y_1! y_2! \ldots y_k!} \pi_1^{y_1} \pi_2^{y_2} \ldots \pi_k^{y_k}, \quad \sum_{i=1}^{k} y_i = n \qquad (3.3.1)$$

which is called the multinomial distribution with parameters $n, k, \pi_1 \ldots, \pi_k$ and is here denoted by $M(n, k, \boldsymbol{\pi})$. Usually, k and n are known and it is the probability π_i that are unknown. These *fundamental parameters* will usually be related or restricted in some way by the statistical model assumed. Thus π_i's may depend on a smaller number $p < k - 1$ of *model parameters*, or *basic parameters* denoted by $\theta = (\theta_1, \ldots, \theta_p)'$. The joint distribution of the data under the model is then Eq. (3.3.1) with the expressions $\pi_i(\theta)$ substituted for π_i. Hence the log-likelihood function is

$$\mathcal{L}_M(\theta; y) = \log P_M(y_1, \ldots, y_n) = c + \sum_{i=1}^{k} y_i \log \pi_i(\theta) \tag{3.3.2}$$

where c is a constant, independent of θ. We denote derivatives of π_i with respect to components of θ by superscripts. For instance, π_3^{12} is the mixed derivative of $\pi_3(\theta)$ with respect to θ_1 and θ_2. Hence the score function whose jth component is

$$U_{Mj}(\theta) = \frac{\partial \mathcal{L}_M}{\partial \theta_j} = \sum_{i=1}^{k} y_i \frac{\pi_i^j(\theta)}{\pi_i(\theta)}. \tag{3.3.3}$$

It is readily seen that $E(U_{Mj}) = 0$. In most regular cases equating the score function to zero, one gets a unique solution which maximizes the likelihood. The MLE $\hat{\theta}$ determines the *fitted value* $\hat{e}_i = n\pi_i(\hat{\theta})$. The maximized log-likelihood is therefore

$$\mathcal{L}_M(\hat{\theta}) = c + \sum_{i=1}^{k} y_i \log(\hat{e}_i) - n\log(n). \tag{3.3.4}$$

Again, the (l, m)th component of the observed information matrix is

$$J_{lm} = -\frac{\partial \mathcal{L}_M}{\partial \theta_l \partial \theta_m} = \sum_{i=1}^{k} y_i \frac{\pi_i^l \pi_i^m - \pi_i^{lm} \pi_l}{(\pi_i)^2}. \tag{3.3.5}$$

Using the fact that the π_i^{lm} sum to zero and $E(Y_i) = n\pi_i$, we find that

$$I_{lm} = E\left(-\frac{\partial \mathcal{L}_M}{\partial \theta_l \partial \theta_m}\right) = n \sum_{i=1}^{k} \frac{\pi_i^l \pi_l^m}{\pi_i}, \tag{3.3.6}$$

This can also be obtained by finding the covariance of U_l and U_m using Var $(Y_i) = n\pi_l(1 - \pi_l)$ and $\text{Cov}(Y_l, Y_m) = -n\pi_l\pi_m$.

In the full model, the fundamental parameters are also the basic parameters and we take them to be

$$\theta = (\pi_1, \ldots, \pi_{k-1})$$

without loss of generality, though their values are unknown. Thus

$$\pi_k(\theta) = 1 - \pi_1 - \cdots \pi_{k-1}$$

and $\pi_i(\theta) = \pi_i$ for $i = 1, \ldots, k-1$ and so

$$\pi_k^j = -1, \quad \pi_i^j = \delta_{ij}, i < k, \quad \pi_i^{lm} = 0$$

where the Kronecker delta δ_{ij} equals 1 when $i = j$ and zero otherwise. Substituting this into (3.3.4) gives

$$U_{Mj} = \frac{\partial \mathcal{L}_M(\pi)}{\partial \pi_j} = \frac{y_j}{\pi_j} - \frac{y_k}{\pi_k}, \; j = 1, \ldots, k-1 \tag{3.3.7}$$

and so the estimates of the π_i are proportional to the y_i. Thus, $\hat{\pi}_i = y_i/n$, the sample proportion. The fitted values are $\hat{e}_i = n\hat{\pi}_i = y_i$, i.e., the data value themselves. Therefore, from Eq. (3.3.4) the maximized log-likelihood is

$$\mathcal{L}(\hat{\theta}) = c + \sum_{i=1}^{k} y_i \log y_i - n \log n. \tag{3.3.8}$$

Substitution into Eqs. (3.3.5) and (3.3.6) for the observed and expected information functions gives

$$J_{lm} = \frac{y_k}{\pi_k^2} + \delta_{lm}\frac{y_l}{\pi_l^2}, \quad I_{lm} = n\left(\frac{1}{\pi_k} + \frac{\delta_{lm}}{\pi_l}\right). \tag{3.3.9}$$

The inverse of the expected information matrix has entries $\pi_i(1-\pi_i)/n$ on the diagonal and entries $-\pi_l\pi_m/n$ off the diagonal. According to asymptotic theory, these are the approximate variances and covariances of the estimates $\hat{\pi}_i$. In fact, they are actually the exact variance covariances of the estimated parameters which follow easily from the known covariances of the Yi.

When the fundamental parameters $\pi = (\pi_1, \ldots, \pi_k)$ have known values

$$\pi_0 = (\pi_{10}, \pi_{20}, \ldots, \pi_{k0})$$

everything about the behavior of the categorical variable X and the counts Y_i is known and summarized in the known multinomial distribution Eq. (3.3.1). Here all the parameters are specified and therefore no parameter to estimate; such a model is called a *simple model*. Since there is no parameter to estimate, there is no score function or information matrices. The estimates of the π_i are the known values π_{i0}. The fitted values will be $\hat{e}_i = n\pi_{i0}$, which will generally disagree with the data values. The log-likelihood function at these "estimates" equals

$$l(\pi_0) = c - \sum_{i=1}^{k} y_i \log \pi_{i0}.$$

3.3.2 Poisson Models

The most common distribution for modeling counts in time (space) is the Poisson distribution. Since the values of a Poisson variables are nonnegative integers, the distribution is generally used to describe experiments in which the observed variable is a count. The Poisson distribution may also be used to to describe number of events occurring randomly and independently in time (space), e.g., number of α-particles emitted by a radioactive substance reaching a given portion of space during a given period of time.

Suppose events of k different types occur singly and independently during a sampling interval $[0, T]$. The probability of an event of type i occurring during time interval $[t, t+\delta]$ is supposed to be $\beta_i\delta$ for small δ. Let Y_i denote the total number of events of type i that occur. Then it can be shown that $Y_1, \ldots, Y_k$ are sufficient for the parameters $\beta_1, \ldots, \beta_k$ and so by sufficiency principle we may ignore the precise time of events and simply analyze the total counts. The distribution of each Y_i will be Poisson with mean $\mu_i = \beta_i T$ and since the Y_i's are independent their joint distribution is

$$P_P(y_1, \ldots, y_k; k, \mu) = e^{-\mu}\frac{\mu_1^{y_1}\ldots\mu_k^{y_k}}{y_1!\ldots y_k!}. \tag{3.3.10}$$

Here, unlike Multinomial distribution, the k *fundamental parameters* $\mu_1, \ldots, \mu_k$ are functionally independent. These parameters will usually be restricted or related somehow by $p \le k$ *basic parameters*, $\theta = (\theta_1, \ldots, \theta_p)$. The statistical model for the data $Y_1, \ldots, Y_k$ comprises the independent Poisson distribution Eq. (3.3.10) with $\mu_i(\theta)$ substituted for μ_i, $i = 1, \ldots, p$.

From (3.3.10) the log-likelihood function

$$\mathcal{L}_P(\theta; y) = c - \sum_{i=1}^{k}\mu_i(\theta) + \sum_{i=1}^{k} y_i \log\mu_i(\theta) \tag{3.3.11}$$

where $\theta = (\theta_1, \ldots, \theta_p)$ are now the main parameters of interest. As before, we denote the derivative of $\mu(\theta)$ wrt components of θ by superscripts. Thus, μ_2^{14} is the mixed derivative of μ_2 wrt θ_1 and θ_4. Unlike the multinomial model, the μ_i are not constrained in their total and so the sums of these derivatives need not equal zero. Taking the derivative of Eq. (3.3.11) wrt θ_j we get the jth component of the score function,

$$U_{Pj}(\theta) = \frac{\partial\mathcal{L}_P}{\partial\theta_j} = \sum_{i=1}^{k}\left\{\frac{y_i}{\mu_i(\theta)} - 1\right\}\mu_i^j(\theta).$$

It is readily seen that $E(U_{Pj}) = 0$. Equating the score function to zero, one gets a unique solution which maximizes the likelihood. The ML estimate $\hat{\theta}$ determines fitted values $\hat{e}_i = \mu_i(\hat{\theta})$. The maximized log-likelihood is then

$$\mathcal{L}_P(\hat{\theta}) = c + \sum_{i=1}^{k} y_i \log(\hat{e}_i) - \sum_{i=1}^{k} \hat{e}_i. \qquad (3.3.12)$$

Expressions for the information follow upon further differentiating U_j. The (l, m)th component of the observed information matrix is

$$J_{lm} = \sum_{i=1}^{k} y_i \frac{\mu_i^l \mu_i^m}{\mu_i^2} + \mu_i^{lm}\left(\frac{y_i}{\mu_i} - 1\right) \qquad (3.3.13)$$

and on taking expectation the expected information matrix has (l, m)th entry

$$I_{lm} = \sum_{i=1}^{k} \frac{\mu_i^l \mu_i^m}{\mu_i} = T \sum_{i=1}^{k} \frac{\beta_i^l \beta_i^m}{\beta_i}. \qquad (3.3.14)$$

The information thus increases proportionately with the sampling time T which, for Poisson sampling, is analogous to the sample size.

In the *full* Poisson model, the fundamental parameters $\mu_1, \ldots, \mu_k$ are also the basic parameters, thus $\mu_i = \theta_i, i = 1, \ldots, k$, though their values are unknown. Hence, $\mu_i^j = \delta_{ij}$ and the score function for θ_j is

$$U_{Pj}(\theta) = \frac{\partial \mathcal{L}_P}{\partial \theta_j} = \frac{y_j}{\mu_j} - 1 \qquad (3.3.15)$$

with solution $\mu_j = y_j$ and so the fitted values are the observed data values. The maximized log-likelihood is

$$\mathcal{L}_P(\hat{\theta}) = c + \sum_{i=1}^{k} y_i \log y_i - \sum_{i=1}^{k} y_i. \qquad (3.3.16)$$

If, however, the μ_i are assumed to have known values μ_{i0} then everything about the distribution of the data is known and the model is *simple*. The fitted values are the assumed model parameter values μ_{i0} and so the log-likelihood is

$$\mathcal{L}_P(\mu_0) = c + \sum_{i=1}^{k} y_i \log \mu_{i0} - \sum_{i=1}^{k} \mu_{i0}.$$

An intermediate model will neither assume values for all the fundamental parameters (as in a simple model) nor leave them entirely unrestricted (as in a full model), but will impose some r restrictions among the μ_i's where $0 < r < k$. The maximized likelihood will then be less than the maximized likelihood for the full model and the difference between the two forms the basis of a goodness-of-fit test discussed in Sect. 3.4.

3.3.3 *The Multinomial-Poisson Connection*

Suppose we observe a sample of k independent Poisson variables Y_i each with mean μ_i. The total number of observed counts $N = Y_1 + \cdots + Y_k$ will therefore be considered as a random variable whose mean value is $\mu_1 + \cdots + \mu_k$ and whose observed value is n. However, if N is fixed at n, the Y_i's would no longer be Poisson, since there would be a restriction on maximum value of each $Y_i (< n)$ and they would not be independent, since a larger value of Y_i would imply a smaller value for the other Y_j. Now,

$$P(Y_1 = y_1, \ldots, Y_k = y_k | N = n) = \frac{P(Y_i = y_i, \ldots, Y_k = y_k \text{ and } \sum_i y_i = n)}{P(\sum_i Y_i = n)}$$

$$= \frac{\Pi_i e^{-\mu_i} \mu_i^{y_i} / y_i!}{e^{-\sum_i \mu_i} (\sum_i \mu_i)^n / n!}$$

$$= \left(\frac{n!}{\Pi_i y_i!}\right) \Pi_i \pi_i^{y_i},$$

where $\pi_i = \mu_i / (\sum_j \mu_j)$. This probability function is, of course, the multinomial probability function (3.3.1). Thus Poisson sampling conditional on the total sample size is therefore equivalent to the multinomial sampling.

3.4 Goodness-of-Fit

A model once fitted, forms the basis for statements about parameters of interest, their structural relations and for predictions of future behavior. A goodness-of-fit test of model $\mathcal{M}$ is a test of

$$H_0 : \mathcal{M} \text{ is true against } H_1 : \text{ full model.}$$

Goodness-of-fit tests are based on comparing the observed number of observations falling in each cell to the number that would have been expected if the hypothesized model was true. If the observed and the expected numbers differ greatly, it is evidenced that the hypothesized model is not correct, that is, it does not fit well. We now consider different tests for goodness-of-fit.

3.4.1 *Likelihood Ratio Statistic*

The likelihood ratio (LR) test statistic is

$$LR = 2(\mathcal{L}_1 - \mathcal{L}_0)$$

where $\mathcal{L}_1$ is the maximized log-likelihood under the more general hypothesis H_1 and $\mathcal{L}_0$ is the maximized log-likelihood under the restrictive hypothesis H_0. Clearly, $\mathcal{L}_0$ is necessarily less than $\mathcal{L}_1$. For multinomial model we have from (3.3.2) and (3.3.8) the LR statistic

$$LR_M = 2\sum_{i=1}^{k} y_i \log\left(\frac{y_i}{\hat{e}_i}\right), \tag{3.4.1}$$

where the $\hat{e}_i$ are the fitted value $n\pi_i(\hat{\theta}_0)$ under the null hypothesis. This statistic may be used to test the goodness-of-fit of any multinomial model. If the number of parameters in the model is p (and the number of parameters in the full model is $k-1$) then the degrees of freedom of the approximating chi-square distribution will be $r = k - 1 - p$.

For Poisson models, Eqs. (3.3.12) and (3.3.16) give the LR statistic

$$LR_P = 2\sum_{i=1}^{k} y_i \log\left(\frac{y_i}{\hat{e}_i}\right) - 2\sum_{i=1}^{k}(y_i - \hat{e}_i) \tag{3.4.2}$$

where $\hat{e}_i$ are the fitted values $\mu_i(\hat{\theta}_0)$ under the null hypothesis. This statistic may be used to test the hypothesis of goodness-of-fit of any Poisson model. If the number of parameters in the model is p (and the number of parameters in the full model is k), then the degrees of freedom of the approximating chi-square distribution will be $k - p$.

The LR goodness-of-fit statistic is a measure of how close the vector of expected values is to the vector of observed values, and it will equal zero if, and only if, the two vectors agree exactly. Naturally, there are many other ways of measuring the agreement of two vectors giving alternative statistics (vide Sect. A.4). However, only the LR statistic measures the relative likelihood of the observed data under the two hypotheses.

3.4.2 Pearson's Goodness-of-Fit Statistic

For either a multinomial model or Poisson formulated categorical data model, the Pearsonian goodness-of-fit statistic is

$$X_P^2 = \sum_{i=1}^{k} \frac{(y_i - \hat{e}_i)^2}{\hat{e}_i} \tag{3.4.3}$$

where y_i is the frequency in the class i and $\hat{e}_i = e_i(\hat{\theta}) = E(Y_i; \hat{\theta})$ under the given model. The distribution of (3.4.3) is approximately $x^2_{(r)}$ where r is the difference in the number of free parameters under the model to be tested and under the full model.

The approximation applies as n increases for fixed k, or more accurately, as the smallest of the fitted values $\hat{e}_i$ grows unbounded.

The large sample properties of both G^2 and X_P^2 in the classical situation, that is, when the independent samples are drawn from an identical population, have been stated explicitly in the Appendix.

Both LR statistic G^2 and Person's statistic X_P^2 are special cases of a general class of statistics called the *power divergence* statistics, defined by Cressie and Read (1984) and Read and Cressie (1988). The statistic takes the form

$$I(\lambda) = \frac{2}{\lambda(\lambda+1)} \sum_{i=1}^{k} y_i \left[\left(\frac{y_i}{\hat{e}_i} \right)^{\lambda} - 1 \right], \quad -\infty < \lambda < \infty, \tag{3.4.4}$$

which measures the distance of the vector of fitted values $\hat{e}_i$ from the vector of observed y_i values. While (3.4.4) is undefined for $\lambda = 0$ or -1, these forms can be defined as the limits of (3.4.4) as $\lambda \to 0$ and $\lambda \to -1$, respectively. When $\lambda = 1$, (3.4.4) is Pearson's X_P^2 statistic. When $\lambda \to 0$, $I(\lambda)$ converges to the LR G^2. As $\lambda \to -1$, I converges to Kullback's minimum discrimination information statistic GM,

$$GM = 2 \sum_i \hat{e}_i \log \left(\frac{\hat{e}_i}{y_i} \right);$$

when $\lambda = 1/2$, it is the Freeman-Turkey (1950) statistic $(FT)^2$,

$$(FT)^2 = 4 \sum_i \left(\sqrt{y_i} - \sqrt{\hat{e}_i} \right)^2;$$

when $\lambda = -2$, it is the Neyman or modified Pearson statistic X_N^2, as proposed by Neyman (1949),

$$X_N^2 = \sum_i \frac{(y_i - \hat{e}_i)^2}{y_i}.$$

The statistic with $\lambda = 2/3$ has some commendable properties when $\hat{e}_i$ values are all greater than one and $n \geq 10$. All members of the power divergence family are asymptotically $\chi^2_{(k-1)}$ under the null hypothesis and under some regularity conditions when $n \to \infty$.

Besides, there exist a number of other goodness-of-fit tests, like the Kolmogorov–Smirnoff test, the Cramer–von Mises test, run tests. We shall however mostly concentrate on X_P^2 and G^2 tests, as these are of frequent use.

3.5 Binomial Data

We will now have a relook at the models for binomial data as it explores the potentiality of modeling the probability of an event, so-called π or p as a regression function on some auxiliary variable x. This concept will be taken up in Sect. 3.7. For the time being, we consider the form of binomial data.

3.5.1 *Binomial Data and the Log-Likelihood Ratio*

We assume that the data $Y_1, \ldots, Y_k$ are independent binomial variables, $Y_i \sim B(n_i, \pi_i)$ with respective parameters (n_i, π_i), The likelihood for k independent binomial random variables $Y_1, \ldots, Y_k$ is then the product of the individual binomial probability functions, i.e.,

$$P_B(y_1, \ldots, y_k; \pi_1, \ldots, \pi_k) = \Pi_{i=1}^k \binom{n_i}{y_i} \pi_i^{y_i}(1-\pi_i)^{n_i - y_i}. \tag{3.5.1}$$

The MLE of π_i from this likelihood is the observed proportion Y_i/n_i.

One is often interested in models that restrict the π_i's in some way. For example, a common null hypothesis of interest is that the π_i have equal value π. The above probability distribution then reduces to

$$\pi^y(1-\pi)^{n-t}\Pi_{i=1}^k \binom{n_i}{y_i} \tag{3.5.2}$$

where $n = \sum_i n_i$, $t = \sum_i y_i$. The MLE of π is $\hat{\pi} = T/n$ and T is the sample total. The statistic T is sufficient for π with $B(n, \pi)$ distribution. The fitted values for this model are $\hat{e}_i = n_i\hat{\pi} = n_i(t/n)$. For a hypothetical model H_0 with fitted values $\hat{e}_i$, and for the saturated or full model, where $\hat{\pi}_i = y_i/n_i$, the LR statistic

$$LR = 2\sum_{i=1}^{k}\left\{y_i \log\left(\frac{y_i}{\hat{e}_i}\right) + (n_i - y_i)\log\left(\frac{n_i - y_i}{n_i - \hat{e}_i}\right)\right\}, \tag{3.5.3}$$

is used for testing fit of H_0 to the data. When $y_i = 0$ or n_i, we define $0\log 0$ as 0.

It is often advantageous to transform the probabilities π_i to

$$\text{logit}\ \pi_i = \log\left(\frac{\pi_i}{1-\pi_i}\right) = \nu_i\ \text{(say)}, \tag{3.5.4}$$

which is also called as the *logistic, logit,* or *log-odds* transformation. Note that while π_i necessarily takes values in [0, 1], its transformation ν_i can take any value between $-\infty$ to $+\infty$. Also,

$$\text{logit}\ (1-\pi) = -\ \text{logit}\ (\pi) = -\nu_i.$$

Again, we can write

$$\pi_i = \exp\{\nu_i\}/(1 + \exp\{\nu_i\}). \tag{3.5.5}$$

In terms of the parameters $\nu_1, \nu_2, \ldots, \nu_k$, the log-likelihood function (3.5.1) becomes

$$\begin{aligned} \mathcal{L}_B(\nu_1, \ldots, \nu_k) &= \sum_{i=}^{k}\{y_i \log \pi_i + (n_i - y_i) \log(1 - \pi_i)\} \\ &= \sum_{i=1}^{k}\{y_i\nu_i - n_i \log(1 + e^{\nu_i})\} \end{aligned} \tag{3.5.6}$$

In Sect. 3.7, we will consider different forms of ν functions.

In considering models for categorical data we noted that the distribution of count data $Y_1, \ldots, Y_k$ depend on k or $k - 1$ cell- probabilities, which themselves depend on p parameters $\theta = (\theta_1, \ldots, \theta_p)$, $p < k$ (or $\leq k$). Therefore in the likelihood function, we replaced π_i by $\pi_i(\theta)$. We considered estimation of θ and hence of π_i as $\pi_i(\hat{\theta})$. We have not considered any special form of θ so far. Of course, the form of θ will depend on the choice of model. In Sects. 3.6 and 3.7, we shall consider two special forms of θ and hence two special types of models.

3.6 Log-Linear Models

3.6.1 Log-Linear Models for Two-Way Tables

In the log-linear models, natural logarithm of cell-probabilities is expressed in a linear model analogous to the analysis of variance (ANOVA) models.

For a 2×2 contingency table with π_{ij} denoting the probability of an element belonging to the (i, j)th cell, we write the model as

$$\ln \pi_{ij} = u + u_{1(i)} + u_{2(j)} + u_{12(ij)},\ i, j = 1, 2, \tag{3.6.1}$$

where u is the general mean effect, $u + u_{1(i)}$ is the mean of the logarithms of probabilities at level i of the first variable, $u + u_{2(j)}$ is the mean of the logarithm of probabilities at level j of the second variable. Thus,

$$\begin{aligned} u &= \frac{1}{4}\sum_i\sum_j \ln \pi_{ij}, \\ u + u_{1(i)} &= \frac{1}{2}(\ln \pi_{i1} + \ln \pi_{i2}),\ i = 1, 2, \\ u + u_{2(j)} &= \frac{1}{2}(\ln \pi_{1j} + \ln \pi_{2j}),\ j = 1, 2. \end{aligned} \tag{3.6.2}$$

Since $u_{1(i)}$ and $u_{2(j)}$ represent deviations from the grand mean u,

$$u_{1(1)} + u_{1(2)} = 0,$$

$$u_{2(1)} + u_{2(2)} = 0.$$

Similarly, $u_{12(ij)}$ represents deviation from $u + u_{1(i)} + u_{2(j)}$, so that

$$u_{12(11)} + u_{12(12)} = 0, \quad u_{12(21)} + u_{12(22)} = 0,$$

$$u_{12(11)} + u_{12(21)} = 0, \quad u_{12(11)} + u_{12(22)} = 0.$$

The general log-linear model for a $2 \times 2 \times 2$ table can be written as

$$\ln \pi_{ijk} = u + u_{1(i)} + u_{2(j)} + u_{3(k)} + u_{12(ij)} + u_{13(ik)} + u_{23(jk)} + u_{123(ijk)}, \quad i, j, k = 1, 2, \tag{3.6.3}$$

where $u_{1(i)}$ means the effect of factor 1 at level i, $u_{12(ij)}$, the interaction between level i of factor 1 and level j of factor 2, $u_{123(ijk)}$, the three-factor interaction among level i of factor 1, level j of factor 2, level k of factor 3, all the effects being expressed in terms of log-probabilities. We need

$$\sum_{i=1}^{2} u_{1(i)} = 0, \sum_{j=1}^{2} u_{2(j)} = 0, \sum_{k=1}^{2} u_{3(k)} = 0,$$

$$\sum_{j(\neq i)=1,2} u_{12(ij)} = 0, \quad \sum_{k(\neq i)=1,2} u_{13(ik)} = 0, \text{ etc..}$$

Suppose now there is a multinomial sample of size n over the $M = I \times J$ cells of an $I \times J$ contingency table; the first factor has I levels represented by rows and the second factor has J levels represented by columns. The cell-probabilities π_{ij} for that multinomial distribution form the joint distribution of the two categorical variables. If we define

$$\mu_{ij} = \log \pi_{ij},$$

then we can write

$$\mu_{ij} = u + u_{1(i)} + u_{2(j)} + u_{12(ij)}$$

by an analog with the ANOVA model. In this formulation,

(i) The first term, u is the grand mean of the logs of the probabilities,

$$u = \frac{\mu_{00}}{IJ}$$

where the zero sign (0) denotes the total when summing across levels over the corresponding factor.

(ii) $u + u_{1(i)}$ is the mean of the logs of the probabilities of the first factor when it is at level i,

$$\frac{\sum_j \log(\pi_{ij})}{J} = \frac{\mu_{i0}}{J}$$

so that $u_{1(i)}$ the deviation from the grand mean u:

$$u_{1(i)} = \frac{\mu_{i0}}{J} - \frac{\mu_{00}}{IJ}.$$

Thus, it satisfies

$$\sum_i \mu_{1(i)} = 0.$$

There are $(I - 1)$ such terms. These terms are often of no intrinsic interest, and represent only the *main effects* of that factor. Similarly, the main effects of the other factor are represented by $u_{2(j)}$, and there are $(J - 1)$ of these terms.

(iii) The remaining component,

$$u_{12(ij)} = \mu_{ij} - \frac{\mu_{i0}}{J} - \frac{\mu_{0j}}{I} + \frac{\mu_{00}}{IJ},$$

can be regarded as measures of departures from the independence of two factors. Since

$$\sum_i u_{12(ij)} = \sum_j u_{12(ij)} = 0,$$

there are $(I - 1)(J - 1)$ of these *interaction terms*. The number is called *the degrees of freedom* in testing for the null hypothesis of independence.

In the case of a general two-way table, numerical values of these $u_{12(ij)}$ terms indicate at which level i of factor 1 and level j of factor 2 the interaction is strong; the negative or positive sign of $u_{12(ij)}$ is not important, it reflects only the arbitrary coding. The results may seem trivial because one can reach the same conclusion by simply inspecting the cell-probabilities. However, when we generalize the models for use with higher dimensional tables, this result may be extremely useful.

3.6.2 *Log-Linear Models for Three-Way Tables*

In a typical study, even if we are interested only in the relationship between a response and an explanatory variable, we still have to control for at least one confounder variable that can influence the relationship under investigation. For example, while studying the relationship between incidence of cancer and smoking habit, we may also have to consider the nature of job of the subjects, as it may affect both the variables. Therefore, we generally end up studying at least three factors simultaneously. Of course, the relationship among the three factors is far more complicated than in the case of two factors. In the log-linear modeling approach, we model the cell-probabilities or, equivalently, *cell counts* or *frequencies* in a contingency table in terms of association among the variables. Suppose there is a multinomial sample of size n over the $M = IJK$ contingency table (I, J,and K are the number of categories for the factors X_1, X_2, X_3, respectively.). Then the log-linear model (3.6.1) for the two-way table can be generalized and expressed for three-way tables as follows.

$$\mu_{ijk} = u + u_{1(i)} + u_{2(j)} + u_{3(k)} + u_{12(ij)} + u_{13(ik)} + u_{23(jk)} + u_{123(ijk)}, \quad (3.6.4)$$

subject to similar constraints (i.e., summing across indices to zero), where $\mu_{ijk} = \ln \pi_{ijk}$. The full or saturated model decomposes the observed frequency n_{ijk}, because the expected cell frequency can be expressed as $m_{ijk} = n\pi_{ijk}$. The model (3.6.4) consists of:

(i) A constant u.
(ii) Terms representing main effects $u_{1(i)}, u_{2(j)}, u_{3(k)}$. There are $(I-1)$, $(J-1)$, and $(K-1)$ of these main effects terms for the three factors, respectively; these terms are only influenced by the marginal distributions of these three factors and, are often of no intrinsic interests.
(iii) Terms representing two-factor interactions, $u_{12(ij)}, u_{13(ik)}$, and $u_{23(jk)}$; such terms are $(I-1)(J-1)$, $(I-1)(K-1)$, and $(J-1)(K-1)$ in number, respectively.
(iv) Terms used as measures of three-factor interactions, $u_{123(ijk)}$. There are $(I-1)(J-1)(K-1)$ of these three-factor interaction terms.

If the terms in the last group are not zero, i.e., three-factor interaction is present, the presence or absence of a factor would modify the relationship between the other two factors.

The models of independence

We may want to see if:

(i) The three factors are *mutually independent*: That is, whether

$$P(X_1 = i, X_2 = j, X_3 = k) = P(X_1 = i)P(X_2 = j)P(X_3 = k)$$

or if

(ii) One factor, say X_3, is *jointly independent* of the other two factors. That is, whether

$$P(X_1 = i, X_2 = j, X_3 = k) = P(X_1 = i, X_2 = j)P(X_3 = k)$$

or if

(iii) Two factors, say X_1 and X_2, are *conditionally independent* given the third factor. That is, whether

$$P(X_1 = i, X_2 = j, X_3 = k) = P(X_1 = i|X_3 = k)P(X_2 = j|X_3 = k).$$

The last item, the concept of conditional independence, is very important and is often the major aim of an epidemiological study.

Starting with the saturated model (3.6.4), we can translate a null hypothesis of independence into a log-linear model by setting certain of the above u-terms equal to zero. The various concepts of independence for three-way tables can be grouped as follows:

Types of Independence	Symbol	Log-linear model
Mutual Independence	(X_1, X_2, X_3)	$u + u_1 + u_2 + u_3$
Joint Independence	(X_1, X_2X_3)	$u + u_1 + u_2 + u_3 + u_{23}$
from two factors	(X_2, X_1X_3)	$u + u_1 + u_2 + u_3 + u_{13}$
	(X_3, X_1X_2)	$u + u_1 + u_2 + u_3 + u_{12}$
Conditional	(X_1X_3, X_2X_3)	$u + u_1 + u_2 + u_3 + u_{13} + u_{23}$
Independence	(X_1X_2, X_2X_3)	$u + u_1 + u_2 + u_3 + u_{12} + u_{23}$
	(X_1X_2, X_1X_3)	$u + u_1 + u_2 + u_3 + u_{12} + u_{13}$
No Three-Factor	(X_1X_2, X_1X_3, X_2X_3)	$u + u_1 + u_2 + u_3 + u_{12} + u_{13} + u_{23}$

(subscripted factor-levels are dropped for simplicity, e.g., u_1 is used for $u_{1(i)}$.) We now consider relationship between terms of a log-linear model and hierarchy of models.

Relationships between terms and hierarchy of models

A log-linear model term u_A is a lower order term relative of term u_B if A is a subset of B. For example, u_2 is a lower order term relative of u_{23} and u_{12} is a lower order term relative of u_{123}. If u_A is a lower order relative of u_B, then u_B is a higher order relative of u_A. For example, u_2 is a lower order relative of u_{23}, and u_{123} is a higher order relative of u_{23}. A log-linear model is a *hierarchical model* under the following conditions:

(i) If a u-term is zero, then all of its higher order relatives are zero; and
(ii) if a u-term is not zero, then all of its lower order relatives are not zero.

For example, all eight models of independence for three-way tables given in the above table are hierarchical. However, the model

$$H : u + u_1 + u_2 + u_3 + u_{123}$$

is a non-hierarchical model.

Testing a specific model

Given the data in a three-way table, we have two different types of statistical inferences:

(i) To test a specific model, for example, we may want to know whether two specific factors are conditionally independent given the third factor.
(ii) To search for a model that can best explain the relationship(s) found in the observed data.

We shall consider the first issue here, as the log-linear model under investigation is generally the result of an enquiry concerning a relationship between factors, which leads to testing a null hypothesis.

Expected Frequencies
Expected frequencies are cell counts obtained under the null hypothesis. Given the null hypothesis, these expected frequencies can be easily determined for the hierarchical model. For example, if we consider the model of conditional independence between X_1 and X_2 given X_3, i.e., independence between (X_1X_3, X_2X_3), or

$$P(X_1 = i, X_2 = j|X_3 = k) = P(X_1 = i|X_3 = k)P(X_2 = j|X_3 = k),$$

then it can be shown that expected frequency

$$\hat{m_{ijk}} = \frac{x_{i0k}x_{0jk}}{x_{00k}}$$

where the x's are observed cell counts and the zero sign indicates a summation across the index. However, in practice these tedious jobs are generally left to computer packaged programs, such as SAS.

Test Statistic
When measuring goodness-of-fit in two-way tables, we often rely on the Pearson's chi-square statistic:

$$\sum_{i,j,k} \frac{(x_{ijk} - \hat{m_{ijk}})^2}{\hat{m_{ijk}}}. \tag{3.6.5}$$

However, for a technical reason, we compare the frequencies and the expected frequencies in higher dimensional tables using the likelihood ratio chi-square statistic:

$$G^2 = 2\sum_{i,j,k} x_{ijk} \log \frac{x_{ijk}}{\hat{m_{ijk}}}. \tag{3.6.6}$$

The degrees of freedom for the above likelihood ratio chi-square statistic is equal to the number of u-term which are set equal to zero in the model being tested. For example, if we want to test the model of conditional independence between X_1 and X_2,

$$H_0 : u + u_1 + u_2 + u_3 + u_{13} + u_{23},$$

for which we set u_{12} and u_{123} equal to zero, the degrees of freedom is

$$df = (I - 1)(J - 1) + (I - 1)(J - 1)(K - 1).$$

Similarly, if we want to test for the model of no three-factor interaction,

$$H_0 : u + u_1 + u_2 + u_3 + u_{12} + u_{13} + u_{23},$$

the degrees of freedom is

$$d.f. = (I - 1)(J - 1)(K - 1).$$

Searching for the best model

We first define when a log-linear model is said to be *nested.* A log-linear model H_2 is *nested* in model H_1 if every nonzero term u in H_2 is also contained in $H_1 (H_2 < H_1)$. For example, if we denote

$$\begin{aligned} H_1 &= u + u_1 + u_2 + u_3 + u_{12} + u_{13} + u_{23} + u_{123} \\ H_2 &= u + u_1 + u_2 + u_3 + u_{12} + u_{13} + u_{23} \\ H_3 &= u + u_1 + u_2 + u_3 + u_{12} + u_{23}, \end{aligned}$$

then

(i) $H_3 < H_2$, and
(ii) $H_2 < H_1$.

In this nested hierarchy of models, it can be shown that if

$$H_3 < H_2 < H_1,$$

then the likelihood ratio chi-square statistics satisfy the reversed inequality

$$\chi^2(H_1) < \chi^2(H_2) < \chi^2(H_3).$$

For Pearson goodness-of-fit statistic X_P^2, this property does not necessarily hold for every set of nested models. Furthermore, it can be shown that if a model H_2 is *nested* in a model H_1, then

$$\chi^2 = \chi^2(H_2) - \chi^2(H_1)$$

is distributed as chi-square with

$$df = df(H_2) - df(H_1).$$

This property also does not hold good for Person's chi-squares.

3.7 Logistic Regression Analysis

The purpose of most research projects is to assess relationships among a set of variables and regression techniques which are often suitable instruments for the statistical analysis of such relationships. Research designs may be classified as experimental or observational. Regression analysis is applicable to both types. In most cases, one variable is usually taken to be the response or dependent variable, that is, a variable to be predicted from or explained by other variables. The other variables are called predictors, or explanatory or independent variables. Choosing an appropriate model and analytical technique depends on the type of the dependent variable under investigation. In a variety of applications, the dependent variable of interest may have only two possible outcomes, and therefore can be represented by an indicator variable taking on values 0 and 1. Consider a study designed to investigate risk factors for cancer. Attributes of people are recorded, including age, gender, smoking pattern, and so on. The target variable is whether or not the person has lung cancer (a 0/1) variable, with 0 for no lung cancer and 1 for the presence of lung cancer. The above example and others show a wide range of applications in which the dependent variable is dichotomous, and hence may be represented by a variable taking the value 1 with probability π and the value o with probability $1 - \pi$. Such a variable is a binomial variable and the model often used to express the probability π as a function of potential independent variables under investigation is the logistic regression model.

3.7.1 The Logistic Regression Model

The goal of usual regression analysis is to describe the 'mean' of a dependent variable Y given the value (s) of the independent variable (s). This quantity is called the conditional mean and will be denoted as $E(Y|x)$, where Y denotes the outcome variable and x a value of the independent variable. In linear regression, we assume that this mean may be expressed as a linear equation in x, such as $E(Y|x) = \beta_0 + \beta_1 x$.

In logistic regression, we use the quantity $\pi(x)$ to represent the conditional mean of Y given x. We use the following specific form of $\pi(x)$,

$$\pi(x) = \frac{e^{\beta_0+\beta_1 x}}{1 + e^{\beta_0+\beta_1 x}}. \tag{3.7.1}$$

A transformation of $\pi(x)$ that is central to our study is the *logit* transformation

$$\nu(x) = \text{logit}\,\{\pi(x)\} = \ln\left\{\frac{\pi(x)}{1-\pi(x)}\right\} = \beta_0 + \beta_1 x. \tag{3.7.2}$$

The function $\nu(x)$ has many of the desirable properties of a linear regression model. The logit $\nu(x)$ is linear in β's and may range from $-\infty$ to ∞, depending on the range of x.

As noted before, the other important difference between the linear and logistic regression model concerns the conditional distribution of Y given x. In linear regression model, we assume that the conditional value of the outcome variable may be expressed as $y = E(Y|x) + \epsilon$ where the error $\epsilon \sim N(0, \sigma^2)$ generally. It then follows that the conditional distribution of Y will be normal with mean $E(Y|x)$ and a constant variance. In case of a dichotomous outcome variable, the situation is completely different. In this situation, we may express the conditional value of the outcome variable as $y = \pi(x) + \epsilon$, where the error ϵ can take only two possible values. If $y = 1$, then ϵ takes the value $1 - \pi(x)$ with probability $\pi(x)$; if $y = 0$, then ϵ takes the value $-\pi(x)$ with probability $1 - \pi(x)$. Thus the conditional distribution of Y follows a binomial distribution with mean $\pi(x)$. Such a random variable is called a *point-binomial* or *Bernoulli variable* and it has the simple discrete probability distribution

$$P(Y = y) = \pi^y (1-\pi)^{1-\pi}; \quad y = 0, 1.$$

3.7.2 Fitting the Logistic Regression Model

Suppose we have a sample of n independent observations of the pair $(x_i, y_i), i = 1, \ldots, n$, where y_i denotes the value of a dichotomous outcome variable and x_i is the value of the independent variable for the ith subject. We also assume that the outcome variable has been coded as 0 or 1, representing the absence or presence of the characteristic, respectively. To fit the logistic regression model in Eq. (3.7.1) to a set of data, we have to estimate the values of β_0 and β_1, the unknown parameters.

In linear regression, we generally use the method of least squares for estimating the unknown parameters, β_0 and β_1. Under the usual assumptions for linear regression, the method of least squares yields estimators with a number of desirable statistical properties. Unfortunately, when the method of least squares is applied to a model with a dichotomous outcome, the estimators no longer have these properties.

The general method of estimation that leads to the least square function under the linear regression model (when the error terms are normally distributed) is the *maximum likelihood* method. We shall use the same method for estimation in the logistic regression model. Under the above simple logistic regression model, the likelihood function is given by

$$
\begin{aligned}
L(\beta) &= \Pi_{i=1}^{n} P(Y_i = y_i) \\
&= \Pi_{i=1}^{n} y_i^{\pi} (1 - y_i)^{1-\pi} \\
&= \Pi_{i=1}^{n} \pi(x_i)^{y_i} [1 - \pi(x_i)]^{1-y_i}; \; y_i = 0, 1,
\end{aligned} \tag{3.7.3}
$$

from which we can obtain 'maximum likelihood estimates' of the parameters β_0 and β_1. However, it is easer mathematically to work with the log of Eq. (3.7.3). Therefore, we have

$$\mathcal{L}(\beta) = ln[L(\beta)] = \sum_{i=1}^{n} \{y_i ln[\pi(x_i)] + (1 - y_i) ln[1 - \pi(x_i)]\}. \tag{3.7.4}$$

To find the MLE of β we differentiate $L(\beta)$ wrt β_0 and β_1 and set the resulting expressions to zero. The likelihood equations are:

$$\sum [y_i - \pi(x_i)] = 0, \tag{3.7.5}$$

and

$$\sum x_i [y_i - \pi(x_i)] = 0. \tag{3.7.6}$$

The expressions in Eqs. (3.7.5) and (3.7.6) are nonlinear in β_0 and β_1, and thus requires special methods for their solution. These methods are iterative in nature and have been programmed into available logistic regression software. The interested reader may see the text by McCullagh and Nelder (1989) for a general discussion of the methods used by most programs. In particular, they show that the solution to Eqs. (3.7.5) and (3.7.6) may be obtained using an iterative weighted least squares procedure.

3.7.3 *The Multiple Logistic Regression Model*

Suppose we have a vector of p independent variables $\mathbf{x} = (x_1, x_2, \ldots, x_p)'$. For the time being, we shall assume that each of these variables is a continuous variable in its own range. The response variable Y is a two-value categorical variable, value 1 for presence and value 0 for absence of certain characteristic. Of course, the value of Y depends on the values of the quantitative and continuous variables $\mathbf{x}$. Let the conditional probability that the outcome is present be denoted by $P(Y = 1|\mathbf{x}) = \pi(\mathbf{x})$. Therefore $P(Y = 0|\mathbf{x}) = 1 - \pi(\mathbf{x})$. In the multiple logistic regression model, we assume that the logit $(\pi(\mathbf{x}))$ is give by the equation

$$\text{logit}(\pi(\mathbf{x})) = \nu(\mathbf{x}) = \log \frac{(\pi(\mathbf{x}))}{1 - \pi(\mathbf{x})} = \beta_0 + \beta_1 x_1 + \cdots + \beta_p x_p, \tag{3.7.7}$$

and the logistic regression model is

$$\pi(\mathbf{x}) = \frac{e^{\nu(\mathbf{x})}}{1 + e^{\nu(\mathbf{x})}}. \tag{3.7.8}$$

If some of the independent variables are discrete, nominal scale variables such as race, sex, age-groups, and so on, they should not be included in the model as if they were interval-scale—variables. In this situation, the method is to use a collection of *design variables* (or *dummy variables*). Suppose, for example, that one of the independent variables is marital status, which has three categories: unmarried, married, and widowed. In this case, two design variables are necessary. One possible coding strategy is that when the respondent is "unmarried", the two design variables D_1 and D_2 would both be set equal to zero; when the respondent is "married," D_1 would be set equal to 1 while D_2 would still equal 0; when the status of the respondent is "widowed", we would use $D_1 = 0$ and $D_2 = 1$.

In general, if a nominal scaled variable has k possible values, then $k - 1$ design variables will be needed. As an illustration of using the notation for these design variables, suppose that the jth independent variable x_j is a categorical variable having k_j categories. These $k_j - 1$ design variables will be denoted as D_{jr} and the coefficients for these design variables will be denoted as $\beta_{jr}, r = 1, 2, \ldots, k_j - 1$. Thus, the logit for a model with p auxiliary variables of which the jth variable is discrete would be

$$\nu(\mathbf{x}) = \beta_0 + \beta_1 x_1 + \cdots + \beta_{j-1} x_{j-1} + \sum_{r=1}^{k_j - 1} \beta_{jr} D_{jr} + \beta_{j+1} x_{j+1} + \cdots + \beta_p x_p. \tag{3.7.9}$$

3.7.4 *Fitting the Multiple Logistic Regression Model*

Suppose that we have a sample of n independent observations $(\mathbf{x}_i, y_i), i = 1, \ldots, n$. As in the univariate case, for fitting the model we have to obtain estimates of the vector $\beta' = (\beta_0, \beta_1, \ldots, \beta_p)$, for which we will use the maximum likelihood method as in the univariate case. The likelihood function will be almost identical to that given in Eq. (3.7.4), with the only difference that $\pi(\mathbf{x})$ is now defined as in Eq. (3.7.8). There will be $p + 1$ likelihood equations. These are obtained by differentiating the log-likelihood function with respect to the $p + 1$ coefficients $\beta_0, \ldots, \beta_p$. The resulting likelihood equations may be expressed as follows:

$$\sum_{i=1}^{n} [y_i - \pi(\mathbf{x}_i)] = 0 \tag{3.7.10}$$

and

$$\sum_{i=1}^{n} x_{ij} [y_i - \pi(\mathbf{x}_i)] = 0 \tag{3.7.11}$$

for $j = 1, 2, \ldots, p$. Here again, the solution of the likelihood equations requires special software which is available in most statistical packages. Let $\hat{\boldsymbol{\beta}}$ denote the solution to these equations. Thus, the fitted values for the multiple logistic regression model are $\hat{\pi}(\mathbf{x}_i)$, the value of the expression in Eq. (3.7.8) computed using $\hat{\boldsymbol{\beta}}$ and $\mathbf{x}_i$.

The variance–covariance matrix of $\hat{\boldsymbol{\beta}}$ is estimated following the theory of maximum likelihood equation (see, for example, Rao 1947). According to this theory, the estimators are obtained from the matrix of second-order partial derivatives of the log-likelihood function. Now,

$$\frac{\partial^2 \mathcal{L}(\boldsymbol{\beta})}{\partial \beta_j^2} = -\sum_{i=1}^{n} x_{ij}^2 \pi_i (1 - \pi_i) \tag{3.7.12}$$

and

$$\frac{\partial^2 \mathcal{L}(\boldsymbol{\beta})}{\partial \beta_j \partial \beta_l} = -\sum_{i=1}^{n} x_{ij} x_{il} \pi_i (1 - \pi_i) \tag{3.7.13}$$

for $j, l = 0, 1, \ldots, p$ where π_i denotes $\pi(\mathbf{x_i})$. The *observed information matrix* $J(\boldsymbol{\beta})$ is the $(p + 1) \times (p + 1)$ matrix whose terms are the negative of the terms in (3.7.12) and (3.7.13). Then, dispersion matrix of $\hat{\boldsymbol{\beta}}$ is $\text{Var}(\hat{\boldsymbol{\beta}}) = \mathbf{J}^{-1}(\boldsymbol{\beta})$. It is not possible to write down an explicit expression for the elements in the matrix, except for special cases.

A formulation of the information matrix which will be very useful in discussing model-fitting is

$$\hat{\mathbf{J}}(\hat{\boldsymbol{\beta}}) = \mathbf{X}'\mathbf{V}\mathbf{X} \tag{3.7.14}$$

where $\mathbf{X}$ is an $n \times (p+1)$ matrix containing the data for each subject (the first column is $\mathbf{1}_n$) and $\mathbf{V}$ is an $n \times n$ diagonal matrix with general diagonal element $\hat{\pi}_i(1 - \hat{\pi}_i)$. Writing $V\hat{a}r(\hat{\beta}_j)$ as the estimated Var $(\hat{\beta}_j)$ the univariate Wald test statistics for testing the significance of β_j is

$$W_j = \hat{\beta}_j / \sqrt{V\hat{a}r(\hat{\beta}_j)}. \tag{3.7.15}$$

The multivariate Wald test for the significance of $\boldsymbol{\beta}$ is

$$\begin{aligned} W &= \hat{\boldsymbol{\beta}}'[V\hat{a}r(\hat{\boldsymbol{\beta}})]^{-1}\hat{\boldsymbol{\beta}} \\ &= \hat{\boldsymbol{\beta}}'(\mathbf{X}'\mathbf{V}\mathbf{X})\hat{\boldsymbol{\beta}}, \end{aligned} \tag{3.7.16}$$

which will be distributed as $\chi^2_{(p+1)}$ under the null hypothesis $H_0(\boldsymbol{\beta} = \mathbf{0})$.

3.7.5 Polytomous Logistic Regression

The logistic regression models stated above apply only to binary data where Y can take only two values 0 and 1, i.e., it has only two levels. We now consider the situations where Y has $J(\geq 2)$ levels.

Consider the situation when there is no natural ordering to the levels of the target variable Y. Multinomial (or polytomous) logistic regression models or more simply, multinomial logit models are derived in the following way. Let π_j, $j = 1, \ldots, J$ be the probability that a unit falls in the jth level of a multinomial target variable Y. Our goal is to construct a model for π_j as a function of the set of values of the predictor variables $\mathbf{x} = (x_1, \ldots, x_p)'$, $\sum_j \pi_j = 1$. The model is again based on logits, but the difficulty here is that there is no single 'success' (or 'failure') on which the model has to be based. We can however construct all of the logits relative to one of the levels, which we shall term *baseline* level. If one of the target levels is seen to be of different type than the other levels, it is natural to choose this level as 'baseline' level. For instance, in a clinical trial the level corresponding to the control group may be taken as baseline.

Suppose the Jth category is the baseline category. The logistic regression model is then

$$\log\left(\frac{\pi_j}{\pi_J}\right) = \beta_{0j} + \beta_{1j}x_1 + \cdots + \beta_{pj}x_p,\ j = 1, \ldots, J - 1. \tag{3.7.17}$$

Thus there are $J - 1$ separate regression equations, each of which is based on distinct set of parameters $\boldsymbol{\beta}_j = (\beta_{0j}, \beta_{1j}, \ldots, \beta_{pj})'$. The choice of the baseline is arbitrary. Thus, if the category I is taken to be the baseline category, then from (3.7.17),

$$\begin{aligned}\log\left(\frac{\pi_j}{\pi_I}\right) &= \log\left(\frac{\pi_j/\pi_J}{\pi_I/\pi_J}\right)\\ &= \log\left(\frac{\pi_j}{\pi_J}\right) - \log\left(\frac{\pi_I}{\pi_J}\right)\\ &= (\beta_{0j} - \beta_{0I}) + (\beta_{1j} - \beta_{1I})x_1 + \cdots + (\beta_{pj} - \beta_{pI})x_p\end{aligned} \tag{3.7.18}$$

We note that there is nothing in this derivation which requires category I to be a baseline category. Thus the choice of a baseline category is arbitrary.

Model (3.7.17) implies a single functional form for these probabilities. The form is a logistic relationship

$$\pi_j = \frac{\exp(\beta_{0j} + \beta_{1j}x_1 + \cdots + \beta_{pj}x_{pj})}{\sum_{k=1}^{J} \exp(\beta_{0k} + \beta_{1k}x_1 + \cdots + \beta_{pk}x_p)}. \tag{3.7.19}$$

The estimators $(\boldsymbol{\beta}_1, \ldots, \boldsymbol{\beta}_{J-1})'$ are obtained using maximum likelihood, where the log-likelihood is

$$\mathcal{L} = \sum_{j=1}^{J} \sum_{y_i=j} \log \pi_{j(i)}, \tag{3.7.20}$$

where the second summation is over all observations i whose response level is j and $\pi_{j(i)}$ is the probability (3.7.19) with values of the predictors corresponding to the ith observation substituted.

3.8 Fitting the Logistic Regression Models to Data from Complex Surveys

Of late, there have been some developments in logistic regression statistical software by including routines to perform analysis with data obtained from complex sample surveys. These routines may be found in STATA, SUDAAN (1997), and other less well-known special purpose packages. In this section, we shall try to provide a brief introduction to these methods. For more details, the reader may refer to Korn and Graubard (1990), Skinner et al. (1989), Roberts et al. (1987), and Thomas and Rao (1987). Recently, Lumley (2010) has given a comprehensive description of fitting a logistic regression model using R.

The central idea in fitting Eq. (3.7.8) to data from complex surveys is to set up a function that approximates the likelihood function in the finite sampled population with a likelihood function formed with the observed sample and known sampling weights. Suppose that the population is divided into H strata, the hth strata containing N_h primary sampling units (psu) ha, $a = 1, \ldots, N_h$ and the hath psu containing M_{ha} second stage units (ssu) hab, $b = 1, \ldots, M_{ha}$ in the population. Suppose again that from the hth stratum n_h psu's are sampled from the N_h psu's in the population and from the hath sampled psu, m_{ha} ssu's are sampled. Denote the total number of sampled ssu's as $m = \sum_{h=1}^{H} \sum_{a=1}^{n_h} m_{ha}$. For the habth observation denote the known sampling weight as w_{hab}, the vector of covariates as $\mathbf{x}_{hab}$ and the dichotomous outcome as y_{hab}. The approximate log-likelihood function is then

$$\sum_{h=1}^{H} \sum_{a=1}^{n_h} \sum_{b=1}^{m_{ha}} [w_{hab} \times y_{hab}] \times ln[\pi(\mathbf{x}_{hab})] + [w_{hab} \times (1 - y_{hab})] \times ln[1 - \pi(\mathbf{x}_{hab})]. \tag{3.8.1}$$

(Compare it with the log-likelihood function (3.7.4).) Differentiating this equation with respect to the unknown regression coefficients $\boldsymbol{\beta}$, we get the $(p+1) \times 1$ vector of score equations

$$\mathbf{X}'\mathbf{W}(\mathbf{y} - \boldsymbol{\pi}) = \mathbf{0}, \tag{3.8.2}$$

where $\mathbf{X}$ is the $m \times (p+1)$ matrix of covariates, $\mathbf{W}$ is an $m \times m$ diagonal matrix containing the weights w_{hab}, $\mathbf{y}$ is the $m \times 1$ vector of observed outcomes and $\boldsymbol{\pi} = (\pi(\mathbf{x}_{111}), \ldots, \pi(\mathbf{x}_{Hn_Hm_{Ha}}))'$ is the $m \times 1$ vector of logistic probabilities.

In theory, any logistic regression package that allows weights could be used to obtain the solutions to Eq. (3.8.2). However, the problem comes in obtaining the correct estimator of Var($\hat{\beta}$). Naive use of a standard logistic regression package with weight matrix **W** would yield instead estimates of the matrix $(\mathbf{X}'\mathbf{D}\mathbf{X})^{-1}$ where $\mathbf{D} = \mathbf{W}\mathbf{V}$ is an $m \times m$ diagonal matrix with general element $w_{hab} \times (\hat{\pi}(\mathbf{x}_{hab}))[1-\hat{\pi}(\mathbf{x}_{hab})]$. The correct estimator is

$$\hat{Var}(\hat{\beta}) = (\mathbf{X}'\mathbf{D}\mathbf{X})^{-1}\mathbf{S}(\mathbf{X}'\mathbf{D}\mathbf{X})^{-1}, \tag{3.8.3}$$

where **S** is a pooled within-stratum estimator of the covariance matrix of the left-hand side of Eq. (3.8.2).

We denote a general element in the vector in (3.8.2) as $\mathbf{z}'_{hab} = \mathbf{x}'_{hab} w_{hab}(y_{hab} - \pi(\mathbf{x}_{hab}))$, the sum over the m_{ha} sampled units in the ath primary sampling unit in the hth stratum as $\mathbf{z}_{ha} = \sum_{b=1}^{n_{ha}} \mathbf{z}_{hab}$ and their stratum-specific mean as $\bar{\mathbf{z}}_h = \sum_{a=1}^{n_h} \mathbf{z}_{ha}/n_h$. The within-stratum estimator for the hth stratum is

$$\mathbf{S}_h = \frac{n_h}{n_h - 1}\sum_{a=1}^{n_h}(\mathbf{z}_{ha} - \bar{\mathbf{z}}_h)(\mathbf{z}_{ha} - \bar{\mathbf{z}}_h)'.$$

The pooled estimator is $\mathbf{S} = \sum_{h=1}^{H}(1 - f_h)\mathbf{S}_h$. The quantity $(1 - f_h)$ is called the finite population correction factor where $f_h = n_h/N_h$ is the ratio of the number of observed primary sampling units to the total number of primary sampling units in stratum h. In settings where N_h is unknown, it is common practice to assume it is large enough that f_h is quite small and the correction factor is equal to one.

We note that the likelihood function (3.8.1) is only an approximation to the true likelihood. Thus, one would expect that inferences about model parameters should be based on univariable and multivariable Wald statistics as in (3.7.15) and (3.7.16), computed from specific elements of (3.8.3). However, simulation studies in Korn and Graubard (1990) as well as Thomas and Rao (1987) show that when data come from a complex survey from a finite population, use of a modified Wald statistic which follows an F-distribution under null hypothesis (details in Chap. 4) provide tests with better adherence to the stated type I error. Results from these modified Wald tests are reported in STATA and SUDAAN. For further details, the reader may refer to Hosmer and Lemeshow (2000).

The readers interested in classical theory of analysis of categorical data may also refer to Bishop et al. (1975), Lloyd (1999), Le (1998), Von and Mun (2013), among many others.

Chapter 4
Analysis of Categorical Data Under a Full Model

Abstract Nowadays, large-scale sample surveys are often conducted to collect data to test different hypotheses in natural and social sciences. Such surveys often use stratified multistage cluster design. Data obtained through such complex survey designs are not generally independently distributed and as a result multinomial models do not hold in such cases. Thus, the classical Pearson statistic and the related usually used test statistic would not be valid tools for testing different hypotheses in these circumstances. Here we propose to investigate the effect of stratification and clustering on the asymptotic distribution of Pearson statistic, log-likelihood ratio statistic for testing goodness-of-fit (simple hypothesis), independence in two-way contingency tables, and homogeneity of several populations.

Keywords Pearson's statistic (X_P^2) · Log-likelihood ratio statistic · Design-based Wald statistic · Goodness-of-fit tests · Tests of homogeneity · Tests of independence · Rao–Scott corrections to X_P^2 · Fay's jackknifed statistic

4.1 Introduction

The classical analysis of categorical data assumes that the data are obtained through multinomial sampling. In testing for goodness-of-fit, homogeneity of several populations, independence in two-way contingency tables Pearson chi-square test statistic is often used. The log-likelihood ratio and the Wald statistic which are asymptotically equivalent to Pearson statistic (see Appendix), are also used.

Nowadays, large-scale sample surveys are often conducted to collect data to test different hypotheses in natural and social sciences. Such surveys often use stratified multistage cluster design. Data obtained through such complex survey designs are not generally independently distributed and as a result multinomial models do not hold in such cases. Thus the classical Pearson statistic and the related usually used test statistic would not be valid tools for testing different hypotheses in these circumstances.

Here we propose to investigate the effect of stratification and clustering on the asymptotic distribution of Pearson statistic, log-likelihood ratio statistic for testing

P. Mukhopadhyay, *Complex Surveys*, DOI 10.1007/978-981-10-0871-9_4

goodness-of-fit (simple hypothesis), independence in two-way contingency tables and homogeneity of several populations. The model is called a full model, as we assume that the population proportions do not involve any other set of unknown parameters.

Section 4.2 considers test statistics for goodness-of-fit hypothesis and their asymptotic distribution. The concepts of *generalized design effects* (in Kish's sense) have been introduced and the classical test statistics have been modified for better control of the type I error. Section 4.3 considers the tests for homogeneity of several populations with respect to the vector of population proportions. The next section examines the effects of survey designs on classical tests of general linear hypotheses, and subsequently tests of independence in a two-way table. This chapter draws mainly from Rao and Scott (1981), Scott and Rao (1981), and Holt et al. (1980).

4.2 Tests of Goodness-of-Fit

While testing a simple hypothesis that the population proportions of different categories, in which elements in a finite population are divided, are equal to some specific values, we shall consider four statistics: (i) Pearson chi-square statistic; (ii) design-based Wald statistic; (iii) Neyman's (1949) statistic; and (iv) likelihood ratio statistic. These are introduced below.

4.2.1 *Pearsonian Chi-Square Statistic*

Consider a finite population of size N whose members are divided into t classes with unknown proportions $\pi_1, \ldots, \pi_t$ $(\sum_{k=1}^{t} \pi_k = 1)$, $\pi_k = N_k/N$, N_k being the unknown number of units in the population in class k. A sample of size n is drawn from this population following a sampling design $p(s)$; for example, a stratified multistage design. Let $n_1, \ldots, n_t$ $(\sum_k n_k = n)$ denote the observed cell frequencies in the sample.

Under these circumstances, the conventional Pearson chi-squared statistic for testing the simple hypothesis $H_0 : \pi_k = \pi_{k0}, k = 1, \ldots, t$ is given by

$$\begin{aligned} \tilde{X}^2 &= \sum_{k=1}^{t} \frac{(n_k - n\pi_{k0})^2}{n\pi_{k0}} \\ &= \frac{n\sum_{k=1}^{t}(\tilde{p}_k - \pi_{k0})^2}{\pi_{ko}} \end{aligned} \tag{4.2.1}$$

where $\tilde{p}_k = n_k/n(k = 1, \ldots, t)$. Under *srs* $\tilde{X}^2$ will be distributed asymptotically as $\chi^2_{(t-1)}$, a central chi-square with $(t - 1)$ degrees of freedom (d.f.).

Now for general non-*srs* design, $\tilde{p}_k$ is not a consistent estimator of π_k unless the design is self-weighting (i.e., has equal sample weights). For general sampling designs, we therefore consider a more general statistic

$$X_P^2 = n \sum_{k=1}^{t} \frac{(\hat{\pi}_k - \pi_{k0})^2}{\pi_{k0}} \tag{4.2.2}$$

where $\hat{\pi}_k$ is an unbiased (or consistent) estimator of π_k under $p(s)(\sum_k \hat{\pi}_k = 1)$. If $n\hat{\pi}_k = n_k$, the statistic X_P^2 in (4.2.2) reduces to $\tilde{X}^2$ in (4.2.1). From now on we shall consider X_P^2 rather than $\tilde{X}^2$. We shall write $\hat{n}_k$ as the adjusted (sample) cell frequencies after adjusting for non-response, if any, and unequal inclusion probabilities, such that $\sum_k \hat{n}_k = n$. In case, no adjustment is required $\hat{n}_k = n_k$. The $\hat{\pi}_k$ are usually ratio estimators if n is not fixed in advance.

Here, we have assumed that $\pi_{10}, \ldots, \pi_{k0}$ are given quantities and there is nothing unknown about them.

4.2.2 Design-Based Wald Statistic

The *design-based Wald statistic* for testing $H_0 : \pi = \pi_0$ where $\pi = (\pi_1, \ldots, \pi_{t-1})'$, $\pi_0 = (\pi_{10}, \ldots, \pi_{t-10})'$ is given by

$$X_W^2 = n(\hat{\pi} - \pi_0)'\hat{\mathbf{V}}^{-1}(\hat{\pi} - \pi_0) \tag{4.2.3}$$

where $\hat{\pi} = (\hat{\pi}_1, \ldots, \hat{\pi}_{t-1})'$ and $\hat{\mathbf{V}}/n$ denotes a consistent estimator of the actual design-covariance matrix $\mathbf{V}/n$ of $\hat{\pi}$. An estimator $\hat{\mathbf{V}}$ can be obtained by the linearization method, balance repeated replication, sample-reuse methods, such as jackknife or any other non-parametric method of variance-estimation.

The statistic X_W^2 is unique for any choice of $(t-1)$ classes and is distributed approximately as a $\chi^2_{(t-1)}$ random variable under H_0 if n is sufficiently large (vide Sect. 4.2.5). The statistic X_W^2 provides an asymptotically exact size α test when applied to $\chi^2_{(t-1)}(\alpha)$. This approach has been well illustrated by Koch et al. (1975), though one of its major disadvantages is that we need to calculate $\hat{\mathbf{V}}$ and this may be difficult if the design is complex.

Fay (1985) has shown that if the degrees of freedom (d.f.) for $\hat{\mathbf{V}}$ is not large relative to the d.f. of χ^2, X_W^2 is often unreliable due to instability in the estimated covariance matrix. Monte Carlo results (Thomas and Rao 1984) also indicate that the Wald statistics although asymptotically valid, does not control the type I error satisfactorily under the above situation unlike the Satterthwaite adjusted X_P^2 (discussed in Sect. 4.2.7) or the jackknife chi-square of Fay (1985) (Sect. 4.2.8).

In practice, X_W^2 can be expected to work reasonably well, if the number of sample clusters m in a stratified multistage sampling design with H strata, is large and t is relatively small, because in that case one can expect $\hat{\mathbf{V}}$ to be stable.

If the number of sample clusters m is small, the number of degrees of freedom available to calculate $\hat{\mathbf{V}}$, $m - H = f$ (say) becomes small and consequently $\hat{\mathbf{V}}$ becomes unstable. To overcome this problem, a d.f.-correction to Wald statistic is applied. Two alternative F-corrected Wald statistic are:

(i) $F_1(X_W^2) = \frac{f-t+2}{f(t-1)} X_W^2 \sim F_{(t-1, f-t+2)}$;

(ii) $F_2(X_W^2) = \frac{X_W^2}{t-1} \sim F_{(t-1, f)}$.

Thomas and Rao (1987) made a comparative study of the performance of various test statistics of a simple goodness-of-fit hypothesis under the situation of instability. Their simulation study indicated that in cases when instability is not too severe, the F-corrected Wald statistic $F_1(X_W^2)$ behaves relatively well in comparison to its competitors.

Now, X_P^2 can be written as

$$X_P^2 = n(\hat{\pi} - \pi_0)'\mathbf{P}_0^{-1}(\hat{\pi} - \pi_0) \tag{4.2.4}$$

where $\mathbf{P}_0$ is the value of $\mathbf{P} = \mathbf{D}_\pi - \pi\pi'$ for $\pi = \pi_0$, $\mathbf{D}_\pi =$ Diag. $(\pi_1, \ldots, \pi_{t-1})'$ and $\mathbf{P}/n$ is the $(t-1)\times(t-1)$ covariance matrix of $\hat{\mathbf{p}} = \frac{\mathbf{n}}{n}$ for multinomial sampling where $\mathbf{n} = (n_1, \ldots, n_{t-1})'$. In case of multinomial sampling $\hat{\pi}_k = n_k/n, k = 1, \ldots, t-1$, the result (4.2.4) can be seen as follows. We have

$$\mathbf{P}_0 = \mathbf{D}_{\pi_0} - \pi_0\pi_0'.$$

Using the relation

$$(\mathbf{A} + \mathbf{uv})^{-1} = \mathbf{A}^{-1} - \frac{\mathbf{A}^{-1}\mathbf{uv}'\mathbf{A}^{-1}}{1 + \mathbf{v}'\mathbf{A}^{-1}\mathbf{u}}$$

where $\mathbf{A}$ is $p \times p$, $\mathbf{u}$ is $p \times 1$, $\mathbf{v}'$ is $1 \times p$, we get

$$\mathbf{P}_0^{-1} = \mathbf{D}_{\pi_0}^{-1} + (1/\pi_{t0})\mathbf{1}_{t-1}\mathbf{1}_{t-1}'.$$

Now

$$\begin{aligned} X_P^2 &= n\sum_{k=1}^{t-1}(\hat{\pi}_k - \pi_{k0})^2/\pi_{k0} + n(\hat{\pi}_t - \pi_{t0})^2/\pi_{t0} \\ &= n(\hat{\boldsymbol{\pi}} - \pi_0)'\mathbf{D}_{\hat{\pi}_0}^{-1}(\hat{\pi} - \pi_0) \\ &\quad + n(\hat{\pi} - \pi_0)'\mathbf{1}_{t-1}\mathbf{1}_{t-1}'(1/\pi_t)(\hat{\boldsymbol{\pi}} - \pi_0) \\ &= n(\hat{\pi} - \pi_0)'(\mathbf{P}_0)^{-1}(\hat{\pi} - \pi_0). \end{aligned} \tag{4.2.5}$$

Comparing with (4.2.3) it is seen that X_P^2 is a special case of X_W^2.

4.2.3 Neyman's (Multinomial Wald) Statistic

An alternative to Pearson statistic X_P^2 is Neyman's statistic

$$\begin{aligned} X_N^2 &= n\sum_{k=1}^{t-1}\frac{(\hat{\pi}_k-\pi_{k0})^2}{\hat{\pi}_k} \\ &= n(\hat{\pi}-\pi_0)'(\hat{\mathbf{P}})^{-1}(\hat{\pi}-\pi_0) \end{aligned} \tag{4.2.6}$$

where $\hat{\mathbf{P}} = \text{Diag.}\,(\hat{\pi}) - \hat{\pi}\hat{\pi}'$ and $\hat{\mathbf{P}}/n$ is the estimated (empirical) multinomial covariance matrix. Under multinomial sampling, $X_N^2 \sim \chi^2_{(t-1)}$ asymptotically; but for more complex designs, the statistic needs adjustment similar to those used for Pearson statistic (vide Sect. A.4.1).

4.2.4 Log-Likelihood Ratio Statistic

The log-likelihood ratio statistic for testing H_0 is given by

$$\begin{aligned} G^2 &= -2\log\left\{\frac{\Pi_{k=1}^t(\pi_{k0})^{\hat{n}_k}}{\Pi_{k=1}^t(\hat{\pi}_k)^{\hat{n}_k}}\right\} \\ &= 2n\sum_{k=1}^{t}\hat{\pi}_k\log\left(\frac{\hat{\pi}_k}{\pi_{k0}}\right), \end{aligned} \tag{4.2.7}$$

provided $\hat{n}_k = n\hat{\pi}_k, k = 1,\ldots,t$. Under multinomial sampling, it is well known that both X_P^2 and G^2 are distributed asymptotically as $\chi^2_{(t-1)}$, when H_0 holds.

4.2.5 Asymptotic Distribution of X_W^2 and X_P^2

Consider any $t-1$ categories, labeled, without any loss of generality, as $1,\ldots,t-1$. Assume that with each unit i in the population, there is a $(t-1)$-dimensional vector $\mathbf{Z}_i = (Z_i^1, Z_i^2,\ldots,Z_i^{t-1})'$ such that $Z_i^k = 1(0)$ if the unit i belongs to category k (otherwise). Now $\sum_{i=1}^N Z_i^k = N_k, \pi_k = N_k/N$ $(k = 1,\ldots,t-1)$ and $N_t = N - \sum_{k=1}^{t-1}N_k$, $\pi_t = 1 - \sum_{k=1}^{t-1}\pi_k$, an estimate of π_k would be a weighted function of $Z_i^k, i \in s$. Hence, a design-based estimator of π_k is

$$\hat{\pi}_k = \sum_{i\in s} w_i(s)Z_i^k \tag{4.2.8}$$

for some weight $w_i(s)$ which may depend upon both the sample s and the unit-label i in the population. Note that we are considering here any arbitrary sampling design $p(s)$. An estimator of a linear function $\mathbf{a}'\pi$, where $\mathbf{a} = (a_1, \ldots, a_{t-1})'$ is a vector of constants, is

$$\begin{aligned}\mathbf{a}'\hat{\pi} &= \sum_{k=1}^{t-1} a_k \sum_{i \in s} w_i(s) Z_i^k \\ &= \sum_{i \in s} w_i(s) \sum_{k=1}^{t-1} a_k Z_i^k \\ &= \sum_{i \in s} w_i(s) y_i \text{ (say).} \end{aligned} \tag{4.2.9}$$

If the sampling design is such that it is amenable to a central limit theorem (CLT) then the statistic $\mathbf{a}'\hat{\pi}$, for any $\mathbf{a}$, is approximately normally distributed with mean $\mathbf{a}'\pi$ and variance $\mathbf{a}'\mathbf{V}\mathbf{a}/n$, where $\mathbf{V} = ((v_{ij}))$ and cov.$(\hat{\pi}_i, \pi_j) = v_{ij}/n$, for sufficiently large n. Hence $(\hat{\pi} - \pi)$ is asymptotically $(t-1)$ variate normal $N_{t-1}(\mathbf{0}, \mathbf{V}/n)$ for large n.

The sample survey literature is replete with examples where sampling design accommodates some CLT; see for example, Madow (1948) and Ha'jek (1960) for *srswor*; Scott and Wu (1981) for results on expansion, ratio and regression estimators for *srs*; Isaki and Fuller (1982), Francisco and Fuller (1991) for asymptotic normality of the Horvitz–Thompson estimator (HTE) in unequal-probability sampling designs; Krewski and Rao (1981), Rao and Wu (1985) for asymptotic normality of the estimators of means and totals when the number of strata approaches infinity in multistage sampling design with two primary units in the sample per stratum.

If $\hat{\mathbf{V}}$ is a consistent estimator $\mathbf{V}$, the generalized Wald statistic X_W^2 given in (4.2.3) will be asymptotically distributed as $\chi^2_{(t-1)}$. This result follows from the general result on the distribution of quadratic forms in the multivariate normal distribution (also see Corollary 4.2.2.2).

The asymptotic distribution of X_P^2 under any general sampling design has been given in Theorem 4.2.2. This theorem follows from the form (4.2.4) of X_P^2, asymptotic multivariate normality of $\hat{\pi}$ (which holds under a class of sampling designs, as noted above) and the standard results on quadratic forms of multivariate normal vector, as noted in Lemma A.3.2. The same lemma is stated below in a slightly different form.

Theorem 4.2.1 (Jhonson and Kotz 1970) *Let the random vector $\mathbf{Z} = (Z_1, \ldots, Z_m)'$ have a multivariate normal distribution $N_m(\mathbf{0}, \mathbf{U})$. Then the distribution of the quadratic form $Q(\mathbf{Z}) = \mathbf{Z}'\mathbf{A}\mathbf{Z} = \sum_{i=1}^m \sum_{j=1}^m a_{ij} Z_i Z_j$ is the same as that of $\sum_{i=1}^m \lambda_i \tau_i^2$, where the variables τ_i are independent $N(0, 1)$ variables and the numbers $\lambda_1 \geq \lambda_2 \geq \cdots \geq \lambda_m$ are the eigenvalues of* $\mathbf{AU}$.

If we take $m = t - 1, \mathbf{Z} = \sqrt{n}(\hat{\pi} - \pi_0)$, then for certain sampling designs as stated before, $\mathbf{Z} \sim N_{t-1}(\mathbf{0}, \mathbf{V}_0)$ under H_0 where $\mathbf{V}_0$ is the value of $\mathbf{V}$ for $\pi = \pi_0$. Also from (4.2.5), $X_P^2 = \mathbf{Z}'\mathbf{P}_0\mathbf{Z}$. Hence, we have the following theorem.

Theorem 4.2.2 *Under $H_0 : \pi = \pi_0$, X_P^2 given in (4.2.2) is approximately distributed as*

$$X_P^2 \sim \sum_{k=1}^{t-1} \lambda_{0k}\tau_k^2 \tag{4.2.10}$$

where $\tau_1, \ldots, \tau_{t-1}$ are independent $N(0, 1)$ variables and λ_{0k}'s are the eigenvalues of $\mathbf{D}_0 = \mathbf{P}_0^{-1}\mathbf{V}_0$ with $\lambda_{01} \geq \lambda_{02} \geq \cdots \lambda_{0t-1} \geq 0$ and $\mathbf{V}_0/n$ denoting the covariance matrix of $\hat{\pi}$ for $\pi = \pi_0$.

Corollary 4.2.2.1 *It follows from (4.2.10) that $X_P^2/\lambda_{01} \leq \sum_{k=1}^{t-1}\tau_k^2$ where $\sum_{k=1}^{t-1}\tau_k^2$ is distributed asymptotically as $\chi_{(t-1)}^2$ under H_0.*

Therefore, if it is known that $\lambda_{01} \leq \lambda^*$, we can obtain an asymptotic conservative test for H_0 by treating X_P^2/λ^* as a $\chi_{(t-1)}^2$ variable.

Corollary 4.2.2.2 *Asymptotic distribution of $X_W^2 = \mathbf{Z}'\hat{\mathbf{V}}_0^{-1}\mathbf{Z}$ given in (4.2.3) is $\sum_{k=1}^{t-1}\tilde{\lambda}_{0k}\tau_k^2$, where $\tilde{\lambda}_{0k}$'s are the eigenvalues of $\hat{\mathbf{V}}_0^{-1}\mathbf{V}_0 \approx \mathbf{I}_{t-1}$. Clearly $\tilde{\lambda}_{0k}$'s should be all close to unity. Hence, asymptotic distribution of X_W^2 under H_0 is $\chi_{(t-1)}^2$.*

Corollary 4.2.2.3 *If for any sampling design $\mathbf{V} = \lambda\mathbf{P}$ for some constant λ for any π_0, then $X_P^2/\lambda \sim \chi_{(t-1)}^2$ asymptotically. It can be shown that the above condition is also necessary.*

Examples 4.2.1 (a) and 4.2.2 show some situations where this condition is fulfilled.

Note 4.2.1 If $\lambda_{01} < 1$, $X_P^2 < \sum_{k=1}^{t-1}\tau_k^2 = \chi_{(t-1)}^2$, then the Pearson chi-square test will be asymptotically conservative. For some examples of this situation see Exercises 4.3 and 4.4.

Note 4.2.2 Solomon and Stephens (1977) considered distribution of quadratic forms $Q_k = \sum_{j=1}^{k} c_j(X_j + a_j)^2$, where the X_j's are independently and identically distributed standard normal variables and c_j and a_j are nonnegative constants. Exact significance points for Q_k for selected values of c_j and all $a_j = 0$ have been published for $k = 2(1)6(2)10$. They also proposed two new approximations to Q_k: (i) fitting a Pearson curve with the same first four moments as Q_k; (ii) Fitting $Q_k = Aw^r$, where w has the $\chi_{(p)}^2$ distribution and a, r, and p are determined by the first three moments of Q_k.

For any given values of λ's, we can use their approximations to evaluate the correct percentage points of the distribution of X_P^2.

Example 4.2.1 (a) *srswor*: If the sampling is by simple random sampling without replacement, $\mathbf{V} = (1 - n/N)\mathbf{P}$ where N is the finite population size. Hence, by Corollary 4.2.2.3, $X_P^2 \sim (1 - n/N)\chi_{(t-1)}^2$ as both N and $n \to \infty$ in such a way that $(N - n) \to \infty$. Here $\lambda_k = \lambda_{0k} = 1 - n/N$ for any π_0 and

$$(1 - n/N)^{-1}X_P^2 \sim \chi_{(t-1)}^2. \tag{4.2.11}$$

Thus the Pearson statistic will be conservative in this case. An asymptotically valid test can be obtained for this sampling scheme by referring to $X_P^2/(1-\frac{n}{N})$ to $\chi^2_{(t-1)}$.

(b) *Stratified Random Sampling with Proportional Allocation*: Suppose that a population of size N is divided into H strata, the hth stratum being of size N_h and a stratified sample $s = (s_1, \dots, s_H)$, where s_h is a sample of size n_h drawn from the hth stratum by *srswr*, is selected ($\sum_h n_h = n$). Let $W_h = N_h/N$, ρ_{hk}, the population proportion in the kth category and n_{hk}, the observed cell frequency in the kth category in the hth stratum, $h = 1, \dots, H$. Then $\rho_k = \pi_k = \sum_h W_h \rho_{hk}$, $\hat{\rho}_{hk} = n_{hk}/n_h$ and $\hat{\rho}_k = \hat{\pi}_k = \sum_h W_h n_{hk}/n_h$. Under proportional allocation where $n_h = nW_h$, $\hat{\rho}_k = \hat{\pi}_k$ reduces to m_k/n where $m_k = \sum_h n_{hk}$.

Under stratified random sampling design with large n_h, $\hat{\pi}$ is approximately distributed as a $(t-1)$-variate normal variable with mean π and covariance matrix $\mathbf{V}/n$ where

$$\begin{aligned}\mathbf{V} &= \mathbf{P} - \sum_{h=1}^{H} W_h(\boldsymbol{\rho}_h - \pi)(\boldsymbol{\rho}_h - \pi)' \\ &= \mathbf{P} - \mathbf{L} \text{ (say)},\end{aligned} \tag{4.2.12}$$

where $\boldsymbol{\rho}_h = (\rho_{h1}, \dots, \rho_{ht-1})'$. Now

$$0 \le \frac{\mathbf{c}'\mathbf{V}\mathbf{c}}{\mathbf{c}'\mathbf{P}\mathbf{C}} = 1 - \frac{\sum_h W_h[\mathbf{c}'(\boldsymbol{\rho}_h - \pi)]^2}{\mathbf{c}'\mathbf{P}\mathbf{c}} \le 1. \tag{4.2.13}$$

Therefore, the first eigenvalue of $\mathbf{D}_0 = \mathbf{P}_0^{-1}\mathbf{V}_0$, $\lambda_{01} \le 1$ for any π_0 and hence

$$X_P^2 \le \sum_{k=1}^{t-1} \tau_k^2 \sim \chi^2_{(t-1)}. \tag{4.2.14}$$

Hence in case of stratified random sampling with proportional allocation, Pearson chi-square statistic X_P^2 is always asymptotically conservative.

However, for this sampling design, $\hat{\mathbf{V}}$ can be easily calculated if the primary data on $n_{hk}(h = 1, \dots, H; k = 1, \dots, t)$ sampled units are available. Therefore, in this case, one should use Wald's statistic in lieu of Pearson's statistic.

If $H \ge t$ and all elements in the same stratum belong to the same category so that the stratification is perfect, value of X_P^2 is zero.

Let us now consider the case $H < t$. Now the rank of $\mathbf{L}$ in (4.2.12) is at most $(H-1)$. Also, all the eigenvalues of $\mathbf{I}_{t-1}$ equal one. Hence, at least $(t-H)$ eigenvalues of $\mathbf{P}^{-1}\mathbf{V}$ must be each equal to one. Therefore, $X_P^2 \ge \sum_{k=1}^{t-H} \tau_k^2 = \chi^2_{(t-H)}$. Therefore, X_P^2 is asymptotically well-approximated by $\chi^2_{(t-1)}$ if $(t-H) \approx (t-1)$ or if t is large and H relatively small.

When $H = 2$,

$$\begin{aligned}\mathbf{D} &= \mathbf{P}^{-1}\mathbf{V}\\ &= \mathbf{I} - W_1 W_2 \mathbf{P}^{-1}(\pi_1 - \pi_2)(\pi_1 - \pi_2)'\\ &= \mathbf{I} - \mathbf{\Lambda} \text{ (say)},\end{aligned} \tag{4.2.15}$$

where $W_2 = 1 - W_1$. Since $\mathbf{\Lambda}$ is of rank one, $(t-2)$ eigenvalues of $\mathbf{\Lambda}$ are each zero and the only nonzero eigenvalue of $\mathbf{\Lambda}$ is tr $(\mathbf{\Lambda}) = W_1 W_2 \sum_{k=1}^{t-1}(\rho_{1k} - \rho_{2k})^2/\pi_k = \delta^*$ (say) , $0 \leq \delta^* \leq 1$. Therefore, $\lambda_1 = \cdots = \lambda_{t-2} = 1$ and $\lambda_{t-1} = 1 - \delta^*$. Therefore,

$$\begin{aligned}X_P^2 &= \sum_{k=1}^{t-2} \tau_k^2 + (1 - \delta_0^*)\tau_{t-1}^2\\ &= \chi_{t-2}^2 + (1 - \delta_0^*)\chi_{(1)}^2,\end{aligned} \tag{4.2.16}$$

where δ_0^* is the value of δ^* for $\pi = \pi_0$. Thus, if t is not small, X_P^2 can be well-approximated by $\chi_{(t-1)}^2$ in this case.

(c) *Two-Stage Sampling*: Suppose we have N primary stage units (*psu*'s) with M_a second stage units (*ssu*'s) in the ath psu in the population ($a = 1, \ldots, N$; $\sum_{a=1}^{N} M_a = M$). A sample of n psu's is selected with probability proportional to size with replacement (*ppswr*), size being M_a, and subsamples each of size m ssu's are selected by *srswr* from each selected psu ($mn = q$).

Let m_{ak} be the observed cell frequency in the kth category in the sampled psu, a. Then $\hat{\pi}_{ak}$ = estimate of the population proportion $\pi_{ak} = M_{ak}/M_a$ in the ath psu, where M_{ak} is the number of ssu's in the kth category in the ath psu, is $m_{ak}/m, k = 1, \ldots, t-1$. Let

$$\begin{aligned}\hat{\rho}_a &= (\hat{\pi}_{a1}, \ldots, \hat{\pi}_{at-1})'\\ \pi &= (\pi_1, \ldots, \pi_{t-1})',\end{aligned}$$

where $\pi_k = \sum_{a=1}^{N} M_a \pi_{ak}/M$, $\hat{\pi}_k = \sum_{a=1}^{n} m_{ak}/q$, $q_k = \sum_{a=1}^{n} m_{ak}$. Therefore, $\hat{\pi} = \mathbf{q}/q$ where $\mathbf{q} = (q_1, \ldots, q_{t-1})'$. Also $E(q_k) = q\pi_k$.

It follows that $\hat{\pi}$ follows approximately $(t-1)$-variate normal distribution with mean vector π and covariance matrix $\mathbf{V}/n$ where

$$\begin{aligned}\mathbf{V} &= \mathbf{P} + (m-1)\sum_{a=1}^{N} W_a(\rho_a - \pi)(\rho_a - \pi)'\\ &= \mathbf{P} + (m-1)\mathbf{\Lambda}_1 \text{ (say)}\end{aligned} \tag{4.2.17}$$

where $W_a = M_a/M$ and $\rho_a = (\pi_{a1}, \ldots, \pi_{at-1})'$. Let $\xi_1 \geq \xi_2 \geq \cdots \geq \xi_{t-1} (\geq 0)$ denote the eigenvalues of $\mathbf{P}^{-1}\mathbf{\Lambda}_1$. Then $\lambda_k = 1 + (m-1)\xi_k$ and

$$X_P^2 = \sum_{k=1}^{t-1} \{1 + (m-1)\xi_{0k}\}\tau_k^2 \tag{4.2.18}$$

where ξ_{0k} is the value of ξ_k for $\pi = \pi_0$. Also

$$\frac{\mathbf{c}' \sum_{a=1}^{N} W_a(\rho_a - \pi)(\rho_a - \pi)'\mathbf{c}}{\mathbf{c}'(\text{Diag.}(\pi) - \pi\pi')\mathbf{c}} = \frac{\mathbf{c}'\mathbf{\Lambda}_1\mathbf{c}}{\mathbf{c}'\mathbf{P}\mathbf{c}} \leq 1 \tag{4.2.19}$$

so that $\xi_{0k} \leq 1 \ \forall\ k$ and we get

$$\begin{aligned}
\sum_{k=1}^{t-1} \tau_k^2 &\leq \{1 + (m-1)\xi_{0,t-1}\} \sum_{k=1}^{t-1} \tau_k^2 \\
&\leq X_P^2 \\
&\leq \{1 + (m-1)\xi_{0,1}\} \sum_{k=1}^{t-1} \tau_k^2 \\
&\leq m \sum_{k=1}^{t-1} \tau_k^2,
\end{aligned} \tag{4.2.20}$$

since $\xi_k \geq 0\ \forall k$. Thus $X_P^2/m \sim \chi^2_{(t-1)}$ approximately asymptotically and this gives an approximate conservative test for the hypothesis $H_0 : \pi = \pi_0$. The ξ_k's may be called the *generalized measures of homogeneity*.

As in the case of stratified random sapling, the computation of $\hat{\mathbf{V}}$ and X_W^2 can be done in a straightforward manner if the primary data on m_{ab} sampled units are available ($a = 1, \ldots, n; b = 1, \ldots, m$). Here

$$\hat{\mathbf{V}} = m \sum_{a=1}^{n} (\hat{\rho}_a - \pi)(\hat{\rho}_a - \pi)'/(n-1). \tag{4.2.21}$$

Example 4.2.2 (*Brier's 1980 model*) There are N clusters in the population from each of which a sample of n units is selected by multinomial sampling. Let $\mathbf{Y}^{(i)} = (Y_1^{(i)}, \ldots, Y_t^{(i)})'$ be the vector of observed cell counts for the ith cluster, $Y_j^{(i)}$ being the number of units in the ith cluster that falls in category j, $\sum_{j=1}^{t} Y_j^{(i)} = n\ \forall\ i = 1, \ldots, N$. Let $p^{(i)} = (p_1^{(i)}, \ldots, p_t^{(i)})'$ be the vector of cell-probabilities for the ith cluster.

We assume that the vectors $p^{(i)} (i = 1, \ldots, N)$ are independently and identically distributed with a Dirichlet distribution defined by the density

$$f(\mathbf{p}|\pi, k) = \frac{\Gamma(k)}{\Pi_{j=1}^{t} \Gamma(k\pi_j)} \Pi_{j=1}^{t} p_j^{k\pi_j - 1} \tag{4.2.22}$$

where $\mathbf{p}$ and $\pi = (\pi_1, \ldots, \pi_t)'$ lie in the $(t-1)$-dimensional space $\xi_t = \{\mathbf{p} = (p_1, \ldots, p_t) : p_j > 0, \sum_j p_j = 1\}$ and k is a parameter of the distribution.

Hence, the unconditional distribution of $\mathbf{Y} = (Y_1, \ldots, Y_t)'$ where $Y_j = \sum_{i=1}^{N} Y_j^{(i)}$ is

$$
\begin{aligned}
f(\mathbf{y}|\pi, k) &= \int \cdots \int_{\xi_t} \binom{n}{y_1, \ldots, y_t} \Pi_{j=1}^{t} p_j^{y_j} \frac{\Gamma(k)}{\Pi_{j=1}^{t}\Gamma(k\pi_j)} \Pi_{j=1}^{t} p_j^{k\pi_j - 1} \\
&= \binom{n}{y_1, \ldots, y_t} \frac{\Gamma(k)}{\Gamma(n+k)\Pi_{j=1}^{t}\Gamma(k\pi_j)} \Pi_{j=1}^{t}\Gamma(y_j + k\pi_j) \qquad (4.2.23)
\end{aligned}
$$

where $\mathbf{y}/n \in \xi_t$. This distribution is referred to as Dirichlet multinomial distribution, $DM_t(n, \pi, k)$. Mosimann (1962) showed that that the mean of the distribution of $\hat{\mathbf{p}} = \mathbf{y}/n$ is π and its covariance matrix is $n^{-1}C(\mathbf{D}_\pi - \pi\pi')$ where $C = (n+k)/(1+k)$. Here, k is the structural parameter representing the cluster effects; if ρ is the intracluster correlation coefficient, then $\rho = 1/(1+k)$.

Since the covariance matrix of $\hat{\mathbf{p}}$ is C times that of a multinomial distribution, the design effect (*deff*) (vide Sect. 4.2.6) will be a constant, C (or zero). For testing the hypothesis of goodness-of-fit $H_0 : \pi = \pi_0$, the asymptotic distribution of X_P^2 will therefore be $C\chi^2_{(t-1)}$. Since $C > 1$, Pearson's statistic will be liberal in this case.

Since a constant deff may not be realistic in many practical situations, Thomas and Rao (1987) extended Brier's model to generate non-constant design effects, as considered in Sect. 5.5.

4.2.6 *Generalized Design Effect of π*

We have already explained in Sect. 2.4, the concept of multivariate design effect of a pair of parameter estimator and covariance matrix estimator (in Skinner's sense) in the context of quantitative variables. Here we extend the concept of design effect (of an estimator) in Kish's sense to the case of categorical variables. From Theorem 4.2.2, we may coin the following definition.

Definition 4.2.1 Generalized Design Effect of π: The matrix $\mathbf{D} = \mathbf{P}^{-1}\mathbf{V}$ where $\mathbf{V}/n$ is the covariance matrix under the sampling design $p(s)$ of $\hat{\pi}$, estimator of π under $p(s)$ and $\mathbf{P}/n$, as defined in (4.2.4) is the covariance matrix of the estimator of π under multinomial sampling design, may be called the *multivariate design effect matrix* or *generalized design effect matrix* of π.

For $t = 2$, $\mathbf{D} = \mathbf{P}^{-1}\mathbf{V}$ reduces to the ordinary design effect (*deff*) of π_1,

$$
deff(\pi_1) = \frac{nV(\hat{\pi}_1)}{\pi_1(1-\pi_1)} \qquad (4.2.24)
$$

where $V(\hat{\pi}_1)$ is the variance of $\hat{\pi}_1$ under the sampling design $p(s)$.

Now, λ_1, the largest eigenvalue of $\mathbf{D} = \mathbf{P}^{-1}\mathbf{V}$ is

$$\begin{aligned}\lambda_1 &= \sup_{\mathbf{c}}\left[\frac{\mathbf{c}'\mathbf{V}\mathbf{c}}{\mathbf{c}'\mathbf{P}\mathbf{c}}\right]\\ &= \sup_{\mathbf{c}}\left[\frac{Var(\sum_{k=1}^{t-1} c_k\hat{\pi}_k)}{Var_{mns}(\sum_{k=1}^{t-1} c_k n_k/n)}\right],\end{aligned} \tag{4.2.25}$$

where Var_{mns} denotes the variance operator under multinomial sampling. The τ_k's in (4.2.10) are special linear combinations of the estimated cell proportions and λ_k may be called the *deff* of τ_k. Thus λ_1 is the largest possible *deff* taken over all individual $\hat{\pi}_k$'s and over all possible linear combinations of the $\hat{\pi}_k$'s; λ_2 is the largest *deff* among all linear combinations of $\hat{\pi}_k$'s, that are orthogonal to τ_1. The other λ_k's also represent *deff*s for special linear combinations of $\hat{\pi}_k$'s. Thus the λ_k's may be termed as the *generalized design effects* in Kish's sense. (Thus, there are two types of generalized design effects, one in Skinner's sense (Sect. 2.4) and other in Kish's sense. Their actual meaning will be clear from the context.)

The Eq. (4.2.10) states that one should use the distribution of X_P^2 as $\sum_{k=1}^{t-1}\lambda_k\tau_k^2$ and not as $\chi^2_{(t-1)}$. However, in practice, one assumes the results for multinomial sampling always and uses $\chi^2_{(t-1)}$. This distorts the significance level of the test. The effect of the sampling design on the change in the significance level depends on the size of the λ's and the degrees of freedom (d.f.).

Table 4.1 adopted from Fellegi (1980) shows the actual size of the X_P^2 test when the significance level of the usually used $\chi^2_{(t-1)}$ test of Pearson chi-square is 5 % for $\lambda_k = \lambda, k = 1, \ldots, t-1$. Clearly, $X_P^2 \sim \lambda\chi^2_{(t-1)}$ in this case. A quick perusal of the table indicates that the use of standard chi-square as a test of goodness-of-fit (or test of independence, etc. in contingency tables) can be misleading. Even with a *deff* of only 1.5, not a large value, the significance tests are practically useless. The null hypothesis will be rejected with a probability of 0.19 if the d.f. are only five, the probability of rejection rising to 0.6 or more with larger values of *deff* and/or larger d.f.

Table 4.1 Actual size of X_P^2 test with nominal 5 % significance level ($\lambda_k = \lambda, k = 1, \ldots, t-1$)

Deff (λ)	$t = 3$	$t = 6$	$t = 10$
1	5	5	5
1.2	8	10	12
1.5	14	19	27
3.0	37	59	81

4.2.7 Modification to X_P^2

Both the Wald procedure and Fay procedure (discussed in Sect. 4.2.8) require detailed survey information from which the covariance matrix $\mathbf{V}$ can be estimated. Such detailed information is not generally available. In secondary analysis from published results, the best one can hope for is the information on deffs for cells of the table and in case of contingency tables (discussed in subsequent sections) perhaps some information on deffs of marginal totals. Thus one important endeavor has been to find methods that can effectively use this limited information and yet provide acceptable tests. With this aim in view various authors, Rao and Scott (1979, 1981, 1987), Fellegi (1980), among others, proposed corrections to Pearson statistic.

The basic idea is to impose a correction on X_P^2 such that the first moment of the corrected X_P^2 is the same as that of $\chi^2_{(t-1)}$, namely, $t-1$. A simple approximation to the asymptotic distribution of X_P^2 that uses very limited information about $\hat{\mathbf{V}}$ would be preferred.

The following adjustments to X_P^2 have been proposed.

(a) *Mean deff adjustment to X_P^2 due to Fellegi (1980) and Holt, et al. (1980)*: This is based on the estimated design effect (deff) of $\hat{\pi}_k$. We have

$$\hat{d}_k = \frac{\hat{V}(\hat{\pi}_k)}{\hat{V}_{mns}(\hat{\pi}_k)} = \frac{\hat{V}(\hat{\pi}_k)}{\hat{\pi}_k(1-\hat{\pi}_k)/n} \tag{4.2.26}$$

where $\hat{V}(\hat{\pi}_k)$ is the estimated true design variance of $\hat{\pi}_k$ and $\hat{V}_{mns}(\hat{\pi}_k)$ is the estimated design variance of $\hat{\pi}_k$ under multinomial sampling. Now, taking $X_P^2 \sim \sum_{k=1}^{t-1}\lambda_k\psi_k \approx \sum_{k=1}^{t-1}d_k\psi_k$, where d_k is the deff, $d_k = E(\hat{d}_k)$, and ψ_k's are independent $\chi^2_{(1)}$ variables,

$$E\left(\sum_{k=1}^{t-1}\lambda_k\psi_k\right) \approx \sum_{k=1}^{t-1}d_k, \quad \text{since } E(\psi_k)=1.$$

Therefore, defining $\bar{d} = \sum_{k=1}^{t-1}d_k/(t-1)$,

$$E(X_P^2/\bar{d}) \approx E\left(\sum_{k=1}^{t-1}\lambda_k\psi_k/\bar{d}\right) \approx \sum_{k=1}^{t-1}d_k/\bar{d} = t-1. \tag{4.2.27}$$

Taking an estimate of $\bar{d}$ as

$$\hat{d}_0 = \sum_{k=1}^{t-1}\hat{d}_k/(t-1), \tag{4.2.28}$$

the mean deff corrected Pearson statistic is

$$X_P^2(\hat{d}_0) = \frac{X_P^2}{\hat{d}_0} \tag{4.2.29}$$

which is asymptotically distributed as $\chi^2_{(t-1)}$. A positive intraclass correlation among the variate-values gives a mean $\bar{d}$ greater than one and so the value of $X_P^2(\hat{d}_0)$ becomes smaller than X_P^2 and this tends to remove the liberality in X_P^2.

The mean deff adjustment can also be executed by calculating the *effective sample size* $n_e = n/\bar{d}$ and using n_e in place of n in the formula for X_P^2 (vide also (2.3.7)).

(b) *Rao–Scott (1979, 1981) first-order adjustment*: The mean deff adjustment to X_P^2 is only approximate, because λ_k's are approximated by d_k's. Since, under H_0, $E(\sum_{k=1}^{t-1}\lambda_k\psi_k) = \sum_{k=1}^{t-1}\lambda_k$, so that $E(\sum_{k=1}^{t-1}\lambda_k\psi_k/\hat{\lambda}_0) = t-1$ where $\hat{\lambda}_0 = \sum_{k=1}^{t-1}\hat{\lambda}_k/(t-1)$, Rao and Scott proposed the first-order adjustment to X_P^2 as

$$X^2_{P(c)} = X_P^2(\hat{\lambda}_0) = X_P^2/\hat{\lambda}_0 \tag{4.2.30}$$

which is distributed asymptotically as a $\chi^2_{(t-1)}$ random variable under H_0. Here $\hat{\lambda}_k$'s are the eigenvalues of $\hat{\mathbf{D}} = \hat{\mathbf{P}}^{-1}\hat{\mathbf{V}}$ and $\hat{\mathbf{P}} =$ Diag. $(\hat{\mathbf{p}}) - \hat{\mathbf{p}}\hat{\mathbf{p}}'$, $\hat{\mathbf{p}} = \mathbf{n}/n$.

We note that $\hat{\lambda}_0$ depends only on the estimated cell variances $\hat{v}_{kk}/n$ (or equivalently, the estimated cell deff's $\hat{d}_1, \ldots, \hat{d}_{t-1}$), since

$$\begin{aligned}\hat{\lambda}_0 &= \frac{tr(\hat{\mathbf{P}}^{-1}\hat{\mathbf{V}})}{t-1}\\ &= \sum_{k=1}^{t-1}\frac{\hat{v}_{kk}}{\hat{\pi}_k(t-1)}\\ &= \sum_{k=1}^{t-1}\frac{(1-\hat{\pi}_k)\hat{d}_k}{(t-1)}\end{aligned} \tag{4.2.31}$$

where $\hat{d}_k = \hat{v}_{kk}/[\hat{\pi}_k(1-\hat{\pi}_k)]$ and $\hat{v}_{kk}/n = \hat{V}(\hat{\pi}_k)$. Some knowledge about $\hat{d}_k$ for ultimate cells in a goodness-of-fit problem is often available. Thus $X^2_{P(c)}$ can be calculated from the knowledge of cell variances alone and no information on covariance terms are required at all.

Alternatively, $\hat{\lambda}_0$ can be obtained from the estimated generalized design effect matrix, $\hat{\mathbf{D}} = \hat{\mathbf{P}}^{-1}\hat{\mathbf{V}}$ by the equation $\hat{\lambda}_0 = tr(\hat{\mathbf{D}})/(t-1)$.

Provided that the variation among $\hat{\lambda}_k$ is small, $X^2_{P(c)}$ is asymptotically distributed as $\chi^2_{(t-1)}$. The statistic $X^2_{P(c)}$ is more exact than $X_P^2(\hat{d}_0)$.

Similar correction applies to the likelihood ratio test statistic,

$$G_c^2 = G^2(\hat{\lambda}_0) = \frac{G^2}{\hat{\lambda}_0}. \tag{4.2.32}$$

(c) *Rao–Scott second-order correction*: The first-order Rao–Scott adjustment stated above aims at correcting X_P^2 so that the asymptotic expectation of the corrected X_P^2 is equal to its d.f. $t - 1$. If the variation in $\hat{\lambda}_k$'s is large, a correction to X_P^2 for this variance is also required. This is achieved by a second-order Rao–Scott adjustment based on Satterthwaite (1946) method. The second-order adjusted Pearson statistic is given by

$$X_{P(cc)}^2 = X_P^2(\hat{\lambda}_0, \hat{a}^2) = \frac{X_{P(c)}^2}{(1 + \hat{a}^2)} \tag{4.2.33}$$

where an estimator of the squared coefficient of variation a^2 of the unknown eigenvalues λ_k's is

$$\hat{a}^2 = \frac{\sum_{k=1}^{t-1}(\hat{\lambda}_k - \hat{\lambda}_0)^2}{(t-1)\hat{\lambda}_0^2}. \tag{4.2.34}$$

It is found that

$$X_P^2(\hat{\lambda}_0, \hat{a}^2) \sim \chi_{(\nu)}^2 \tag{4.2.35}$$

asymptotically, where $\nu = (t - 1)/(1 + \hat{a}^2)$.

We note that

$$\sum_{k=1}^{t-1} \hat{\lambda}_k^2 = tr(\hat{\mathbf{D}}^2) = \sum_{k=1}^{t} \sum_{k'=1}^{t} \frac{\hat{v}_{kk'}^2}{\hat{\pi}_k \hat{\pi}_{k'}} \tag{4.2.36}$$

so that a^2 and ν can be readily calculated from $\hat{\mathbf{V}}$.

Similar correction holds for the likelihood ratio test,

$$G_{cc}^2 = G^2(\hat{\lambda}_0, \hat{a}^2) = \frac{G^2(\hat{\lambda}_0)}{(1 + \hat{a}^2)}. \tag{4.2.37}$$

Thomas and Rao (1987) showed that when $\hat{a}$ is small, that is, when $\hat{\lambda}_k$'s and hence λ_k's are of similar size, the first-order correction (4.2.30) provides good control of the level of the test. In fact, for $a = 0$, when the λ_k's are all equal, the first-order corrections are asymptotically exact. However, when the variations among the λ_k's becomes appreciable, first-order corrections tend to become somewhat liberal. In this case, the second-order corrected tests provide good control of test level as well as good power-performance. On the basis of a Monte Carlo study, Thomas and Rao (1987) recommended using Satterthwaite procedure when the full matrix $\hat{\mathbf{V}}$ is available.

Note 4.2.3 In unstable situations, that is, when the number of sampled clusters m is small, an F-correction to the Rao–Scott (first-order) adjustment (4.2.30) may be helpful. It is given by

$$F(X^2_{P(c)}) = \frac{X^2_{P(c)}}{(t-1)} \tag{4.2.38}$$

which is now treated as a F-variable with d.f. $(t-1)$ and $\nu(t-1)$, respectively, where ν is the d.f. available to estimate $\mathbf{V}$. This statistic is better than the uncorrected first-order adjusted $X^2_{P(c)}$ in unstable conditions in terms of Type I error performance (Thomas and Rao 1987). Similar corrections can be made to $X^2_{P(cc)}$ obtaining

$$F(X^2_{P(cc)}) = \frac{X^2_{P(cc)}}{(t-1)}. \tag{4.2.39}$$

Similarly, one obtains

$$F(G^2_{cc}) = \frac{G^2_{cc}}{(t-1)}. \tag{4.2.40}$$

Note 4.2.4 For Neyman (multinomial Wald) statistic X^2_N in (4.2.6), we have a mean deff adjusted statistic $X^2_N(\hat{d}_0) = X^2_N/\hat{d}_0$, a first-order Rao–Scott adjusted statistic $X^2_{N(c)} = X^2_N/\hat{\lambda}_0$, a second-order Rao–Scott adjusted statistic $X^2_{N(cc)} = X^2_N(\hat{\lambda}_0, \hat{a}^2) = X^2_{N(c)}/(1+\hat{a}^2)$ and an F-corrected first-order Rao–Scott adjusted statistic $F(X^2_{N(c)}) = X^2_N(\hat{\lambda}_0)/(t-1)$.

4.2.8 *Fay's Jackknifed Statistic*

Fay (1979, 1984, 1985) discusses the adjustment of Pearson and likelihood ratio chi-squared statistics for testing the fit of a model to a cross-classified table of counts through a jackknifing approach. The jackknifed tests, denoted as X_{PJ} and G_J are related to the Rao–Scott corrected tests and can be regarded as an alternative method of removing or at least reducing the distortion in the complex survey distribution of X^2_P and G^2 as characterized by the eigenvalues $\lambda_i (i = 1, \ldots, t-1)$. Fay's method, though computationally intensive than the other methods, is remarkable for its simplicity. The technique may generally be applied whenever a standard replication method, such as the jackknife, bootstrap, or repeated half-samples, provide a consistent estimate of the covariance matrix of the sample estimates. Softwares for implementing the jackknife approach is available and simulation studies by Thomas and Rao (1987), Rao and Thomas (2003) have shown that the jackknifed tests are competitive from the point of view of control of type I error and of power.

In the general case, suppose $\mathbf{Y}$ represents an observed cross-classification, possibly in the form of estimated population totals for a finite population derived from a complex sample survey. Fay considered a class of replication methods to be based on (pseudo) replicates $\mathbf{Y} + \mathbf{W}^{(i,j)}, i = 1, \ldots, I; j = 1, \ldots, J_i$ typically based on the same data as $\mathbf{Y}$. The asymptotic theory for the jackknifed tests requires that

$$\sum_j \mathbf{W}^{(i,j)} = \mathbf{0} \tag{4.2.41}$$

for each i. An estimate, $\text{cov}^*(\mathbf{Y})$, of the covariance of $\mathbf{Y}$ should be given by

$$\text{cov}^*(\mathbf{Y}) = \sum_i b_i \sum_j \mathbf{W}^{(i,j)} \mathbf{W}^{(i,j)'}, \tag{4.2.42}$$

where b_i are a fixed set of constants appropriate for the problem.

The standard jackknife may be applied when $\mathbf{Y}$ can be represented as the sum of n iid random variables $\mathbf{Z}^{(i)}$. The standard leave-one-out replicates are $\mathbf{Y}^{(-j)} = \mathbf{Y} - \mathbf{Z}^{(j)}$. This may be weighted by the factor $n/(n-1)$ to get the same expected total as $\mathbf{Y}$ and written as

$$\frac{n\mathbf{Y}^{(-j)}}{n-1} = \mathbf{Y} + \frac{\mathbf{Y} - n\mathbf{Z}^{(j)})}{n-1}. \tag{4.2.43}$$

The second term on the right of (4.2.43) defines $\mathbf{W}^{(i,j)}$ and satisfies (4.2.41). (Here the subscript i is fixed at 1.) The value $(n-1)/n$ represents the usual choice for b_i.

For calculating the jackknifed values of the test statistics, we have to refit the given model to the replicates, $\mathbf{Y} + \mathbf{W}^{(i,j)}$ and recompute the test statistics, $X_P^2(\mathbf{Y} + \mathbf{W}^{(i,j)})$ or $G^2(\mathbf{Y} + \mathbf{W}^{(i,j)})$, for these new tables. Using the b_i introduced in (4.2.42), the jackknifed test statistic X_{PJ} is defined by

$$X_{PJ} = \frac{[(X_P^2(\mathbf{Y}))^{1/2} - (K^+)^{1/2}]}{\{V/(8X_P^2(\mathbf{Y}))\}^{1/2}}, \tag{4.2.44}$$

where

$$P_{ij} = X_P^2(\mathbf{Y} + \mathbf{W}^{(i,j)}) - X_P^2(\mathbf{Y}), \tag{4.2.45}$$

$$K = \sum_i b_i \sum_j P_{ij}, \tag{4.2.46}$$

$$V = \sum_i b_i \sum_j P_{ij}^2, \tag{4.2.47}$$

and K^+ takes the value K for positive K and zero otherwise.

The jackknifed version of G^2, denoted as G_J^2 is defined in an analogous manner. Both $X_P(J)$ and G_J are asymptotically distributed as

$$\sqrt{2}[(\chi^2_{(t-1)})^{1/2} - (t-1)^{1/2}] \tag{4.2.48}$$

and the test is to reject H_0 if the observed value of X_{PJ} (or G_J) exceeds the upper $100(\alpha)\%$ critical value derived from (4.2.48).

Consider now a two-stage cluster sampling in which a t-category sample of m units is drawn independently from each of r sampled clusters. Let $\mathbf{m}_i = (m_{i1}, m_{i2}, \ldots, m_{i,t-1})'$ be the vector of counts in the first $t-1$ categories in the ith sample cluster ($i = 1, \ldots, r$) and $\mathbf{m} = \sum_{i=1}^r \mathbf{m}_i = (m_1, m_2, \ldots, m_{t-1})'$ be the corresponding vector for the whole sample, i.e., $m_k = \sum_{i=1}^r m_{i,k}$, $k = 1, \ldots, t-1$. The total number of observations in the sample is $n = \sum_i m_i$. Let $\hat{\pi}$ be a design-unbiased (or design-consistent estimator) of π with variance, $V(\hat{\pi}) = \mathbf{V}/n$, $\mathbf{V}$ being a suitable $(t-1) \times (t-1)$ matrix. Here, $\hat{\pi} = \sum_{i=1}^r \sum_{i=1}^r \mathbf{m}_i$, sum of r iid random vectors. Our aim is to test the null hypothesis $H_0 : \pi = \pi_0$. Then Rao and Thomas (1987) has shown that Fay's jackknifed statistics takes the following forms. Let

$$\begin{aligned}
\hat{\pi}_k(-i) &= \tfrac{r}{r-1}(\hat{\pi}_k - \tfrac{m_{ik}}{n}),\\
Q^2(-i) &= \sum_{k=1}^{t} \{\hat{\pi}_k(-i) - \pi_{k0}\}^2/\pi_{k0},\\
P(i) &= n(Q^2(-i) - Q^2),\\
K_J &= \tfrac{r-1}{r} \sum_{i=1}^{r} P(i),\\
V_J &= \tfrac{r-1}{r} \sum_{i=1}^{r} P^2(i)
\end{aligned} \tag{4.2.49}$$

where $Q^2 = X_P^2/n$, $\hat{\pi}_t = 1 - (\hat{\pi}_1 + \cdots + \hat{\pi}_{t-1})$. The Jackknife X_P^2 is

$$X_P^2(J) = \frac{(X_P^2)^{1/2} - (K_J)^{1/2}}{(V_J/8X_P^2)^{1/2}}. \tag{4.2.50}$$

Residual analysis Whenever H_0 is rejected, residual analysis can provide an insight into the nature of the deviations from H_0. Standardized residual can be defined as

$$\begin{aligned}
r_k &= \tfrac{\hat{\pi}_k - \pi_{k0}}{\sqrt{v_{kk}/n}}\\
&= \tfrac{\sqrt{n}(\hat{\pi}_k - \pi_{k0})}{\sqrt{d_{kk}\hat{\pi}_k(1-\hat{\pi}_k)}}.
\end{aligned}$$

It is assumed that r_k's are approximately independent $N(0, 1)$ variables. Cells that deviate from H_0 will be indicated by large values of $|r_k|$.

4.3 Testing for Homogeneity

In this section, we test if the given populations share the same vector of population proportions. The results are presented following Holt et al. (1980).

Suppose we have independent samples of sizes n_1 and n_2 from two populations. Let $\pi_i = (\pi_{i1}, \pi_{i2}, \ldots, \pi_{it-1})'$ be the population proportions for the ith population, $i = 1, 2$. The null hypotheses is

$$H_0 : \pi_1 = \pi_2 = \pi. \tag{4.3.1}$$

Suppose that the estimates $\hat{\pi}_i = (\hat{\pi}_{i1}, \ldots, \hat{\pi}_{it-1)})'(i = 1, 2)$ are calculated from the sample data obtained under an arbitrary sampling design $p(s)$ and

$$\sqrt{n_i}(\hat{\pi}_i - \pi) \sim N_{t-1}(\mathbf{0}, \mathbf{V}_i) \ (i = 1, 2), \tag{4.3.2}$$

as sample size $n_i \to \infty$.

If we have consistent estimators $\hat{\mathbf{V}}_1$ and $\hat{\mathbf{V}}_2$ of $\mathbf{V}_1$ and $\mathbf{V}_2$, respectively, then we can use a generalized Wald statistic

$$X^2_{WH} = (\hat{\pi}_1 - \hat{\pi}_2)' \left[\frac{\hat{\mathbf{V}}_1}{n_1} + \frac{\hat{\mathbf{V}}_2}{n_2} \right]^{-1} (\hat{\pi}_1 - \hat{\pi}_2) \tag{4.3.3}$$

which is asymptotically distributed as $\chi^2_{(t-1)}$ under H_0. This approach is due to Koch et al. (1975).

In practice, an estimate of $\mathbf{V}_i$ is often not available. This is specially true in secondary analysis of data that have been collected for other purposes. In such cases, practitioners often resort to the ordinary Pearson statistic for testing homogeneity, using the observed cell counts n_{ij} (number of observations in the jth category from population i) in the case of self-weighting design or, more generally, the estimated cell counts $\hat{n}_{ij} = n_i\hat{\pi}_{ij}$, that is, they use the statistic,

$$X^2_{PH} = \sum_{i=1}^{2}\sum_{j=1}^{t} \frac{n_i(\hat{\pi}_{ij} - \hat{\pi}_j)^2}{\hat{\pi}_j}, \tag{4.3.4}$$

where $\hat{\pi}_j = (n_1\hat{\pi}_{1j} + n_2\hat{\pi}_{2j})/(n_1 + n_2)$, an estimator of π_{ij} under $H_0(i = 1, 2)$. The statistic X^2_{PH} follows $\chi^2_{(t-1)}$ under multinomial sampling.

Now, X^2_{PH} can be written as

$$X^2_{PH} = \frac{n_1 n_2}{n_1 + n_2}(\hat{\pi}_1 - \hat{\pi}_2)'\hat{\mathbf{P}}^{-1}(\hat{\pi}_1 - \hat{\pi}_2), \tag{4.3.5}$$

where $\hat{\mathbf{P}} = \text{Diag.}(\hat{\pi}) - \hat{\pi}\hat{\pi}'$. Thus X^2_{PH} is equivalent to the modified Wald statistic, X^2_{WH} in the multinomial case, since then $\mathbf{V}_1 = \mathbf{V}_2 = \mathbf{P}$ under H_0.

It follows, therefore, from Theorem 4.2.1, that under a general sampling design $p(s)$ (and under H_0)

$$X^2_{PH} \sim \sum_{k=1}^{t-1} \lambda_i \tau_i^2 \tag{4.3.6}$$

where $\tau_1, \ldots, \tau_{t-1}$ are independent $N(0, 1)$ random variables and the λ_i's are the eigenvalues of

$$\begin{aligned} \mathbf{D}_H &= \mathbf{P}^{-1}\left(\frac{n_2\mathbf{V}_1 + n_1\mathbf{V}_2}{n_1 + n_2}\right) \\ &= \frac{n_2\mathbf{D}_1 + n_1\mathbf{D}_2}{n_1 + n_2} \end{aligned} \tag{4.3.7}$$

where $\mathbf{D}_i = \mathbf{P}^{-1}\mathbf{V}_i$ is the design effect matrix for the ith population. As in Sect. 4.2.6, λ_i may be regarded as the deff of τ_i under H_0.

Since X^2_{PH} is not asymptotically $\chi^2_{(t-1)}$ under H_0 for general sampling designs, it is important to find out whether testing X^2_{PH} as $\chi^2_{(t-1)}$ can be seriously misleading, and if so, whether it is possible to modify the statistic in a simple way to give better results.

Scott and Rao (1981) in a numerical study have shown that that the naive use of the Pearson chi-squared statistic with a multistage design can give very misleading results.

For any given set of λ_i's, we can use the approximations given for linear combinations of χ^2 random variables in Solomon and Stephens (1977) to evaluate the correct percentage points of the distribution of X^2_{PH} and hence study the effect of using the percentage points of $\chi^2_{(t-1)}$ in their place.

It follows that

$$\lambda_{r+s-1} \leq \frac{n_2\lambda_{1r} + n_1\lambda_{2s}}{n_1 + n_2} \quad (r + s \leq t), \tag{4.3.8}$$

where $\lambda_{i1} \geq \lambda_{i2} \geq \cdots$ are the ordered eigenvalues of $\mathbf{D}_i (i = 1, 2)$ and λ_i's are as given in (4.3.6). In particular, with the proportionally allocated stratified sampling $\lambda_{ij} \leq 1 \; \forall \; i, j$ so that $\lambda_j \leq 1$ and the Pearson X^2_{PH} test becomes conservative. Again, let

$$\lambda_0 = \frac{tr(\mathbf{D}_H)}{t-1} = \frac{n_2 tr(\mathbf{D}_1) + n_1 tr(\mathbf{D}_2)}{(n_1+n_2)(t-1)}$$
$$= \frac{n_2\bar{\lambda}_1 + n_1\bar{\lambda}_2}{n_1+n_2} \tag{4.3.9}$$

where $\bar{\lambda}_i = \sum_{j=1}^{t-1}\lambda_{ij}/(t-1)$, $(i = 1, 2)$ and λ_0 is the overall average eigenvalue.

As has been shown (Rao and Scott 1979), $tr(\mathbf{D}_i) = \sum_{k=1}^{t-1} n_i V_{ikk}/\pi_k(1-\pi_k)$ where V_{ikk} is the variance of $\hat{\pi}_{ik}$ under arbitrary sampling design $p(s)$; hence, if only the cell variances var $(\hat{\pi}_{ik})$ of the two populations are known, $\bar{\lambda}_1$ and $\bar{\lambda}_2$ and hence λ_0 can be calculated.

Alternatively, as has been shown by Scott and Rao (1981)

$$(t-1)\lambda_0 = \tilde{n}\sum_{i=1}^{2}\sum_{k=1}^{t-1}(1-\pi_k)d_{ik}/n_i$$

where

$$d_{ik} = \frac{n_i Var(\hat{\pi}_{ik})}{\pi_k(1-\pi_k)}$$

is the design effect for the (i, k)th cell and $\tilde{n} = (n_1 n_2)/(n_1+n_2)$. Thus, λ_0 is a weighted average of the individual cell design effects.

As before, a modified X^2_{PH} statistic is $X^2_{PH(c)} = X^2_{PH}/\hat{\lambda}_0$ which is distributed asymptotically as $\chi^2_{(t-1)}$ under H_0. Rao and Scott (1979), Holt et al. (1980) have shown that treating $X^2_{PH(c)}$ as a $\chi^2_{(t-1)}$ random variable under H_0 gives a very good approximation. Note that If one sample is much smaller in size than the other, then $\hat{\lambda}_0$ will be essentially equivalent to the estimated average deff of the smaller sample.

Another possible approximation due to Scott and Rao (1981) is as follows. Let $n_i^* = n_i/\hat{\bar{\lambda}}_i$, where $\bar{\lambda}_i = tr(\mathbf{D}_i)/(t-1)$. n_i^* can be regarded as the approximate sample size that is needed to get the same average accuracy as the actual design for all the t classes. Replacing in X^2_{PH}, the cell counts $\hat{n}_{ij}$ by $\hat{n}^*_{ij} = n_i^*\hat{\pi}_{ij}$ we get another modified Pearson statistic X^{2*}_{PH}. It can be shown that X^{2*}_{PH} has the same asymptotic distribution as $X^2_{PH(c)}$. Note that in this case, we are implicitly estimating π_k by $(n_1^*\hat{\pi}_{1k} + n_2^*\hat{\pi}_{2k})/(n_1^* + n_2^*)$ which may be preferable to $\hat{\pi}_k$ in small samples $(k = 1, \ldots, t)$.

We can directly extend these results to the problem of testing of homogeneity of r populations. Suppose there are r populations and an independent sample of size n_i is selected from the ith population by an arbitrary sampling design $p(s)$. Let $\hat{\pi}_i = (\hat{\pi}_{i1}, \ldots, \hat{\pi}_{it-1})'$ denote the vector of estimated proportions from the ith sample and suppose that

$$\sqrt{n_i}(\hat{\pi}_i - \pi) \rightarrow^L N_{t-1}(\mathbf{0}, \mathbf{V}_i), \tag{4.3.10}$$

as $n_i \rightarrow \infty$, $i = 1, \ldots, r$ and $\pi = (\pi_1, \ldots, \pi_{t-1})'$.

The usual chi-square test of homogeneity is

$$X_{PH}^2 = \sum_{i=1}^{r}\sum_{j=1}^{t}\frac{n_i(\hat{\pi}_{ij}-\hat{\pi}_j)^2}{\hat{\pi}_j} \tag{4.3.11}$$

where $\hat{\pi}_j = \sum_i n_i\hat{\pi}_{ij}/\sum_i n_i$, an estimate of π_j under H_0. With independent multinomial sampling in each population, X_{PH}^2 in (4.3.11) has asymptotically a $\chi^2_{((r-1)(t-1))}$ distribution. We now study the behavior of X_{PH}^2 under a more general sampling design.

Theorem 4.3.1 *Under any sampling design $p(s)$, the asymptotic distribution of X_{PH}^2 is*

$$X_{PH}^2 \sim \sum_{i=1}^{(r-1)(t-1)} \lambda_i\tau_i^2 \tag{4.3.12}$$

where $\tau_1, \tau_2, \ldots$ are independent $N(0,1)$ random variables and λ_i's are the nonzero eigenvalues of the matrix

$$\mathbf{\Lambda} = \begin{bmatrix} (1-f_1)\mathbf{D}_1 & -f_1\mathbf{D}_2 & \ldots & -f_1\mathbf{D}_r \\ -f_2\mathbf{D}_1 & (1-f_2)\mathbf{D}_2 & \ldots & -f_2\mathbf{D}_r \\ . & . & \ldots & . \\ -f_r\mathbf{D}_1 & -f_r\mathbf{D}_2 & \ldots & (1-f_r)\mathbf{D}_r \end{bmatrix} \tag{4.3.13}$$

where $f_i = \lim \frac{n_i}{n}$, $n = \sum_i n_i$ and $\mathbf{D}_i = \mathbf{P}^{-1}\mathbf{V}_i$ is the design effect matrix for the ith population $(i = 1, \ldots, r)$.

Proof Suppose that the sample sizes n_i and n increase simultaneously in such a way that $\lim \frac{n_i}{n} = f_i, 0 < f_i < 1 (i = 1, \ldots, r)$.

Let $\hat{\pi}_0$ be the $(t-1)r \times 1$ vector, $\hat{\pi}_0 = (\hat{\pi}_1', \hat{\pi}_2', \ldots, \hat{\pi}_r')'$. Let us define $\pi_0 = (\pi', \pi', \ldots, \pi')'$ where $\pi = (\pi_1, \ldots, \pi_{t-1})'$. It follows by the assumption (4.3.10) that

$$\sqrt{n}(\hat{\pi}_0 - \pi_0) \to^L N(\mathbf{0}, \mathbf{V}_0) \ \text{ as } n \to \infty \tag{4.3.14}$$

where $n = \sum_i n_i$ and $\mathbf{V}_0 = \oplus_{k=1}^r(\mathbf{V}_i/f_i)$ and the symbol $\oplus$ denote the block-diagonal matrix.

It can be shown that X_{PH}^2 in (4.3.11) has the same asymptotic distribution as

$$\tilde{X}_{PH}^2 = \sum_{i=1}^{r}\sum_{j=1}^{t}\frac{n_i(\hat{\pi}_{ij}-\hat{\pi}_j)^2}{\pi_j} \tag{4.3.15}$$

under the assumption (4.3.10). We can write $\tilde{X}^2_{PH}$ as

$$\tilde{X}^2_{PH} = n(\hat{\pi}_0 - \pi_0)'\mathbf{B}(\hat{\pi}_0 - \pi_0) \tag{4.3.16}$$

where

$$\mathbf{B} = \mathbf{F} \otimes \mathbf{P}^{-1}, \ \mathbf{P} = \text{Diag. } (\pi) - \pi\pi', \mathbf{F} = \text{ Diag.}(\mathbf{f}) - \mathbf{ff}', \tag{4.3.17}$$

and $\mathbf{f} = (f_1, f_2, \ldots, f_r)'$ and the symbol $\otimes$ denotes the direct product. Note that rank $(\mathbf{B})$ = rank $(\mathbf{F}) \times$ rank $(\mathbf{P}) = (r-1)(t-1)$.

It follows from Theorem 4.2.1 that

$$\tilde{X}^2_{PH} \sim \sum_{k=1}^{(t-1)(r-1)} \lambda_i \tau_i^2 \tag{4.3.18}$$

where $\tau_1, \tau_2, \ldots$ are asymptotically independent $N(0, 1)$ random variables and $\lambda_1, \lambda_2, \ldots$ are eigenvalues of $\mathbf{BV}_0 = \mathbf{\Lambda}$. Hence the proof. □

Note 4.3.1 As usual, the Pearson test X^2_{PH} is conservative with proportionally allocated stratified sampling.

The modified chi-square statistic $X^2_{PH(c)} = X^2_{PH}/\bar{\lambda}_0$ where

$$(r-1)(t-1)\bar{\lambda}_0 = tr(\mathbf{\Lambda}) = \sum_{i=1}^{r}(1-f_i)tr(\mathbf{D}_i), \tag{4.3.19}$$

that is,

$$\begin{aligned} \bar{\lambda}_0 &= \sum_{i=1}^{r} \frac{(1-f_i)\bar{\lambda}_{i0}}{r-1} \\ &= \sum_{i=1}^{r}\sum_{k=1}^{t} \frac{(1-f_i)(1-\pi_k)d_{i,k}}{(r-1)(t-1)}, \end{aligned} \tag{4.3.20}$$

where $d_{i.k}$ is the deff of the (i, k)th cell and $\bar{\lambda}_{i0}$ is the average design effect for the ith population. Therefore, $\bar{\lambda}_0$ can be calculated from the information about the cell variances for each population. The corrected statistic $X^2_{PH(c)}$ is again treated as a χ^2 variable with $(r-1)(t-1)$ d.f. under H_0. It may be noted that as r becomes large, $\bar{\lambda}_0$ will tend toward the unweighted average of the $\bar{\lambda}_{i0}$'s provided no single n_i dominates the others. $\bar{\lambda}_0$ is simply a weighted average of population design effects and should stay relatively stable as r increases.

Like the goodness-of-fit case, a second-order correction $X^2_{PH(cc)} = X^2_{PH(c)}/(1+\hat{a}^2)$ can be obtained when $\hat{a}^2$ is given by (4.2.34) and making the appropriate changes. Corresponding modifications to G^2 can be similarly defined.

Residual analysis: Standardized residuals which have an approximately standard normal distribution under H_0 are given by

$$\hat{r}_{ij} = (\hat{\pi}_{ij} - \hat{\pi}_j)/\sqrt{\hat{Var}(\hat{\pi}_{ij} - \hat{\pi}_j)}, i = 1, \ldots, r;\ j = 1, \ldots, t \tag{4.3.21}$$

where

$$\hat{Var}(\hat{\pi}_{ij} - \hat{\pi}_j) = \frac{1}{n^2}(\hat{\pi}_j(1 - \hat{\pi}_j))\left\{\frac{n(n - 2n_i)}{n_i}\hat{d}_{ij} + \sum_{l=1}^{r} n_l \hat{d}_{lj}\right\} \tag{4.3.22}$$

4.3.1 A Simple Method for Binary Data from Cluster Sampling

Suppose there are I populations of clusters each receiving a treatment. From the ith population m_i clusters are selected at random and suppose that among the n_{ij} units in jth cluster so selected x_{ij} units are affected by the treatment and the remaining $(n_{ij} - x_{ij})$ units are not affected. Let π_i be the unknown proportion of affected units in the ith population and π be the unknown overall proportion of affected units in this set-up.

For the ith population a natural estimator of π_i is $\hat{\pi}_i = \bar{x}_i/\bar{n}_i$ where $\bar{x}_i = \sum_{j=1}^{m_i} x_{ij}/m_i$ and $\bar{n}_i = \sum_{j=1}^{m_i} n_{ij}/m_i$. Since $\hat{\pi}_i$ is the ratio of two means, an estimator of variance of $\hat{\pi}_i$ for large m_i is

$$v_i = \frac{m_i}{(m_i - 1)n_i^2}\sum_{j=1}^{m_i} r_{ij}^2 \tag{4.3.23}$$

where $r_{ij} = x_{ij} - n_{ij}\hat{\pi}_i$ (Cochran 1977). Under mild regularity conditions on the population variances of n_{ij}'s and r_{ij}'s, it follows that $(\hat{\pi}_i - \pi_i)/\sqrt{v_i}$ is asymptotically $N(0, 1)$ as $m_i \to \infty$. Also, v_i is a consistent estimator of $V(\hat{\pi})$ in the sense that $m_i[v_i - V(\hat{\pi}_i)] \to 0$ as $m_i \to \infty$. Dividing v_i by the estimated binomial variance, we get the design effect (deff) of $\hat{\pi}_i$,

$$d_i = (n_i v_i)/[\hat{\pi}_i(1 - \hat{\pi}_i)]. \tag{4.3.24}$$

Again, $\tilde{n}_i = n_i/d_i$ is the effective sample size. Let $\tilde{x}_i = x_i/d_i$.

We transform the data from (x_i, n_i) to $(\tilde{x}_i, \tilde{n}_i)$ and treat $\tilde{x}_i$ as a binomial variate with parameters $(\tilde{n}_i, \pi_i)$. Now the estimated binomial variance of $\tilde{\pi}_i = \tilde{x}_i/\tilde{n}_i = \hat{\pi}_i$ is given by $\tilde{\pi}_i(1 - \tilde{\pi}_i)/\tilde{n}_i = v_i$ which is the same as the estimated variance of $\hat{\pi}_i$ and since $(\tilde{\pi}_i - \pi_i)/[\tilde{\pi}_i(1 - \tilde{\pi}_i)/\tilde{n}_i]^{1/2} = (\hat{\pi}_i - \pi_i)/\sqrt{v_i}$, it is asymptotically $N(0, 1)$. Therefore, tests based on $(\tilde{x}_i, \tilde{n}_i)$ leads to the asymptotically (as $m_i \to \infty$) correct results. Now $(\sqrt{n_1}(\tilde{\pi}_1 - \pi_1), \ldots, \sqrt{n_I}(\tilde{\pi}_I - \pi_I))'$ tends to be distributed as $N_{I-1}(\mathbf{0}, \mathbf{\Sigma})$ where $\mathbf{\Sigma}$ is the diagonal matrix with $\pi_i(1 - \pi_i)$ in its ith diagonal. Rao and Scott (1992) therefore conclude that replacing (x_i, n_i) by $(\tilde{x}_i, \tilde{n}_i)$ or equivalently replacing $(n_i, \hat{\pi}_i)$ by $(\tilde{n}_i, \tilde{\pi}_i)$ in any binomial-based procedure gives asymptotically correct results.

Testing homogeneity: The hypothesis of homogeneity is given by $H_0 : \pi_1 = \cdots = \pi_I$. The asymptotic distribution of the standard chi-square statistic

$$X_P^2 = \sum_{i=1}^{I} (x_i - n_i\hat{\pi})^2/[n_i\hat{\pi}(1 - \hat{\pi})] \tag{4.3.25}$$

with $\hat{\pi} = \sum_i x_i / \sum_i n_i$, will be a weighted sum of $(I - 1)$ independent $\chi^2_{(1)}$ random variables with weights depending on the population deff $D_i = n_i V(\hat{\pi}_i)/[\pi_i(1-\pi_i)]$. This weights will be larger than 1 because of positive intraclass correlation among the units within the same cluster so that the actual type I error rate would be larger than the nominal level. Replacing (x_i, n_i) by $(\tilde{x}_i, \tilde{n}_i)$ in (4.3.25), we get the adjusted chi-square statistic

$$X_P'^2 = \sum_{i=1}^{I} (\tilde{x}_i - \tilde{n}_i\tilde{\pi})^2/[\tilde{n}_i\tilde{\pi}(1 - \tilde{\pi})]. \tag{4.3.26}$$

Under H_0, $X_P'^2$ is asymptotically distributed as a $\chi^2_{(I-1)}$ random variable. In the special case when $D_i = D \; \forall \; i$, we may use $\tilde{x}_i = x_i/d$ and $\tilde{n}_i = n_i/d$ where d is pooled estimator given by

$$(I - 1)d = \sum_{i=1}^{I} (1 - f_i)\frac{\hat{\pi}_i(1 - \hat{\pi}_i)}{\hat{p}(1 - \hat{p})}d_i, \tag{4.3.27}$$

$f_i = n_i/n$ and $n = \sum_i n_i$. In this case $X_P'2$ reduces to X_P^2/d.

Earlier Donner (1989) suggested a modified chi-square under a common intraclass correlation model.

We now consider tests of independence in a two-way contingency table. However, since this hypothesis involves a number of hypotheses relating to linear functions of π_k's, we shall first consider tests of general linear hypotheses and effects of survey designs on these tests.

4.4 Effect of Survey Design on Classical Tests of General Linear Hypotheses

Suppose that the hypothesis of interest is

$$H_0 : h_i(\pi) = 0, \; i = 1, \ldots, b. \tag{4.4.1}$$

We assume the following regularity conditions on $h_i(\pi)$.

(i) $\frac{\partial h_i(\pi)}{\partial \pi_j}$ is a continuous function in the neighborhood of true π $\forall$ i, j;
(ii) The matrix $\mathbf{H}(\pi) = ((\frac{\partial h_i(\pi)}{\partial \pi_j}))_{b\times(t-1)}$ has rank b.

Now, by Taylor-series expansion,

$$\mathbf{h}(\hat{\pi}) \approx \mathbf{h}(\pi) + \mathbf{H}(\pi)(\hat{\pi} - \pi), \tag{4.4.2}$$

retaining only the terms up to first-order derivative. Here $\mathbf{h}(\pi) = (h_1(\pi), \ldots, h_b(\pi))'$ and $\mathbf{h}(\hat{\pi}) = (h_1(\hat{\pi}), \ldots, h_b(\pi))'$.

As in Sect. 4.2.5, we assume that $\sqrt{n}(\hat{\pi} - \pi) \sim N_{t-1}(\mathbf{0}, \mathbf{V})$ where $\mathbf{V}/n$ is the covariance matrix of $\hat{\pi}$ under a general sampling design $p(s)$.

Hence, it follows that $\sqrt{n}(\mathbf{h}(\hat{\pi}) - \mathbf{h}(\pi))$ is asymptotically b-variate normal $N_b(\mathbf{0}, \mathbf{HVH}')$. Therefore, if a consistent estimator $\hat{\mathbf{V}}$ of $\mathbf{V}$ is available, we can use the corresponding generalized Wald statistic

$$X_W^2(\mathbf{h}) = n\mathbf{h}(\hat{\pi})'(\hat{\mathbf{H}}\hat{\mathbf{V}}\hat{\mathbf{H}}^{-1})^{-1}\mathbf{h}(\hat{\pi}) \tag{4.4.3}$$

where $\hat{\mathbf{H}} = \mathbf{H}(\hat{\pi})$, which is distributed asymptotically as $\chi^2_{(b)}$ under H_0.

However, no estimate of $\mathbf{V}$ or $\mathbf{HVH}'$ is generally available and the researcher assumes the multinomial sampling and uses the multinomial covariance matrix $\mathbf{P} = \mathbf{D}_\pi - \pi\pi'$ in place of $\mathbf{V}$. The test statistic is then

$$X_P^2(\mathbf{h}) = n\mathbf{h}(\hat{\pi})'(\hat{\mathbf{H}}_0\hat{\mathbf{P}}_0\hat{\mathbf{H}}_0)^{-1}\mathbf{h}(\hat{\pi}) \tag{4.4.4}$$

where $\hat{\mathbf{H}}_0\hat{\mathbf{P}}_0\hat{\mathbf{H}}_0'$ is any consistent estimator of $\mathbf{H}_0\mathbf{P}_0\mathbf{H}_0'$ under the general sampling design when H_0 is true. The asymptotic distribution of $X_P^2(\mathbf{h})$ follows directly from Theorem 4.2.1.

Theorem 4.4.1 *Under the null hypothesis* H_0 : $\mathbf{h}(\pi) = \mathbf{0}$, $X_P^2(\mathbf{h}) = \sum_{k=1}^b \delta_{0k}\psi_k$ *where the* δ_k*'s are the eigenvalues of* $(\mathbf{HPH}')^{-1}(\mathbf{HVH}')$; $\delta_1 \geq \delta_2 \geq \cdots \geq \delta_b > 0$; $\psi_1, \ldots, \psi_b$ *are independent* $\chi^2_{(1)}$ *random variables and* δ_{0k} *is the value of* δ_k *under* H_0.

If $\mathbf{HVH}'$ *or* $\mathbf{HPH}'$ *are not known, their estimates are to be used throughout.* □

As before the δ_k's can be interpreted as the design effects of linear combinations L_k of the components of $\mathbf{H}\hat{\pi}$. Obviously, $\lambda_1 \geq \delta_k \geq \lambda_{t-1}$ for $k = 1, \ldots, b$, where λ's are the eigenvalues of $\mathbf{P}^{-1}\mathbf{V}$, since the L_k's are particular linear combinations of the π_k's.

As before, $X_P^2(\mathbf{h})/\tilde{\delta}$ with $\tilde{\delta} \geq \delta_{01}$ provides an asymptotically conservative test, when we can provide such an upper bound $\tilde{\delta}$. For stratified random sampling with proportional allocation $\delta_k \leq 1$ $\forall$ k and the conventional test $X_P^2(\mathbf{h})$ is asymptotically conservative.

Following Sect. 4.3, we obtain a good approximation test by treating $X_P^2(\mathbf{h})/\hat{\delta}_0$ as a $\chi^2_{(b)}$ random variable, where $\hat{\delta}_0 = \sum_{k=1}^b \hat{\delta}_k/b$ and $\hat{\delta}_k$ is a consistent estimate of δ_k under H_0. However, in general, $\hat{\delta}_0$ requires the knowledge of full matrix $\hat{\mathbf{V}}$. Tests based on $X_P^2(\mathbf{h})/\hat{d}_0$ have been suggested by Fellegi (1980) and others, where $\hat{d}_0$ is the estimated average cell deff.

4.5 Tests of Independence in a Two-Way Table

Suppose a contingency table has r rows and c columns and let $\pi = (\pi_{11}, \pi_{12}, \ldots, \pi_{rc-1})'$ denote the vector of cell-probabilities π_{ij}, $\sum_{i=1}^{r}\sum_{j=1}^{c}\pi_{ij} = 1$. Let $\pi_{i0} = \sum_{j=1}^{c}\pi_{ij}$, $\pi_{0j} = \sum_{i=1}^{r}\pi_{ij}$. As usual, we assume that we have estimated probabilities $\hat{\pi} = (\hat{\pi}_{11}, \hat{\pi}_{12}, \ldots, \hat{\pi}_{rc-1})'$ and that

$$\sqrt{n}(\hat{\pi} - \pi) \to^{L} N_{rc-1}(\mathbf{0}, \mathbf{V}). \tag{4.5.1}$$

The hypothesis of interest is

$$H_0 : h_{ij}(\pi) = \pi_{ij} - \pi_{i0}\pi_{0j} = 0,\ i = 1, \ldots, r-1;\ j = 1, \ldots, c-1. \tag{4.5.2}$$

The usual Pearson statistic for testing H_0 is

$$X_{PI}^2 = n\sum_{i=1}^{r}\sum_{j=1}^{c}\frac{(\hat{\pi}_{ij} - \hat{\pi}_{i0}\hat{\pi}_{0j})^2}{(\hat{\pi}_{i0}\hat{\pi}_{0j})}. \tag{4.5.3}$$

Here $\hat{\pi}_{ij}$ is the estimate of π_{ij} in the unrestricted case and $\tilde{\pi}_{ij} = \hat{\pi}_{i0}\hat{\pi}_{0j}$ is the estimate of π_{ij} under the null hypothesis. Now, (4.5.3) can be written as

$$X_{PI}^2 = n\mathbf{h}(\hat{\pi})'(\hat{\mathbf{P}}_r^{-1} \otimes \hat{\mathbf{P}}_c^{-1})\mathbf{h}(\hat{\pi}). \tag{4.5.4}$$

Here the symbol $\otimes$ denotes the direct product operator, $\mathbf{h}(\hat{\pi}) = (h_{11}(\hat{\pi}), \ldots, h_{(r-1)(c-1)}(\hat{\pi}))'$, $\hat{\mathbf{P}}_r$ = value of $\mathbf{P}_r$ = Diag. $(\pi_r) - \pi_r\pi_r'$ for $\pi_r = \hat{\pi}_r = (\hat{\pi}_{10}, \ldots, \hat{\pi}_{r-10})'$; $\hat{\mathbf{P}}_c$ = value of $\mathbf{P}_c$ = Diag. $(\pi_c) - \pi_c\pi_c'$ for $\pi_c = \hat{\pi}_c = (\hat{\pi}_{01}, \ldots, \hat{\pi}_{0c})'$.

The generalized Wald statistic for testing H_0 is given by

$$X_{WI}^2 = n\mathbf{h}(\hat{\pi})'\hat{\mathbf{V}}_h^{-1}\mathbf{h}(\hat{\pi}) \tag{4.5.5}$$

where $\mathbf{V}_h/n$ is the estimate of the covariance matrix $\mathbf{V}_h/n = \mathbf{HVH}/n$ of $\mathbf{h}(\hat{\pi})$. The statistic X_{WI}^2 is approximately distributed as $\chi^2_{(b)}$ under H_0 for sufficiently large n, where $b = (r-1)(c-1)$. The estimator $\hat{\mathbf{V}}_h/n$, if the sampling design does not permit a direct estimate, can be obtained by the familiar linearization method, the balanced repeated replication method, jackknife method, or any other sample-reuse method.

The hypothesis (4.5.2) is a special case of (4.4.1) with $t = rc$ and $b = (r-1)(c-1)$. It is straightforward to show that when H_0 is true and the sampling distribution is multinomial, $\mathbf{HPH}'$ reduces to $\mathbf{P}_r \otimes \mathbf{P}_c$. Thus X^2_{PI} is of the form $X^2_P(\mathbf{h})$ given by (4.4.4) and hence by Theorem 4.2.1, $X^2_{PI} \sim \sum_{i=1}^{b} \delta_{0i}\psi_i$ under H_0 where the δ_i's are the eigenvalues of the generalized deff matrix

$$(\mathbf{P}_r^{-1} \otimes \mathbf{P}_c^{-1})(\mathbf{HVH}') = (\mathbf{P}_c^{-1} \otimes \mathbf{P}_r^{-1})\mathbf{V}_h = \mathbf{D}_I \text{ (say)} \tag{4.5.6}$$

and δ_{0i} is the value of δ_i under H_0 and ψ's are independent $\chi^2_{(1)}$ variables. Again, the δ_i's can be interpreted as the design effects of the components of $\mathbf{H}\hat{\pi}$. As usual, this means that the ordinary Pearson chi-square test is conservative for proportionally allocated stratified sampling.

A first-order correction to the Pearson statistic can be obtained by dividing X^2_{PI} by $\hat{\delta}_0$, the mean of the estimated eigenvalues of the generalized deff matrix $\mathbf{D}_I$ and the first-order corrected statistic is

$$X^2_{PI(c)} = \frac{X^2_{PI}}{\hat{\delta}_0} \tag{4.5.7}$$

where

$$\begin{aligned}\hat{\delta}_0 &= \sum_{i=1}^{r}\sum_{j=1}^{c} \hat{v}_{ij}(\mathbf{h})/(b\hat{\pi}_{i0}\hat{\pi}_{0j}) \\ &= \sum_{i=1}^{r}\sum_{j=1}^{c}(1-\hat{\pi}_{i0})(1-\hat{\pi}_{0j})\hat{\delta}_{ij}/b,\end{aligned} \tag{4.5.8}$$

$\hat{v}_{ij}(\mathbf{h})/n$ is the estimator of variance of $h_{ij}(\hat{\pi})$ and $\hat{\delta}_{ij}$ is the estimated deff of $h_{ij}(\hat{\pi})$; that is,

$$\hat{\delta}_{ij} = \hat{v}_{ij}(\mathbf{h})/[\hat{\pi}_{i0}\hat{\pi}_{0j}(1-\hat{\pi}_{i0})(1-\hat{\pi}_{0j})]. \tag{4.5.9}$$

Hence, $X^2_{PI(c)}$ requires only the knowledge of the deff's of $h_{ij}(\hat{\pi})$'s, which may not be generally available. Information on $\hat{\mathbf{V}}_h$ is never available with published data. We can only hope to have information on the diagonal elements of $\mathbf{V}$ at best.

Scott and Rao (1981) have stated that if the design effects of $\hat{\pi}_{ij}$ are not too variable, the $\hat{\delta}_0$ might be expected to be close to

$$\begin{aligned}\hat{\lambda}_0 &= \sum_{i=1}^{r}\sum_{j=1}^{c} v_{ij}/[(rc-1)\hat{\pi}_{ij}] \\ &= \sum_{i=1}^{r}\sum_{j=1}^{c}(1-\hat{\pi}_{ij})\hat{d}_{ij}/(rc-1)\end{aligned} \tag{4.5.10}$$

where $v_{ij} = \hat{V}(\hat{\pi}_{ij})$ and $\hat{d}_{ij}$ is the estimated deff of the individual cell-estimators $\hat{\pi}_{ij}$. However, if b is large, even v_{ij}'s would be difficult to come by.

Estimates of $\hat{\delta}_0$, the mean of the eigenvalues of $\hat{\mathbf{D}}_I$, which depend only on the partial information on $\hat{\mathbf{V}}$ has been given by Rao and Scott (1982, 1984), Bedrick (1983) and Gross (1984). This is given by

$$\hat{\delta}_0 = \frac{1}{(r-1)(c-1)} \sum_{i=1}^{r} \sum_{j=1}^{c} \frac{\hat{\pi}_{ij}(1-\hat{\pi}_{ij})}{\hat{\pi}_{i0}\hat{\pi}_{0j}} \hat{d}_{ij} - \sum_{i=1}^{r} (1-\hat{\pi}_{i0})\hat{d}_{A(i)} - \sum_{j=1}^{c} (1-\hat{\pi}_{0j})\hat{d}_{B(j)} \tag{4.5.11}$$

where

$$\hat{d}_{ij} = \hat{V}(\hat{\pi}_{ij})n/\hat{\pi}_{ij}(1-\hat{\pi}_{ij}) \tag{4.5.12}$$

is the (i, j)th cell deff and

$$\begin{aligned} \hat{d}_{A(i)} &= \hat{V}(\hat{\pi}_{i0})n/\hat{\pi}_{i0}(1-\hat{\pi}_{i0}) \\ \hat{d}_{B(j)} &= \hat{V}(\hat{\pi}_{0j})n/(\hat{\pi}_{0j}(1-\hat{\pi}_{0j})) \end{aligned} \tag{4.5.13}$$

are the deffs of the ith row and jth column totals, respectively. The first-order corrected test now consists of referring $X^2_{PI(c)} = X^2_{PI}/\hat{\delta}_0$ to $\chi^2_{(r-1)(c-1)}(\alpha)$.

A second-order corrected statistic is given by $X^2_{PI(cc)} = X^2_{PI(c)}/(1+\hat{a}^2)$ where $\hat{a}^2$ is given by (4.2.34) with $\hat{\delta}_0$ replaced by (4.5.11), $\sum \hat{\delta}_i^2$ replaced by

$$\sum_{k=1}^{(r-1)(c-1)} \hat{\delta}_k^2 = \sum_{i,i'}^{r-1} \sum_{j,j'}^{c-1} \hat{v}^2_{h,ij,i'j'}/(\hat{\pi}_{i0}\hat{\pi}_{0j})(\hat{\pi}_{i'0}\hat{\pi}_{0j'}), \tag{4.5.14}$$

and $(t-1)$ replaced by $(r-1)(c-1)$, when $\hat{v}_{h,ij,i'j'}$ is the element of $\hat{\mathbf{V}}_h$ corresponding to the covariance of $\hat{\pi}_{ij}$ and $\hat{\pi}_{i'j'}$. The second-order procedure refers to $X^2_{PI(cc)}$ to χ^2_ν where $\nu = (r-1)(c-1)/(1+\hat{a}^2)$.

Similarly an F-based version of the Wald test given in (4.5.5) is given by

$$F_1(X^2_{WI}) = \left(\frac{f-(r-1)(c-1)+1}{f(r-1)(c-1)} X^2_{WI} \right) \tag{4.5.15}$$

to an F-distribution on $(r-1)(c-1)$ and $(f-(r-1)(c-1)+1)$ degrees of freedom.

Residual Analysis Standardized residuals that have approximate standard normal distributions under H_0 are given by

$$\hat{r}_{ij} = r_{ij}/\sqrt{\hat{d}_{h,ij}} \tag{4.5.16}$$

where the

$$r_{ij} = \hat{h}_{ij}/\{\hat{\pi}_{i0}\hat{\pi}_{0j}(1-\hat{\pi}_{i0})(1-\hat{\pi}_{0j})\}^{1/2} \qquad (4.5.17)$$

are the standardized residuals defined under the assumption that the sampling design is *srswr* and

$$\hat{d}_{h,ij} = \hat{v}_{h,ij}/\{n^{-1}\hat{\pi}_{i0}\hat{\pi}_{0j}(1-\hat{\pi}_{i0})(1\hat{\pi}_{0j})\}^{1/2} \qquad (4.5.18)$$

are the estimated deffs of the residuals $\hat{h}_{ij}$ under H_0.

Note 4.5.1 In the case of multinomial sampling, it is well-known that the test statistic for independence and homogeneity are identical; but this property does not carry over to more complex sampling designs and the effect on the asymptotic distribution of Pearson statistic can be very different in the two situations (Holt et al. 1980).

Note 4.5.2 Bedrick (1983) considered an asymptotic mean correction to X^2_{PI} and the corresponding G^2_I under a log-linear model. The log-linear model has been considered in detail in Sect. 5.2.

Consider the problem of testing for independence in a two-way cross-classification of variables R and C having r and c levels, respectively. The model of interest is $\pi_{ij} = \pi_{i0}\pi_{0j}$ $(i = 1, \ldots, r;\ j = 1, \ldots, c)$. The corresponding log-linear model is

$$\log \pi_{ij} = u + u_{R(i)} + u_{C(j)}\ (i = 1, \ldots, r;\ j = 1, \ldots, c). \qquad (4.5.19)$$

Let $\hat{\pi}_{ij}$ and $\tilde{\pi}_{ij} = \hat{\pi}_{i0}\hat{\pi}_{0j}$ denote the estimate of the cell-proportion π_{ij} in the unrestricted case and under null hypothesis, respectively. The Pearson statistic for testing independence is given by X^2_{PI} in (4.5.3). The log-likelihood ratio statistic G^2 is

$$\begin{aligned} G^2_I &= 2n\sum_{i=1}^{r}\sum_{j=1}^{c}\hat{\pi}_{ij}\log(\frac{\hat{\pi}_{ij}}{\tilde{\pi}_{ij}}) \\ &= 2n\left[\sum_{i=1}^{r}\sum_{j=1}^{c}\hat{\pi}_{ij}\log\hat{\pi}_{ij} - \sum_{i=1}^{r}\hat{\pi}_{i0}\log\hat{\pi}_{i0}\right. \\ &\qquad \left.\sum_{j=1}^{c}\hat{\pi}_{0j}\log\hat{\pi}_{0j}\right]. \end{aligned} \qquad (4.5.20)$$

The asymptotic mean of X^2_{PI} and G^2_I can be calculated by calculating the asymptotic mean of the first nonzero term in a Taylor-series expansion of $G^2_I(\hat{\pi})$ about $G^2(\pi)$. Assuming that the model of independence is correct and $\sum_{i,j}\tilde{\pi}_{ij} = 1$, it can be shown that the constant and linear terms are zero. It follows that the quadratic terms reduce to

$$n\left[\sum_i\sum_j \frac{(\hat{\pi}_{ij}-\pi_{ij})^2}{\pi_{ij}} - \sum_i \frac{(\hat{\pi}_{i0}-\pi_{i0})^2}{\pi_{i0}} - \sum_j \frac{(\hat{\pi}_{0j}-\pi_{0j})^2}{\pi_{0j}}\right]. \tag{4.5.21}$$

Therefore,

$$\begin{aligned} E[G_I^2(\hat{\pi})] &= \sum_{i,j} \frac{Var(\sqrt{n}\hat{\pi}_{ij})}{\pi_{ij}} - \sum_i \frac{Var(\sqrt{n}\hat{\pi}_{i0})}{\pi_{i0}} - \sum_j \frac{Var(\sqrt{n}\hat{\pi}_{0j})}{\pi_{0j}} \\ &= (r-1)(c-1)\bar{\lambda} \text{ (say).} \end{aligned} \tag{4.5.22}$$

Let $d_{RC}(i,j)$, $d_R(i)$ and $d_C(j)$ denote the (i,j)th cell design effect, the ith cell design effect for the row variable R, and the jth cell design effect for the column variable C, respectively. Then

$$\begin{aligned} E\{G_I^2(\hat{\pi})\} &= \sum_{i,j}(1-\pi_{ij})d_{RC}(i,j) - \sum_i(1-\pi_{i0})d_R(i) - \sum_j(1-\pi_{0j})d_C(j) \\ &= (rc-1)\bar{d}_{RC} - (r-1)\bar{d}_R - (c-1)\bar{d}_C \end{aligned} \tag{4.5.23}$$

where

$$\begin{aligned} \bar{d}_{Rc} &= \sum_{i,j}(1-\pi_{ij})d_{RC}(i,j)/(rc-1), \\ \bar{d}_R &= \sum_i(1-\pi_{i0})d_R(i)/(r-1), \\ \bar{d}_C &= \sum_j(1-\pi_{0j})d_C(j)/(c-1) \end{aligned} \tag{4.5.24}$$

are weighted averages of cell and marginal cell design effects. If as in Rao and Scott (1981) we need first moment adjustment to X_{PI}^2 (or G_I^2), consistent estimates of these design effects are required. If $r=c=2$, $X_{PI}^2/\bar{\lambda}$ (or $G_I^2/\bar{\lambda}$) $\sim \chi^2_{(1)}$.

Note: Nandram et al. (2013) considered a Bayesian approach to the study of independence in a two-way contingency table which is obtained from a two-stage cluster sampling design. If a procedure based on single-stage simple random sampling is used to test for independence, the resulting p-value may be too small, thus inflating the type I error of the test. In such situations Rao–Scott corrections to the standard chi-square or likelihood ratio tests provide appropriate inference. For smaller surveys, however, Rao–Scott corrections may not be accurate, mainly due to the fact that the chi-square tests are inaccurate in such cases. The authors use a hierarchical Bayesian model to convert the observed cluster sample into a simple random sample. This procedure provides surrogate samples which can be used to derive the distribution of the Bayes factor. The authors demonstrate their procedure using an example and also provide a simulation study to establish that their methodology is a viable alternative to the Rao–Scott approximation for relatively small two-stage cluster sampling.

4.6 Some Evaluation of Tests Under Cluster Sampling

Thomas et al. (1996) made a comparative study of different tests of independence on two-way tables under cluster sampling. Let $\pi = (\pi_{11}, \ldots, \pi_{rc-1})', \pi_R = (\pi_{10}, \ldots, \pi_{r-10})', \pi_C = (\pi_{01}, \ldots, \pi_{0c-1})'$.

The independence hypothesis can be expressed in two equivalent forms:

(i) $H_0 : h_{ij} = \pi_{ij} - \pi_{i0}\pi_{0j} = 0, i = 1, \ldots, (r-1); j = 1, \ldots, (c-1)$;

(ii) the log-linear form: $H_0 : \ln(\pi_{ij}) = \mu + \mu_{1(i)} + \mu_{2(j)}, i = 1, \ldots, (r-1); j = 1, \ldots, (c-1)$.

The different formulations give rise to different statistics for testing of independence hypothesis. We consider three different sets of generalized design effects.

(i) $\lambda_{R(k)}, k = 1, \ldots, (r-1)$, the eigenvalues of the design-effect matrix $\mathbf{D}_R = \mathbf{P}_R^{-1}\mathbf{V}_R$ arising from the test of goodness-of-fit on the row-marginals π_R. Here $\mathbf{V}_R$ denotes the covariance matrix of a consistent estimate of π_R, $\mathbf{P}_R$ denotes the corresponding multinomial covariance matrix. The mean of the $\lambda_{R(k)}$'s will be denoted by $\bar{\lambda}_R$.

(ii) $\lambda_{C(k)}, k = 1, \ldots, c-1$ the eigenvalues of the design-effect matrix arising from the test of goodness-of-fit on column marginals π_C. The mean of the $\lambda_{C(k)}$'s will be denoted as $\bar{\lambda}_C$.

(iii) $\delta_k, k = 1, \ldots, (r-1)(c-1)$, the eigenvalues of the generalized design-effects matrix $\mathbf{D}_I$ corresponding to the test of independence. $\mathbf{D}_I$ may be expressed in the form

$$\mathbf{D}_I = n(\mathbf{C}'\mathbf{D}_\pi^{-1}\mathbf{C})^{-1}(\mathbf{C}'\mathbf{D}_\pi^{-1}\mathbf{V}\mathbf{D}_\pi^{-1}\mathbf{C}) \tag{4.6.1}$$

where n is the sample size, $\mathbf{C}$ is the completion of the design matrix $\mathbf{X}$ for the logarithmic form of the independence hypothesis (expressed in the matrix form), $\mathbf{V}$ is the covariance matrix of a consistent estimate of π, $\mathbf{D}_\pi$ is a diagonal matrix with elements $\pi_{ii} = \pi_{i0}\pi_{0i}$ on its diagonal. The mean of the δ_k's will be denoted by $\bar{\delta}$. A measure of variation among δ_k's will be denoted as $a(\delta) = [\sum_{i=1}^{\nu}\delta_i^2/\nu\bar{\delta}^2 - 1]^{1/2}$ where $\nu = (r-1)(c-1)$. (For a proof of (4.6.1), see (5.3.3).)

Several models of two-stage cluster sampling were considered, including Brier's (1980) Dirichlet multinomial model and its extensions (Thomas and Rao 1987; Scott and Rao 1981). A new model, based on a 'modified logistic normal' distribution was developed for use in the study.

In their Monte Carlo study Thomas et al. (1996) considered, among others, the following statistics for row-column independence.

(i) The Pearson X_P^2 and the log-likelihood G^2 tests;
(ii) First-order Rao–Scott correction to X_P^2 and G^2, denoted by $X_{P(c)}^2$ (vide (4.2.30)) and G_c^2 (vide (4.2.32));
(iii) F-corrected Rao–Scott statistic $X_{P(c)}^2$ and G_c^2, denoted as $F(X_{P(c)}^2)$ (vide (4.2.38)) and $F(G_c^2)$.
(iv) The second-order Rao–Scott corrected statistics $X_{P(cc)}^2$ (vide (4.2.33)) and G_{cc}^2 (vide (4.2.37)).
(v) F-corrected tests $F(X_{P(cc)}^2)$ and $F(G_{cc}^2)$;
(vi) Fellegi-corrected X_P^2, denoted as $X_P^2(\hat{d}_0)$ (vide (4.2.29));
(vii) Fay jackknifed procedures applied to X_P^2 and G^2, denoted as X_J^2 and G_J^2, respectively (Fay 1985);
(viii) Two Wald procedures based, respectively, on the residual and log-linear form of the independence hypothesis;
(ix) F-corrected versions of the above two Wald procedures (vide Sect. 4.2.2).

Thomas and Rao (1987) in their Monte Carlo study compared the performances of $X_P^2, G^2, X_W^2, F_1(X_W^2), X_{P(c)}^2, F(X_{P(c)}^2), X_P^2(\hat{d}_0), F(G_c^2), X_{P(cc)}^2, G_{(cc)}^2, X_P^2(J)$ and G_J^2 with respect to control of Type I error and the power of the test. The modified statistic $F(X_{P(c)}^2)$ performs better than $X_{P(c)}^2$ in respect of control of type I error. Since both the statistics require the same amount of information, namely the estimated cell *deff*'s, one may use $F(X_{P(c)}^2)$ in preference to $X_{P(c)}^2$.

Among the statistics $X_{P(c)}^2$, X_J^2 and $F_1(X_W^2)$ which require detailed information, both $X_{P(c)}^2$ and X_J^2 perform better than $F_1(X_W^2)$. Their Monte Carlo study clearly shows that the F-version of the Wald statistic $F_1(X_W^2)$ is much better than the original statistic X_W^2. The performance of X_J^2 and $X_{P(cc)}^2$ are similar, although $X_{P(cc)}^2$ seems to have a slight advantage in the case of varying *deff*.

As is well known by now, analysis of categorical data from complex surveys, using classical statistical methods without taking into account the complex nature of the data may lead to asymptotically invalid statistical inferences. The methods that have been developed to account for the survey design require additional information such as survey weights, design effects or cluster identification. Benhim and Rao (2004) developed alternative approaches that undo the complex data structures using repeated inverse sampling so that standard methods can be applied to the generalized inverse sampling data. The authors proposed a combined estimating equation approach to analyze such data in the context of categorical survey data. For simplicity, they focussed on the goodness-of-fit statistic under cluster sampling.

4.7 Exercises and Complements

4.1 Let $\pi = (\pi_1, \ldots, \pi_t)'$ and define $\mathbf{P}$ and $\mathbf{V}$ accordingly. Prove the following results.

(i) $\mathbf{1}'\pi = \mathbf{1}\hat{\pi} = 1$;
(ii) $\mathbf{D}_\pi \mathbf{1} = \pi, \mathbf{D}_\pi^{-1}\pi = \mathbf{1}$;
(iii) $\mathbf{V1} = \mathbf{0} = \mathbf{VD}_\pi^{-1}\pi$ since $\mathbf{V}$ is a singular matrix. In particular, show that $\mathbf{P1} = \mathbf{PD}_\pi^{-1}\pi = \mathbf{0}$;
(iv) Observe that

$$\begin{aligned}\mathbf{PD}_\pi^{-1}\mathbf{V} &= (\mathbf{D}_\pi - \pi\pi')\mathbf{D}_\pi^{-1}\mathbf{V} = \mathbf{V} - \pi\pi'\mathbf{D}_\pi^{-1}\mathbf{V}\\ &= \mathbf{V} - \pi\mathbf{1}'\mathbf{V}\\ &= \mathbf{V} - \pi(\mathbf{V1})' = \mathbf{V}.\end{aligned}$$

Similarly, note that $\mathbf{VD}_\pi^{-1}\mathbf{P} = \mathbf{V}$; in particular, $\mathbf{PD}_\pi^{-1}\mathbf{P} = \mathbf{P}$ so that $\mathbf{D}_\pi^{-1}$ is a generalized inverse of $\mathbf{P}$.

4.2 Show that in the case of stratified random sampling,

$$E(X_P^2) = t - 1 - \sum_{k=1}^{t} \frac{1}{n\pi_k} \sum_h n_h(\pi_{hk} - \pi_k)^2 \neq t - 1$$

where n_h and π_{hk} are, respectively, the sample size and the proportion of units in category k within stratum h and π_k is the population proportion in category k. Hence, show that the asymptotic distribution of X_P^2 as $\chi^2_{(t-1)}$ does not hold in this case.
(Fellegi 1980)

4.3 Prove the relation (4.2.12).

4.4 Prove the relation (4.2.15).

4.5 Prove the relation (4.2.17).

4.6 Prove the relation (4.2.31).

4.7 Prove the relation (4.3.19).

4.8 Suppose that the survey population consists of R psu's (clusters) with M_i secondaries in the ith psu ($i = 1, \ldots, R$; $\sum_i M_i = N$). Let $Z_{i\lambda k} = 1(0)$ if the λth population unit in the ith psu is in category k (otherwise), $\lambda = 1, \ldots, M_i$; $i = 1, \ldots, R$; $k = 1, \ldots, t$.

Following Altham (1976) assume that the random variable $Z_{i\lambda k}$'s in different clusters are independent and

$$\mathcal{E}(Z_{i\lambda k}) = \pi_k$$
$$\mathcal{E}(Z_{i\lambda k} - \pi_k)(Z_{i\mu k'} - \pi_{k'}) = b_{kk'},\ \lambda \neq \mu \tag{i}$$

where $\mathcal{E}$ denotes expectation with respect to model.

A two-stage sample s is denoted by $(s_1, \ldots, s_r)$ where s_l is a subsample of size m_l from the lth sampled psu and r is the number of sampled psu's ($\sum_{l=1}^{r} m_l = n$). Show that the sample cell frequencies can be written as

$$n_k = \sum_{l=1}^{r}\sum_{h=1}^{m_l} Z_{lhk} = \sum_{l=1}^{r} m_{lk}. \tag{ii}$$

Then show the following results:

$\epsilon(n_k/n) = \pi_k$ for every s, i.e., the estimator $\hat{\pi}_k = n_k/n$ is model-unbiased for π_k. Also, the model-covariance matrix of $\mathbf{n} = (n_1, \ldots, n_{t-1})'$ for a given s is

$$\mathbf{V}_s = \mathbf{\Delta} + (m_{0s} - 1)\mathbf{B} \tag{iii}$$

where $\mathbf{\Delta} =$ Diag. $(\pi) - \pi\pi'$, $\mathbf{B} = (b_{kk'})$, $\pi = (\pi_1, \ldots, \pi_{t-1})'$ and $m_{0s} = \sum_{\in s} m_l^2/n$.

A general hypothesis on model parameters π is given by

$$H_0 : h_i(\pi) = 0,\ i = 1, \ldots, b \tag{iv}$$

and the chi-square test is similar to (4.4.4). Since the m_{lk} observations s in different clusters are independent, it follows that $\sqrt{n}(\hat{\pi} - \pi) \sim N(\mathbf{0}, \mathbf{V}_s)$ for large r and hence $\sqrt{n}(\mathbf{h}(\hat{\pi}) - \mathbf{h}(\pi)) \sim N(\mathbf{0}, \mathbf{H}\mathbf{V}_s\mathbf{H}')$ where $\mathbf{H} = \mathbf{H}(\pi)$. Hence, show by Theorem 4.2.1 that $X_P^2(\mathbf{h}) \sim \sum_{i=1}^{b} \hat{\delta}_{0i}\psi_i$ where $\hat{\delta}_i = 1 + (m_{0s} - 1)\hat{\rho}_i(\mathbf{h})$ and $\hat{\rho}_i(\mathbf{h})$'s are the eigenvalues of $(\mathbf{H}\Delta\mathbf{H}')^{-1}(\mathbf{HBH}')$. It follows that $\mathbf{H}\Delta\mathbf{H}' - \mathbf{HBH}'$ is nonnegative definite so that $\hat{\rho}_i(\mathbf{h}) \leq 1\ \forall\ i = 1, \ldots, b$. Hence, obtain a conservative test by treating $\chi^2(\mathbf{h})/m_{0s}$ as a $\chi^2_{(b)}$ variable under H_0.

(Rao and Scott 1981)

4.9 Consider the usual chi-squared statistic for testing independence in a $r \times c$ table,

$$X_{PI}^2 = n\sum_i\sum_j \frac{(\hat{\pi}_{ij} - \hat{\pi}_{i0}\hat{\pi}_{0j})^2}{\hat{\pi}_{i0}\hat{\pi}_{0j}} \tag{i}$$

where the notations have the usual meanings. Assume that

$$\sqrt{n}(\hat{\pi} - \pi) \rightarrow^L N(\mathbf{0}, \mathbf{V}) \tag{ii}$$

where $\pi = (\pi_{11}, \pi_{12}, \ldots, \pi_{rc})'$ and $\hat{\pi} = (\hat{\pi}_{11}, \hat{\pi}_{12}, \ldots, \hat{\pi}_{rc})'$ are the $rc \times 1$ vectors of cell proportions and their estimates. Here $\mathbf{V}$ is a positive semi-definite matrix of rank $rc - 1$ with each row and column summing to zero. Show that X_{PI}^2 can be written as

$$X^2_{PI} = n\mathbf{h}(\hat{\boldsymbol{\pi}})'(\text{Diag. }(\hat{\pi}))^{-1}\mathbf{h}(\hat{\pi}_I), \qquad (iii)$$

where $\mathbf{h}(\pi)$ and π_I are the $rc \times 1$ vectors whose co-ordinates are given by

$$h_{ij}(\pi) = \pi_{ij} - \pi_{i0}\pi_{0j}, \qquad (iv)$$

$$(\pi_I)_{ij} = \pi_{i0}\pi_{0j}. \qquad (v)$$

If $\mathbf{H} = \partial\mathbf{h}(\pi)/\partial\pi$ is the $rc \times rc$ matrix of partial derivatives, show that

$$\sqrt{n}(\mathbf{h}(\hat{\pi}) - \mathbf{h}(\pi)) \rightarrow^L N(\mathbf{0}, \mathbf{HVH}'). \qquad (vi)$$

From (iii) and (vi) show that

$$X^2_{PI} \sim \sum_{i=1}^{(r-1)(c-1)} \delta_i\tau_i^2 \qquad (vii)$$

where the τ_i's are asymptotically independent $N(0, 1)$ variables and the δ_i's are the nonzero eigenvalues of

$$\mathbf{D}_I = (\text{Diag. }(\pi_I))^{-1}\mathbf{HVH}'. \qquad (viii)$$

Show that $\mathbf{D}$ has rank $(r-1)(c-1)$. Holt et al. (1980) suggested a modified test

$$X^2_{PI(c)} = \frac{X^2_{PI}}{\bar{\delta}} \qquad (ix)$$

where $\bar{\delta} = \sum_i \delta_i/(r-1)(c-1)$. Holt et al. suggested that to calculate $\bar{\delta}$ an expression in terms of variances of proportional estimates only will be desirable. Show that

$$\bar{\delta} = \bar{d}_k = \frac{1}{(c-1)}\bar{d}_r - \frac{1}{(r-1)}\bar{d}_c \qquad (ix)$$

where

$$\bar{d}_k = \frac{1}{(r-1)(c-1)}\sum_i\sum_j \frac{v_{ijij}}{\pi_{i0}\pi_{0j}} \qquad (x)$$

and $\bar{d}_r, \bar{d}_c$ are the average design effects for the rows and column proportions and are given by

$$\begin{aligned}
\bar{d}_r &= \frac{1}{c-1}\sum_{i=1}^{r}\frac{v_{ii}^{(r)}}{\pi_{i0}},\\
\bar{d}_c &= \frac{1}{c-1}\sum_{j=1}^{c}\frac{v_{jj}^{(c)}}{\pi_{0j}},\\
v_{ii}^{(r)} &= \sum_{k,l} v_{ikil},\ v_{jj}^{(c)} = \sum_{k,l} v_{kjlj}.
\end{aligned}$$

(Gross 1984)

Chapter 5
Analysis of Categorical Data Under Log-Linear Models

Abstract This chapter considers analysis of categorical data from complex surveys using log-linear models for cell probabilities π in contingency tables. Noting that appropriate ML equations for the model parameter θ and hence of $\pi(\theta)$ are difficult to obtain for general survey designs, 'pseudo-MLE's have been used to estimate the cell probabilities. The asymptotic distributions of goodness-of-fit (G-o-F) statistic X_P^2, and likelihood ratio (LR) statistic G^2 have been derived and these test statistics have been modified using Rao-Scott (J Amer Stat Assoc 76: 221–230, 1981, Ann Stat 12: 46–60, 1984) first- and second-order corrections, F-based corrections, and Fellegi's (J Amer Stat Assoc 75: 261–268, 1980) correction. Wald's test statistic has been looked into. All these modified statistics have been examined in G-o-F tests, homogeneity tests, and independence tests. Fay's Jackknifed versions to these statistics have been considered. Brier's model has also been looked into. Lastly, nested models have been considered and all the above results have been examined in its light.

Keywords Log-linear models · Pseudo-MLE · Goodness-of-fit tests · Homogeneity tests · Independence tests · Rao-Scott corrections · F-based corrections · Fay's Jackknifed version · Wald test · Brier's model · Nested model

5.1 Introduction

In general, the cell probabilities $\pi_1, \ldots, \pi_t$ will involve some unknown parameters $\theta_1, \ldots, \theta_s$. In this chapter (and the next chapter) we will consider testing of different hypotheses under this setup, for which we will have to generally take recourse to the theory developed in Sects. A.4 and A.5 of the Appendix.

This chapter considers analysis of categorical data under log-linear models. Textbooks in the analysis of categorical data through log-linear models, such as Bishop et al. (1975), Goodman (1978), and Fienberg (1980), discuss these models in terms of classical sampling distributions, namely the Poisson, multinomial, and product-multinomial (with occasional use of hypergeometric distribution). In this context, the Pearson, the likelihood ratio, the Freeman–Tukey chi-squared tests, and Wald

P. Mukhopadhyay, *Complex Surveys*, DOI 10.1007/978-981-10-0871-9_5

statistic are often used. Fineberg (1979) reviewed the literature and properties of these statistics under these models.

However, as is well known at present that these simple test statistics may give extremely erroneous results when applied to data arising from a complex survey design.

5.2 Log-Linear Models in Contingency Tables

We have already introduced log-linear models in Sect. 3.6. In the log-linear models, natural logarithm of cell probabilities is expressed in a linear model analogous to the analysis-of-variance models.

Thus for a 2×2 contingency table with π_{ij} denoting the probability of an element belonging to the (i, j)th cell, we write the model as

$$\ln \pi_{ij} = u + u_{1(i)} + u_{2(j)} + u_{12(ij)}, \ i, j = 1, 2, \tag{5.2.1}$$

where u is the general mean effect, $u + u_{1(i)}$ is the mean of the logarithms of probabilities at level i of the first variable, $u + u_{2(j)}$ is the mean of the logarithm of probabilities at level j of the second variable. Here

$$\begin{aligned} u &= \tfrac{1}{4}\sum_i \sum_j \ln \pi_{ij}, \\ u + u_{1(i)} &= \tfrac{1}{2}(\ln \pi_{i1} + \ln \pi_{i2}), \ i = 1, 2, \\ u + u_{2(j)} &= \tfrac{1}{2}(\ln \pi_{1j} + \ln \pi_{2j}), \ j = 1, 2. \end{aligned} \tag{5.2.2}$$

Since $u_{1(i)}$ and $u_{2(j)}$ represent deviations from the grand mean u,

$$\begin{aligned} u_{1(1)} + u_{1(2)} &= 0, \\ u_{2(1)} + u_{2(2)} &= 0. \end{aligned}$$

Similarly, $u_{12(ij)}$ represents deviation from $u + u_{1(i)} + u_{2(j)}$, so that

$$\begin{aligned} u_{12(11)} + u_{12(12)} &= 0, \quad u_{12(21)} + u_{12(22)} = 0, \\ u_{12(11)} + u_{12(21)} &= 0, \quad u_{12(11)} + u_{12(22)} = 0. \end{aligned}$$

The general log-linear model for a $2 \times 2 \times 2$ table can be written as

$$\ln \pi_{ijk} = u + u_{1(i)} + u_{2(j)} + u_{3(k)} + u_{12(ij)} + u_{13(ik)} + u_{23(jk)} + u_{123(ijk)}, \ i, j, k = 1, 2, \tag{5.2.3}$$

where $u_{1(i)}$ means the effect of factor 1 at level i, $u_{12(ij)}$, the interaction between level i of factor 1 and level j of factor 2, $u_{123(ijk)}$, the three-factor interaction among level i of factor 1, level j of factor 2, level k of factor 3, all the effects being expressed in terms of log probabilities. We need

$$\sum_{i=1}^{2} u_{1(i)} = 0, \ \sum_{j=1}^{2} u_{2(j)} = 0, \ \sum_{k=1}^{2} u_{3(k)} = 0,$$
$$\sum_{j(\neq i)=1,2} u_{12(ij)} = 0, \quad \sum_{k(\neq i)=1,2} u_{13(ik)} = 0, \text{ etc..}$$

The parameters involved in $\mathbf{ln}\,\pi = \boldsymbol{\mu} = (\mu_{111}, \mu_{112}, \mu_{121}, \mu_{122}, \mu_{211}, \mu_{212}, \mu_{221}, \mu_{222})'$ where $\mu_{ijk} = \ln \pi_{ijk}$, are therefore u's. Therefore

$$\theta = (u_{1(1)}, u_{2(1)}, u_{3(1)}, u_{12(11)}, u_{13(11)}, u_{23(11)}, u_{123(111)})'.$$

Bishop et al. (1975), (p. 33) have given expressions for μ_{ijk} terms in terms of these parameters in their sign table of u terms of fully saturated mode for three dichotomous variables. Consider therefore $\mathbf{X}$ an 8×7 matrix consisting of 1's and -1's such that each column of $\mathbf{X}$ add to zero, that is, $\mathbf{X}'\mathbf{1} = \mathbf{0}$.

If $\boldsymbol{\mu}$ is as stated above, then the last column of $\mathbf{X}$ is $(1, -1, -1, 1, -1, 1, 1, -1)'$. Under $H_0 : u_{123(111)} = 0$, the last column of $\mathbf{X}$ will be deleted and $\mathbf{X}$ becomes an 8×6 matrix.

In general, a log-linear model can be written as

$$\begin{bmatrix} \ln \pi_1 \\ \ln \pi_2 \\ \cdot \\ \cdot \\ \cdot \\ \ln \pi_T \end{bmatrix} = \begin{bmatrix} \mathbf{x}_1'\theta \\ \mathbf{x}_2'\theta \\ \cdot \\ \cdot \\ \cdot \\ \mathbf{x}_T'\theta \end{bmatrix} + \ln \left\{ \frac{1}{\exp(\mathbf{x}_1'\theta) + \exp(\mathbf{x}_2'\theta) + \ldots + \exp(\mathbf{x}_T'\theta)} \right\} \mathbf{1}_T \quad (5.2.4)$$

where

$$\mathbf{X} = \begin{bmatrix} \mathbf{x}_1' \\ \mathbf{x}_2' \\ \cdot \\ \cdot \\ \mathbf{x}_T' \end{bmatrix}$$

is a known $T \times r$ matrix of $+1$'s and -1's and is of full rank $r(\leq T - 1)$ and $\theta = (\theta_1, \theta_2, \ldots, \theta_r)'$ is a r-vector of unknown parameters. Writing $\ln \pi_k = \mu_k$, $\mu = (\mu_1, \ldots, \mu_T)'$ and

$$\ln \left\{ \frac{1}{\exp(\mathbf{x}_1'\theta) + \ldots + \exp(\mathbf{x}_T'\theta)} \right\} = \ln \left\{ \frac{1}{\mathbf{1}_T' \exp(\mathbf{X}\theta)} \right\} = \tilde{u}(\theta) \quad (5.2.5)$$

the log-linear model (5.2.4) can be written as

$$\mu = \tilde{u}(\theta)\mathbf{1}_T + \mathbf{X}\theta, \quad (5.2.6)$$

where $\tilde{u}$ is the normalizing factor which ensures that $\sum \pi_i(\theta) = 1$. If $r = T - 1$, we get the general or saturated log-linear model. Under multinomial sampling the likelihood equations for π are given by

$$\mathbf{X}'\hat{\pi} = \mathbf{X}'(\mathbf{n}/n) \tag{5.2.7}$$

where $\mathbf{n} = (n_1, \ldots, n_T)'$ is the T-vector of observed frequencies, $\sum_k n_k = n$ and $\hat{\pi} = \hat{\pi}(\hat{\theta})$ is the maximum likelihood estimate (MLE) of π under the model (5.2.4). The MLE $\hat{\pi}$ is easily obtained under such hierarchical models. The method of iterative proportional fitting (IPF) gives the fitted proportions $\pi(\hat{\theta})$ from Eq. (5.2.7) without estimating MLE $\hat{\theta}$ of θ (Bishop et al. 1975, p. 83). The Newton–Raphson method is often used as an alternative since $\hat{\theta}$ is then obtained and the iteration has quadratic convergence unlike the IPF method.

For general survey design, appropriate likelihood equations for π_k's are difficult to obtain. Hence we use a 'pseudo-MLE' $\hat{\pi}$ obtainable from the modified Eq. (5.2.7) by replacing $(\mathbf{n}/n)$ by $\hat{\mathbf{p}} = (\hat{p}_1, \ldots, \hat{p}_T)'$, a design-consistent estimate of π. Under standard regularity conditions the consistency of $\hat{\mathbf{p}}$ follows from the consistency of $\hat{\pi}$.

It may be noted that for any survey design a consistent estimate of π_i is $\hat{N}_i/\hat{N}$ where $\hat{N}_i$ represents a consistent estimate of population cell count N_i and $\hat{N} = \sum_i \hat{N}_i$. We shall denote $(\hat{N}_1, \ldots, \hat{N}_T)'$ as $\hat{\mathbf{N}}$.

5.3 Tests for Goodness of Fit

The general Pearson statistic for testing goodness of fit of the model (5.2.6) with $r < T - 1$ is given by

$$X_P^2 = n \sum_{i=1}^{T} \frac{\left(\hat{p}_i - \pi_i(\hat{\theta})\right)^2}{\pi_i(\hat{\theta})}. \tag{5.3.1}$$

Similarly, the log-likelihood ratio statistic may be written as

$$G^2 = 2n \sum_{i=1}^{T} \hat{p}_i \log\left[\frac{\hat{p}_i}{\pi(\hat{\theta})}\right]. \tag{5.3.2}$$

Rao and Scott (1984) have shown that X_P^2(or G^2) is distributed asymptotically as a weighted sum $\sum_{i=1}^{T-r-1} \delta_i \psi_i$ of $T - r - 1$ independent $\chi^2_{(1)}$ variables $\psi_i (\delta_1 \geq \delta_2 \geq \cdots \geq \delta_{T-r-1} > 0)$. The weights δ_i's are the eigenvalues of a complex design effects matrix given by

$$\mathbf{\Delta} = (\mathbf{C}'\mathbf{D}_\pi^{-1}\mathbf{C})^{-1}(\mathbf{C}'\mathbf{D}_\pi^{-1}\mathbf{V}\mathbf{D}_\pi^{-1}\mathbf{C}) \tag{5.3.3}$$

where $\mathbf{V} = Var(\hat{\pi})$ under the actual sampling design $p(s)$ (or true model) and $\mathbf{C}$ is any $T \times (T - r - 1)$ matrix of full rank such that $\mathbf{C}'\mathbf{X} = \mathbf{0}$ and $\mathbf{C}'\mathbf{1}_T = \mathbf{0}$. (For a proof see Note 5.6.1). In particular, under the standard parametrization of a log-linear model, $\mathbf{C}$ may be chosen as the matrix complement of $\mathbf{X}$ to form a model matrix in the saturated case.

The δ_i's may again be interpreted as generalized design effects (deff's), δ_1 being the largest possible deff taken over all linear combinations of the elements of the vector $\mathbf{C}' \ln \hat{\mathbf{p}}$.

In the case of multinomial sampling, the deff $\delta_i = 1 \; \forall \, i$ and hence the standard X_P^2 or G^2 is distributed asymptotically as $\chi^2_{(T-r-1)}$ variable which follows as a special case.

As in Chap. 4 (vide Examples 4.2.1 (a), (b); 4.2.2) the following results hold for special designs.

(i) All $\delta_i = 1 - \frac{n}{N}$ for *srswor*;
(ii) All $\delta_i \leq 1$ for stratified random sampling with proportional allocation;
(iii) $\delta_i = \delta(> 1) \; \forall \, i$ under Brier's model for two-stage sampling. For a proof see Theorem 5.5.1.

5.3.1 *Other Standard Tests and Their First- and Second-Order Corrections*

First-order Rao-Scott procedure: The first-order Rao-Scott (1979, 1981, 1984) strategy is based on the observations that X_P^2/δ_0, G^2/δ_0 (where δ_0 is the mean of the generalized design effects δ_i) have the same first moment as $\chi^2_{(T-r-1)}$, the asymptotic distribution of Pearson statistic under multinomial sampling. Thus a first-order correction to X_P^2 or G^2 is obtained by referring to

$$X_P^2(\hat{\delta}_0) = \frac{X_P^2}{\hat{\delta}_0} \text{ or } G^2(\hat{\delta}_0) = \frac{G^2}{\hat{\delta}_0} \tag{5.3.4}$$

as $\chi^2_{(T-r-1)}$ random variable, where $\hat{\delta}_0$ is a consistent estimate of δ_0 given by

$$(T - r - 1)\delta_0 = \sum_{i=1}^{T-r-1} \delta_i = \text{tr}(\Delta). \tag{5.3.5}$$

The tests based on $X_P^2(\delta_0)$ and $G^2(\delta_0)$ are asymptotically correct when the individual design effects are nearly equal as has been confirmed by many simulation studies including the one discussed in Rao and Thomas (2003). The second-order correction indicated below is designed for cases when the variation among the design effects is expected to be appreciable.

As indicated before, one of the main advantages of the first-order procedures is that knowledge of the full estimate $\hat{\mathbf{V}}$ of the covariance matrix is not always required. The trace function in Eq. (5.3.5) can often be expressed in closed form as a function of the estimated design effects of the specific cells and marginal probabilities which may be available from published data. In fact, for most of the special cases described below, it is not necessary to have the full matrix $\hat{\mathbf{V}}$, but only variance estimates (or equivalently deffs) for individual cell estimates, that is, the diagonal estimates of the $\hat{\mathbf{V}}$ and for certain marginals.

(a) *Simple goodness of fit*: Here $H_0 : \pi = \pi_0 = (\pi_{10}, \ldots, \pi_{T0})'$ and

$$(T-1)\hat{\delta}_0 = \sum_{i=1}^{T} \frac{\hat{p}_i}{\pi_{i0}}(1-\hat{p}_i)\hat{d}_i \tag{5.3.6}$$

where

$$\hat{d}_i = \frac{n\hat{V}(\hat{p}_i)}{\hat{p}_i(1-\hat{p}_i)}$$

and $\hat{V}ar(\hat{p}_i)$ is the estimated variance of a design-consistent estimator $\hat{p}_i$ of π_i under the true design $p(s)$.

(b) *Test of independence in a $r \times c$ table*: Here

$$X_P^2 = n\sum_{i=1}^{r}\sum_{j=1}^{c} \frac{(\hat{\pi}_{ij} - \hat{\pi}_{i0}\hat{\pi}_{0j})^2}{\hat{\pi}_{i0}\hat{\pi}_{0j}}$$

and the formula (5.3.5) reduces to Rao and Scott (1984), Bedrick (1983), Gross (1984)

$$(r-1)(c-1)\hat{\delta}_0 = \sum_i\sum_j \frac{\hat{\pi}_{ij}}{\hat{\pi}_{i0}\hat{\pi}_{0j}}(1-\hat{\pi}_{ij})\hat{d}_{ij} - \sum_i (1-\hat{\pi}_{i0})\hat{\pi}_{i0}(1) - \sum_j (1-\hat{\pi}_{0j})\hat{\pi}_{0j}(2) \tag{5.3.7}$$

where

$$\hat{d}_{ij} = \frac{n\hat{V}ar(\hat{\pi}_{ij})}{\hat{\pi}_{ij}(1-\hat{\pi}_{ij})}$$

is the estimated design effect of $\hat{\pi}_{ij} = \pi_{ij}(\hat{\theta})$ and $\hat{d}_i(1), \hat{d}_j(2)$ are the estimated deffs of row and column marginals $\hat{\pi}_{i0}$ and $\hat{\pi}_{oj}$, respectively. The corrected statistics $X_P^2(\hat{\delta}_0)$ or $G^2(\hat{\delta}_0)$ is related to χ^2 variable with $(r-1)(c-1)$ d.f. under H_0.

Some alternative corrections to X_P^2 due to Fellegi (1980) are

$$\begin{aligned} X_P^2(\hat{d}_{00}) &= \frac{X_P^2}{\hat{d}_{00}} \\ X_P^2(\hat{d}_m) &= \frac{X_P^2}{\hat{d}_m} \end{aligned} \tag{5.3.8}$$

where $\hat{d}_{00} = \sum_i \sum_j \hat{d}_{ij}/rc$ is the average deff and $\hat{d}_m = \min .(\sum_i \hat{d}_i(1)/r, \sum_j \hat{d}_j(2)/c)$. The corrected statistic $X_P^2(\hat{d}_m)$ is particularly useful, when only marginal deffs are published.

(c) *Test of homogeneity*: Denoting by π_{ij} the population proportion of the jth category in the i th population ($i = 1, \ldots, r$; $j = 1, \ldots, c$), we want to test $H_0(\tilde{\pi}_i = \pi \; \forall \; i)$ where $\tilde{\pi}_i = (\pi_{i1}, \ldots, \pi_{ic})'$ and $\pi = (\pi_1, \ldots, \pi_c)'$

The general Pearson statistic can be written as

$$X_P^2 = \sum_{i=1}^{r} n_i \sum_{j=1}^{c} \frac{(\hat{\pi}_{ij} - \hat{\pi}_j)^2}{\hat{\pi}_j} \tag{5.3.9}$$

with $\hat{\pi}_j = \sum_{i=1}^{r} n_i \hat{\mu}_{ij}/n$, n_i being the sample size for the ith population, $\sum_i n_i = n$. The formula (5.3.5) reduces to Rao and Scott (1981)

$$(r-1)(c-1)\hat{\delta}_0 = \sum_{i=1}^{r} \sum_{j=1}^{c} \frac{\hat{\pi}_{ij}}{\hat{\pi}_j} (1 - \hat{\pi}_{ij}) \left(1 - \frac{n_i}{n}\right) \hat{d}_{ij} \tag{5.3.10}$$

where

$$\hat{d}_{ij} = \frac{n_i \hat{V}ar(\hat{\pi}_{ij})}{\hat{\pi}_{ij}(1 - \hat{\pi}_{ij})}.$$

The corrected statistic $X_P^2(\hat{\delta}_0)$ can therefore be implemented from published tables which report only the estimates π_{ij} and their associated deffs, $\hat{d}_{ij}$. The corrected statistic is again treated as a $\chi^2_{(r-1)(c-1)}$ random variable.

Rao and Scott (1984) derived similar results for three-way tables.

Second-order Rao-Scott procedures: When an estimate $\hat{\mathbf{V}}$ of the full covariance matrix $\mathbf{V}$ is available, a more accurate correction to X_P^2 or G^2 can be obtained by ensuring that the first two moments of the weighted sum $\sum_{i=1}^{T-r-1} \delta_i \psi_i$ are the same as those of $\chi^2_{(T-r-1)}$. Unlike the first-order correction, the second-order correction which is based on Satterthwate (1946) method takes into account the variations among the generalized design effects δ_i. For goodness-of-fit test, it is implemented by referring

$$X_{Pcc}^2 = X_P^2(\hat{\delta}_0, \hat{a}^2) = \frac{X_P^2(\hat{\delta}_0)}{1 + \hat{a}^2}$$

or

$$G_{cc}^2 = G^2(\hat{\delta}_0, \hat{a}^2) = \frac{G^2(\hat{\delta}_0)}{1 + \hat{a}^2} \tag{5.3.11}$$

to the upper 100 (α)% point of a $\chi^2_{(\nu)}$ random variable where $\nu = (T - r - 1)/(1 + \hat{a}^2)$ and $\hat{a}$ is a measure of the variation among design effects, given by

$$\hat{a}^2 = \sum_{i=1}^{T-r-1} (\hat{\delta}_i - \hat{\delta}_0)^2/\{(T-r-1)\hat{\delta}^2\}. \quad (5.3.12)$$

The simulation results of Thomas and Rao (1987) for simple goodness of fit confirm that the second-order corrected tests are more effective for controlling Type I error then the first-order corrected tests when the variation among design effects is appreciable.

Similar corrections can be applied to tests for independence and tests for homogeneity.

F-based corrections to Rao-Scott procedures For a simple goodness-of-fit test Thomas and Rao (1987) gave a heuristic large-sample argument to justify referring F-based version of the first- and second-order Rao-Scott tests to F distribution. Extension of their argument for log-linear model gives the following results.

For goodness-of-fit test an F-corrected $X_P^2(\hat{\delta}_0)$ is

$$F(X_P^2(\hat{\delta}_0)) = \frac{X_P^2(\hat{\delta}_0)}{T-r-1} \quad (5.3.13)$$

which follows an F distribution with $(T-r-1)$ and $(T-r-1)\nu_e$ degrees of freedom under H_0 where ν_e is the d.f. for estimating $\mathbf{V}$. For test of independence

$$F(X_P^2(\hat{\delta}_0)) = \frac{X_P^2(\hat{\delta}_0))}{(r-1)(c-1)} \quad (5.3.14)$$

which follows a F distribution with $(r-1)(c-1)$ and $\nu_e(r-1)(c-1)$ d.f. The statistic $F(G^2(\hat{\delta}_0))$ is defined similarly.

Similar corrections can be made to $X_{P(cc)}^2$ to obtain the second-order F-corrected estimator $F(X_{P(cc)}^2)$.

Wald Statistic

Whenever an estimate of the full covariance matrix of $\hat{\pi}$ under the true design $p(s)$ is available, a Wald test of goodness of fit of the log-linear model can also be constructed. Consider the saturated model corresponding to (5.2.6), written in the form

$$\ln(\pi) = \tilde{u}(\theta)\mathbf{1} + \mathbf{X}\theta + \mathbf{C}\theta_C \quad (5.3.15)$$

where θ_C is a $(T-r-1) \times 1$ vector of parameters. The test of goodness of fit of model (5.2.6) is thus equivalent to

$$H_0^{(1)} : \theta_C = \mathbf{0}.$$

Again, since $\mathbf{C}$ satisfies $\mathbf{C}'\mathbf{1} = \mathbf{0}$ and $\mathbf{C}'\mathbf{X} = \mathbf{0}$, $H_0^{(1)}$ is equivalent to

$$H_0^{(2)} : \mathbf{C}'\ln(\pi) = \phi = \mathbf{0}$$

where ϕ is a $(T-r-1)\times 1$ vector. A Wald statistic for testing H_{02} is then obtained by referring

$$\begin{aligned} X^2_{WI} &= \hat{\phi}'(\hat{\mathbf{D}}(\hat{\phi}))^{-1}\hat{\phi} \\ &= (\mathbf{C}'\ln\hat{\pi})'[\mathbf{C}'\mathbf{D}_{\hat{\pi}}^{-1}\hat{\mathbf{V}}\mathbf{D}_{\hat{\pi}}^{-1}\mathbf{C}]^{-1}(\mathbf{C}'\ln\hat{\pi}) \end{aligned} \tag{5.3.16}$$

as $\chi^2_{(T-r-1)}$ (see Theorem 5.6.2), where $\hat{\phi}$, the estimate of ϕ under the saturated model is given by $\mathbf{C}'\ln(\hat{\pi})$ and $\hat{\mathbf{D}}(\hat{\phi})$ is the estimated covariance matrix of $\hat{\phi}$. Here, $\mathbf{D}_{\hat{\pi}}$ is the diagonal matrix having elements $\hat{\pi}_i$, $(i=1,\ldots,T)$ on the diagonal. Clearly, the Wald statistics (5.3.16) requires that all estimated cell probabilities $\hat{\pi}_i$ must be nonzero. The Wald statistic X^2_W is invariant to the choice of $\mathbf{C}$.

The test may lead to inflated type I error rates in finite samples, as the d.f. ν_e for estimating $\mathbf{V}$ decreases and the number of cells increases, as shown by Thomas and Rao (1987) on the basis of a goodness-of-fit Monte Carlo study. The statistic is not defined if $\nu_e < T-r-1$.

If the d.f. for estimating $\mathbf{V}$ are not large enough relative to $T-r-1$, an improved test statistic can be obtained by treating

$$F_W = \frac{(\nu_e - T + r + 2)}{\nu_e(T-r-1)} X^2_W \tag{5.3.17}$$

as an F random variable with $T-r-1$ and $\nu_e - T + r + 2$ d.f. The Monte Carlo study of Thomas and Rao (1987) has shown that F_W gives a better control of Type I error rate than X^2_W, although it also tends to perform poorly as ν_e approaches $T-r-1$.

5.3.2 *Fay's Jackknifed Tests*

Fay (1979, 1984, 1985) proposed and developed jackknifed versions of X^2_P and G^2, which we shall denote as X_{PJ} and G_J, respectively. We have seen that consistent estimators of T finite population proportions in a general cross-tabulation can be obtained as $(\hat{N})^{-1}\hat{\mathbf{N}}$ where $\hat{N}=\sum_i \hat{N}_i$, $\hat{\mathbf{N}}=(\hat{N}_1,\ldots,\hat{N}_T)'$ and N_i is the ith cell population count $(i=1,\ldots,T)$. Fay (1985) considered the class of replication methods based on (pseudo-) replicates $\hat{\mathbf{N}}+\mathbf{W}^{(i,j)}$, $i=1,\ldots,T$; $j=1,\ldots,J_i$, typically based on the same data as $\hat{\mathbf{N}}$. The asymptotic theory for the jackknifed tests requires that

$$\sum_j \mathbf{W}^{(i,j)} = \mathbf{0} \tag{5.3.18}$$

for each i. An estimate, $\text{cov}^*(\hat{\mathbf{N}})$, of the covariance of $\hat{\mathbf{N}}$ should be given by

$$\text{cov}^*(\hat{\mathbf{N}}) = \sum_i b_i \sum_j \mathbf{W}^{(i,j)} \otimes \mathbf{W}^{(i,j)}, \tag{5.3.19}$$

where $\mathbf{W}^{(i,j)} \otimes \mathbf{W}^{(i,j)}$ represents the outer product of $\mathbf{W}^{(i,j)}$ with itself (the standard cross-product matrix) and the b_i are a fixed set of constants appropriate for the problem.

A number of replication schemes can be represented in this way. In case of stratified cluster sampling suppose the population is divided into $I = T$ strata and n_i samples are selected independently from each stratum. In this setup $\hat{\mathbf{N}}$ may be represented as

$$\hat{\mathbf{N}} = \sum_i \sum_j \mathbf{Z}^{(i,j)}, \tag{5.3.20}$$

where the $\mathbf{Z}^{(i,j)}$, for fixed i, are the n_i iid random variables within stratum i. (These variables are not, however, assumed to be identically distributed across strata.) For each stratum i,

$$\hat{\mathbf{N}} + \mathbf{W}^{(i,j)} = \hat{\mathbf{N}} + \left(\left(\sum_{j'} \mathbf{Z}^{(i,j')} \right) - n_i \mathbf{Z}^{(i,j)} \right) / (n_i - 1), \tag{5.3.21}$$

has the same expected value as $\hat{\mathbf{N}}$ and defines $\mathbf{W}^{(i,j)}$, satisfying (5.3.17). The corresponding choice for b_i is $(n_i - 1)/n_i$.

For calculating the jackknifed values of the test statistics we have to refit the given log-linear model to the replicates, $\hat{\mathbf{N}} + \mathbf{W}^{(i,j)}$ and recompute the test statistics, $X_P^2(\hat{\mathbf{N}} + \mathbf{W}^{(i,j)})$ or $G^2(\hat{\mathbf{N}} + \mathbf{W}^{(i,j)})$, for these new tables. Using the b_i introduced in (5.3.19), the jackknifed test statistic X_{PJ} is defined by

$$X_{PJ} = \frac{[(X_P^2(\hat{\mathbf{N}}))^{1/2} - (K^+)^{1/2}]}{\{V/(8X_P^2(\hat{\mathbf{N}}))\}^{1/2}}, \tag{5.3.22}$$

where

$$P_{ij} = X_P^2(\hat{\mathbf{N}} + \mathbf{W}^{(i,j)}) - X_P^2(\hat{\mathbf{N}}), \tag{5.3.23}$$

$$K = \sum_i b_i \sum_j P_{ij}, \tag{5.3.24}$$

$$V = \sum_i b_i \sum_j P_{ij}^2, \tag{5.3.25}$$

and K^+ takes the value K for positive K and zero otherwise. The jackknifed version of G^2, denoted as G_J, is obtained by replacing X_P^2 by G^2 in Eqs. (5.3.22)–(5.3.25).

It has been shown by Fay (1985) that both X_{PJ} and G_J are asymptotically distributed as a function of weighted sums of $T - r - 1$ independent $\chi^2_{(1)}$ variables, the weights being functions of the eigenvalues δ_i's of the design matrix $\mathbf{\Delta}$ in (5.3.3). When the δ_i's are all equal, the asymptotic distribution of both X_{PJ} and G_J is

$$2^{1/2}[(\chi^2_{(T-r-1)})^{1/2} - (T - r - 1)^{1/2}]. \tag{5.3.26}$$

Numerical investigations have shown that (5.3.26) is a good approximation to the jackknifed statistics even when δ_i's are not all equal. In practice, therefore, the test procedure consists in rejecting the null hypothesis when X_{PJ} (or G_J) exceeds the upper $100(\alpha)\%$ critical point obtained from (5.3.26).

5.4 Asymptotic Covariance Matrix of the Pseudo-MLE $\hat{\pi}$

To derive the asymptotic covariance matrix of $\hat{\pi}$ we need some of the results of Birch (1964) stated in the Appendix for multinomial sampling.

Lemma 5.4.1 *Under the regularity conditions of Sect. A.4.2,*

$$\begin{aligned}\hat{\theta} - \theta &\sim (\mathbf{X}'\mathbf{P}\mathbf{X})^{-1}\mathbf{X}'(\hat{\mathbf{p}} - \pi)\\ \hat{\pi} - \pi &\sim \mathbf{P}\mathbf{X}(\hat{\theta} - \theta)\end{aligned} \tag{5.4.1}$$

where $\mathbf{P} = \mathbf{D}_\pi - \pi\pi'$. *Note that* $\hat{\pi}$ *is the pseudo-MLE of* π *as explained above, while* $\hat{\mathbf{p}}$ *is a consistent estimate of* π *under the sampling design* $p(s)$. *(Here, asymptotic covariance matrix of* $\hat{\mathbf{p}}$ *is* $\mathbf{V}/n$.*)*

Proof From Eqs. (A.4.10) and (A.4.13),

$$\hat{\theta} - \theta \sim (\mathbf{A}'\mathbf{A})^{-1}\mathbf{A}'\mathbf{D}_\pi^{-1}(\hat{\mathbf{p}} - \pi) \tag{5.4.2}$$

and

$$\hat{\pi} - \pi \sim \mathbf{D}_\pi^{1/2}\mathbf{A}(\hat{\theta} - \theta) \tag{5.4.3}$$

where $\mathbf{A}$ is the $T \times r$ matrix whose (i, j)th element is $\pi_i^{-1/2}(\partial\pi_i/\partial\theta_j)$. Under the log-linear model (5.2.4), $\mathbf{A} = \mathbf{D}_\pi^{-1/2}\mathbf{P}\mathbf{X}$. Also,

$$\begin{aligned}\mathbf{A}'\mathbf{A} &= \mathbf{X}'\mathbf{P}'\mathbf{D}_\pi^{-1}\mathbf{P}\mathbf{X}\\ &\quad\mathbf{X}'(\mathbf{I} - \pi\mathbf{1}')\mathbf{P}\mathbf{X} = \mathbf{X}'\mathbf{P}\mathbf{X},\end{aligned}$$

because $\mathbf{1}'\mathbf{P} = \mathbf{0}$. Similarly,

$$\begin{aligned}\mathbf{A}'\mathbf{D}_\pi^{-1}(\hat{\mathbf{p}} - \pi) &= \mathbf{X}'(\mathbf{I} - \pi\mathbf{1}')(\hat{\mathbf{p}} - \pi) \\ &= \mathbf{X}'(\hat{\mathbf{p}} - \pi),\end{aligned}$$

because $\mathbf{1}'(\hat{\mathbf{p}} - \pi) = 0$.

Hence the proof. □

Since by assumption, the asymptotic covariance matrix of $\hat{\mathbf{p}}$ is $\mathbf{V}/n$, we have from (5.4.1), the asymptotic covariance matrix of $\hat{\theta}$ as

$$\mathbf{D}(\hat{\theta}) = n^{-1}(\mathbf{X}'\mathbf{P}\mathbf{X})^{-1}(\mathbf{X}'\mathbf{V}\mathbf{X})(\mathbf{X}'\mathbf{P}\mathbf{X})^{-1}. \tag{5.4.4}$$

Hence the asymptotic covariance matrix of $\hat{\pi}$ is

$$\mathbf{D}(\hat{\pi}) = \mathbf{P}\mathbf{X}\mathbf{D}(\hat{\theta})\mathbf{X}'\mathbf{P}. \tag{5.4.5}$$

In the case of multinomial sampling, we have $\mathbf{V} = \mathbf{P}$ and (5.4.4) reduces to the well-known result $\mathbf{D}(\hat{\theta}) = (\mathbf{X}'\mathbf{P}\mathbf{X})^{-1}/n$.

The asymptotic covariance matrix of the residual $\hat{\mathbf{p}} - \hat{\pi}$ is obtained by noting that

$$\begin{aligned}\hat{\mathbf{p}} - \hat{\pi} &\approx [\mathbf{I} - \mathbf{P}\mathbf{X}(\mathbf{X}'\mathbf{P}\mathbf{X})^{-1}\mathbf{X}'\mathbf{P}'\mathbf{D}_\pi^{-1}](\hat{\mathbf{p}} - \pi) \\ &[\mathbf{I} - \mathbf{P}\mathbf{X}(\mathbf{X}'\mathbf{P}\mathbf{X})^{-1}\mathbf{X}'](\hat{\mathbf{p}} - \pi)]\end{aligned} \tag{5.4.6}$$

since $\mathbf{X}'\mathbf{P}'\mathbf{D}_\pi^{-1}(\hat{\mathbf{p}} - \pi) = \mathbf{X}'(\hat{\mathbf{p}} - \pi)$ so that

$$\mathbf{D}(\hat{\mathbf{p}} - \hat{\pi}) = n^{-1}[\mathbf{I} - \mathbf{P}\mathbf{X}(\mathbf{X}'\mathbf{P}\mathbf{X})^{-1}\mathbf{X}']\mathbf{V}[\mathbf{I} - \mathbf{X}(\mathbf{X}'\mathbf{P}\mathbf{X})^{-1}\mathbf{X}'\mathbf{P}]. \tag{5.4.7}$$

If $\mathbf{V} = \mathbf{P}$, (5.4.7) reduces to

$$n^{-1}[\mathbf{P} - \mathbf{P}\mathbf{X}(\mathbf{X}'\mathbf{P}\mathbf{X})^{-1}\mathbf{X}'\mathbf{P}].$$

5.4.1 Residual Analysis

If a test rejects the hypothesis of goodness of fit of a log-linear model of the form (5.2.6), residual analysis provides an understanding of the nature of deviations from the hypothesis. The standardized residual is

$$r_i = \frac{\hat{p}_i - \pi_i(\hat{\theta})}{s.e.\{\hat{p}_i - \pi(\hat{\theta})\}} \tag{5.4.8}$$

to detect deviations from the hypothesis, using the fact that r_i's are approximately $N(0, 1)$ variables under the hypothesis.

In the case of test of homogeneity of proportions across r regions, the standardized residuals $r_{ij}(i = 1, \ldots, r;\ j = 1, \ldots, c)$ may be expressed in terms of the estimated population proportions $\hat{\pi}_{ij}$ and the associated deffs. $\hat{d}_{ij}$ is as follows:

$$r_{ij} = \frac{\hat{p}_{ij} - \hat{\pi}_j}{\{\hat{V}ar(\hat{p}_{ij} - \hat{\pi}_j)\}^{1/2}} \tag{5.4.9}$$

where

$$\hat{V}ar(\hat{p}_{ij} - \hat{\pi}_j) = n^{-2}\hat{\pi}_j(1 - \hat{\pi}_j)\{\frac{n(n - n_i)}{n_i}\hat{\pi}_{ij} + \sum_{i=1}^{r} n_i\hat{d}_{ij}\},$$

$\hat{d}_{ij}$ being given in (5.3.10).

5.5 Brier's Model

We now consider a case where δ_i's, the eigenvalues of Δ in Eq. (5.3.3) are either zeros or a constant.

Consider Brier's model in Sect. 4.2.5. Suppose we are interested in testing the hypothesis

$$H_0 : \pi = \mathbf{f}(\theta) \ \ (\theta \in \Theta \subseteq \mathcal{R}^r) \tag{5.5.1}$$

versus the alternative $H_A : \pi \in \xi_T(\xi_T$ defined in Eq. (A.4.1)). Under both H_0 and H_A, $\mathbf{Y}^{(i)}(i = 1, \ldots, N)$ are independently distributed as $DM_T(n, \pi, k)$. We assume that the relation $\pi = \mathbf{f}(\theta)$ satisfies the regularity conditions in Sect. A.4.3 given by Birch (1964). These conditions are not very restrictive, and in particular, all hierarchical log-linear models considered by Bishop et al. (1975) satisfy these conditions.

Let $\mathbf{U} = \sum_{i=1}^{N} \mathbf{Y}_i$ be the vector of cell counts. Two statistics that are frequently used are

$$X_P^2 = \sum_{j=1}^{T} \frac{(U_j - M\hat{\pi}_j(\hat{\theta}))^2}{M\hat{\pi}_j(\hat{\theta})} \tag{5.5.2}$$

and

$$G^2 = 2\sum_{j=1}^{T} U_j \log\left(\frac{U_j}{M\hat{\pi}_j}\right) \tag{5.5.3}$$

where $M = nN$ and $\hat{\pi}_j = y_j/n$ is the MLE of π_j assuming that $\mathbf{U}$ is multinomially distributed. It has been noted in Sect. A.4.1 that under H_0, both X_P^2 and G^2 have distributions that are asymptotically $\chi^2_{(T-r-1)}$ if $\mathbf{U}$ has multinomial distribution. We now consider the null distribution of X_P^2 and G^2 when the sample is a cluster sample and therefore, $\mathbf{y}$ follows $DM_T(n, \pi, k)$ distribution.

Theorem 5.5.1 *Let* $\mathbf{Y}^{(1)}, \ldots, \mathbf{Y}^{(N)}$ *be independently and identically distributed as* $DM_T(n, \pi, k)$. *Under* $H_0 : \pi = \mathbf{f}(\theta)$, *if the regularity conditions of Sect. 3.4.2 are satisfied, then* X_P^2 *and* G^2 *are distributed asymptotically as* $C\chi^2_{(T-r-1)}$ *as* $N \to \infty$, *where* X_P^2 *and* G^2 *are defined in (5.5.2) and (5.5.3), respectively, and* $C = \frac{n+k}{1+k}$.

Proof Let $\hat{\mathbf{p}} = \mathbf{U}/M$ be the vector of observed cell proportions and let π_0 and θ_0 be the true values of π and θ, respectively. Applying the central limit theorem to $(\mathbf{Y}^{(1)}, \ldots, \mathbf{Y}^{(N)})'$, we have

$$\sqrt{N}(\hat{\mathbf{p}} - \pi_0) \to^L N(\mathbf{0}, \mathbf{V})$$

where $\mathbf{V} = (\frac{C}{n})(\mathbf{D}_{\pi_0} - \pi_0\pi_0')$. Note that this implies

$$\hat{\mathbf{p}} - \pi_0 = 0_p(N^{-1/2}). \tag{5.5.4}$$

Define

$$\mathbf{A} = ((a_{ij})) = \{\pi_i(\theta)_0\}^{-1/2}\left[\frac{\partial \imath_i(\theta)}{\partial \theta_j}\right]_{\theta=\theta_0}.$$

From Birch's result (A.4.11)

$$\hat{\theta} = \theta_0 + (\mathbf{A}'\mathbf{A})^{-1}\mathbf{A}'\mathbf{D}_{\pi_0}^{-1/2}(\hat{\mathbf{p}} - \pi_0) + o_p(N^{-1/2}) \tag{5.5.5}$$

where $\hat{\theta}$ is the MLE of θ assuming $\mathbf{U}$ is multinomially distributed. This result still holds in our case since the result is based upon the regularity conditions of $\mathbf{f}(\theta)$ and (5.5.4). Using (5.5.4) we have

$$\hat{\pi} = \pi_0 + \mathbf{D}_{\pi_0}^{-1/2}\mathbf{A}(\mathbf{A}'\mathbf{A})^{-1}\mathbf{A}'\mathbf{D}_{\pi_0}^{-1/2}(\hat{\mathbf{p}} - \pi_0) + o_p(N^{-1/2}) \tag{5.5.6}$$

by the delta method. It follows therefore that

$$\sqrt{N}\{(\hat{\mathbf{p}}, \hat{\pi}) - (\pi_0, \pi_0)\} \to^L N(\mathbf{0}, \Gamma) \tag{5.5.7}$$

where

$$\Gamma = (\mathbf{I}, \mathbf{L})'\mathbf{V}(\mathbf{I}, \mathbf{L}) \text{ and } \mathbf{L} = \mathbf{D}_{\pi_0}^{-1}\mathbf{A}(\mathbf{A}'\mathbf{A})^{-1}\mathbf{A}'\mathbf{D}_{\pi_0}^{1/2}.$$

It follows from Theorem A.5.3 that the asymptotic distribution of X_P^2 is $\sum_i \delta_i \psi_i$, where ψ_i's are independent $\chi^2_{(1)}$ variables and δ_i's are the eigenvalues of $C\{\mathbf{I} - \sqrt{\pi}\sqrt{\pi'} - \mathbf{A}(\mathbf{A}'\mathbf{A})^{-1}\mathbf{A}'\}$, where $\sqrt{\pi} = (\sqrt{\pi_1}, \ldots, \sqrt{\pi_T})'$. It can be verified that $T - r - 1$ of the δ_i's are C and the remaining δ_i's are zero. This completes the proof for X_P^2. This together with Theorem A.5.1 establishes the result for G^2. □

5.6 Nested Models

We denote the model (5.2.6) as M_1. Let $\mathbf{X} = (\mathbf{X}_1(T \times s), \mathbf{X}_2(T \times u))$ and $\theta' = (\theta_1'(1 \times s), \theta_2'(1 \times u))$ where $s + u = r$. As before we need $\mathbf{X}_1'\mathbf{1}_T = \mathbf{0}, \mathbf{X}_2'\mathbf{1}_T = \mathbf{0}$. Given the unconstrained model M_1, we are interested in testing the null hypothesis $H_{2|1} : \theta_2 = \mathbf{0}$ so that under $H_{2|1}$ we get the reduced model M_2 as

$$\mu = \tilde{u} \begin{bmatrix} \theta_1 \\ \mathbf{0} \end{bmatrix} \mathbf{1}_T + \mathbf{X}_1\theta_1. \tag{5.6.1}$$

Clearly, M_2 is nested within M_1 (vide Sect. A.6).

Example 5.6.1 As in (5.2.1), consider the saturated log-linear model applicable for two-way cross-classified data according to two categorical variables A and B having r and c categories, respectively. Let π_{ij} be the probability that a unit belongs to the category i of A and category j of B. We want to test the hypothesis of independence

$$H_0 : \pi_{ij} = \pi_{i0}\pi_{0j} \ \forall \ i, j$$

where $\pi_{i0} = \sum_j \pi_{ij}, \pi_{0j} = \sum_i \pi_{ij}$.

The model is

$$\ln \pi_{ij} = \mu_{ij} = u + u_{1(i)} + u_{2(j)} + u_{12(ij)}$$

where the parameters $u_{1(i)}$ and $u_{2(j)}$ are constrained by $\sum_i u_{1(i)} = 0$, $\sum_j u_{2(j)} = 0, \sum_i u_{12(ij)} = 0 \ \forall \ j$ and $\sum_j u_{12(ij)} = 0 \ \forall \ i$ and u is a normalizing factor to ensure that $\sum_i \sum_j \pi_{ij} = 1$. In matrix notation this may be written as

$$\boldsymbol{\mu} = u\mathbf{1} + \mathbf{X}_1\theta_1 + \mathbf{X}_2\theta_2$$

with $\mathbf{X}_2'\mathbf{X}_1 = \mathbf{0}, \mathbf{X}_1'\mathbf{1} = \mathbf{0}$ and $\mathbf{X}_2'\mathbf{1} = \mathbf{0}$ where $\boldsymbol{\mu} = (\mu_{11}, \ldots, \mu_{1r}, \ldots, \mu_{r1}, \ldots, \mu_{rc})', \theta_1$ is the $(r + c - 2)$ vector of parameters $u_{1(1)}, \ldots, u_{1(c-1)}, u_{2(1)}, \ldots, u_{2(c-1)}$ with associated model matrix $\mathbf{X}_1$ consisting of $+1$'s,0's, and -1's; θ_2 is the $(r-1)(c-1)$ vector of parameters $u_{12(11)}, \ldots, u_{12(1,c-1)}, \ldots, u_{12(r-1,1)}, \ldots, u_{12((r-1),(c-1))}$ with associated model matrix $\mathbf{X}_2$. The hypothesis of independence may be expressed as $H_0 : \theta_2 = \mathbf{0}$. □

Let $\hat{\hat{\theta}}_1$ and $\hat{\hat{\pi}} = \pi(\hat{\hat{\theta}}_1)$ denote the 'pseudo-MLE' of θ_1 and π, respectively, under M_2 as obtained from the likelihood equations

$$\mathbf{X}_1'\pi(\hat{\hat{\theta}}_1) = \mathbf{X}_1'\hat{\mathbf{p}} \tag{5.6.2}$$

where $\hat{\mathbf{p}}$ is a design-consistent estimator of π. The consistency of $\hat{\mathbf{p}}$ ensures the consistency of $\hat{\hat{\pi}}$ under M_2.

5.6.1 Pearsonian Chi-Square and the Likelihood Ratio Statistic

The Pearson chi-square statistic for testing $H_{2|1} : \theta_2 = \mathbf{0}$ given the model M_1, i.e., for testing $H_{2|1} : \theta_2 = \mathbf{0}$ under the nested model M_2 is given by

$$\begin{aligned} X_P^2(2|1) &= n\sum_{i=1}^{T} \tfrac{(\hat{\pi}_i - \hat{\hat{\pi}}_i)^2}{\hat{\hat{\pi}}_i} \\ &= n(\hat{\pi} - \hat{\hat{\pi}})'\mathbf{D}_{\hat{\hat{\pi}}}^{-1}(\hat{\pi} - \hat{\hat{\pi}}). \end{aligned} \tag{5.6.3}$$

where we denote the estimate of π under the models M_1 as $\hat{\pi} = \pi(\hat{\theta})$.

Again, writing the likelihood ratio statistic under model M_i as G_i^2, the likelihood ratio statistic for testing $H_{2|1}$ is

$$\begin{aligned} G^2(2|1) &= G_2^2 - G_1^2 \\ &= 2n\sum \hat{\pi}_i \ln(\tfrac{\hat{\pi}_i}{\hat{\hat{\pi}}_i}) - 2n\sum \hat{p}_i \ln(\tfrac{\hat{p}_i}{\hat{\pi}_i}). \end{aligned} \tag{5.6.4}$$

The likelihood ratio statistic $G^2(2|1)$ is usually preferred to $X_P^2(2|1)$ due to its additive properties:

$$G^2(2|1) + G^2(1) = G^2(2)$$

where

$$\begin{aligned} G^2(1) &= 2n\sum_i \hat{p}_i \ln[\tfrac{\hat{p}_i}{\pi_i(\hat{\theta})}] \\ G^2(2) &= 2n\sum_i \hat{p}_i \ln[\tfrac{\hat{p}_i}{\pi_i(\hat{\hat{\theta}})}], \end{aligned} \tag{5.6.5}$$

since $\hat{\mathbf{p}} \approx \hat{\pi}$. However, as in the case of multinomial sampling, when the null hypothesis $H_{2|1}$ is true, both the statistic $X_P^2(2|1)$ and $G^2(2|1)$ are approximately equivalent (vide Theorem A.5.1).

For multinomial sampling, $X_P^2(2|1)$ or $G^2(2|1)$ follow approximately $\chi^2_{(u)}$ (vide Theorem A.6.1). The same asymptotic null distribution holds under a product-multinomial sampling scheme which arises with stratified simple random sampling when the strata correspond to levels of one dimension of the contingency table and the samples are drawn independently from different strata covering all the categories of the other variable. This result, however, does not hold with more complex survey designs involving clustering or stratification based on variables different from those corresponding to the contingency tables.

We shall now derive the asymptotic null distribution of $X_P^2(2|1)$ or $G^2(2|1)$ for any survey design $p(s)$.

Theorem 5.6.1 (Rao and Scott 1984) *Let $\hat{\theta}' = (\hat{\theta}_1', \hat{\theta}_2')$. Under $H_{2|1}$, asymptotic distribution of Pearson chi-square $X_P^2(2|1)$ in (5.6.2) is given by*

$$X_P^2(2|1) \sim n\hat{\theta}_2'(\tilde{X}_2'\mathbf{P}\tilde{\mathbf{X}}_2)\hat{\hat{\theta}}_2 \tag{5.6.6}$$

where

$$\tilde{\mathbf{X}}_2 = (\mathbf{I} - \mathbf{X}_1(\mathbf{X}_1'\mathbf{P}\mathbf{X}_1)^{-1}\mathbf{X}_1'\mathbf{P})\mathbf{X}_2. \tag{5.6.7}$$

Moreover, under $H_{2|1}$,

$$X_P^2(2|1) = \sum_{i=1}^{u} \delta(2|1)_i \psi_i \tag{5.6.8}$$

where ψ_i's are independent $\chi^2_{(1)}$ variables and $\delta(2|1)_i$'s (all >0) are the eigenvalues of the matrix $\Delta(2|1) = (\tilde{\mathbf{X}}_2\mathbf{P}\tilde{\mathbf{X}}_2)^{-1}(\tilde{\mathbf{X}}_2\mathbf{V}\tilde{\mathbf{X}}_2)$.

Proof We have from (5.4.1), $\hat{\pi} - \pi \sim \mathbf{P}\mathbf{X}(\hat{\theta} - \theta)$. Analogously,

$$\hat{\hat{\pi}} - \pi \sim \mathbf{P}\mathbf{X}_1(\hat{\hat{\theta}}_1 - \theta_1). \tag{5.6.9}$$

Hence,

$$\begin{aligned}
\hat{\pi} - \hat{\hat{\pi}} &= (\hat{\pi} - \pi) - (\hat{\hat{\pi}} - \pi) \\
&\sim \mathbf{P}\mathbf{X}(\hat{\theta} - \theta) - \mathbf{P}\mathbf{X}_1(\hat{\hat{\theta}}_1 - \theta_1) \\
&= \mathbf{P}[\mathbf{X}_1\hat{\theta}_1 + \mathbf{X}_2\hat{\theta}_2 - \mathbf{X}_1\theta_1 - \mathbf{X}_2\theta_2 - \mathbf{X}_1\hat{\hat{\theta}}_1 + \mathbf{X}_1\theta_1] \\
&= \mathbf{P}[\mathbf{X}_1(\hat{\theta}_1 - \theta_1) + \mathbf{X}_2\hat{\theta}_2 - \mathbf{X}_1(\hat{\hat{\theta}}_1 - \theta_1)],
\end{aligned} \tag{5.6.10}$$

since under $H_{2|1}, \theta_2 = \mathbf{0}$. Also, from (5.4.1), $\hat{\theta} - \theta \sim (\mathbf{X}'\mathbf{P}\mathbf{X})^{-1}\mathbf{X}'(\hat{\mathbf{p}} - \pi)$. Analogously,

$$\hat{\hat{\theta}}_1 - \theta_1 \sim (\mathbf{X}_1'\mathbf{P}\mathbf{X}_1)^{-1}\mathbf{X}_1'(\hat{\mathbf{p}} - \pi). \tag{5.6.11}$$

Now, $\mathbf{X}'\mathbf{P}\mathbf{X}$ can be expressed as the partitioned matrix

$$\mathbf{X}'\mathbf{P}\mathbf{X} = \begin{bmatrix} \mathbf{X}_1'\mathbf{P}\mathbf{X}_1 & \mathbf{X}_1'\mathbf{P}\mathbf{X}_2 \\ \mathbf{X}_2'\mathbf{P}\mathbf{X}_1 & \mathbf{X}_2'\mathbf{P}\mathbf{X}_2 \end{bmatrix}.$$

Using the standard formula for the inverse of a partitioned matrix (see, e.g., Mukhopadhyay 2008, p. 464) and thereby calculating $\hat{\hat{\theta}}_1 - \theta_1$, it can be shown from (5.4.1) and (5.6.11) that

$$\hat{\hat{\theta}}_1 - \theta_1 \sim (\hat{\theta}_1 - \theta_1) + (\mathbf{X}_1'\mathbf{P}\mathbf{X}_1)^{-1}(\mathbf{X}_1'\mathbf{P}\mathbf{X}_2)\hat{\theta}_2. \tag{5.6.12}$$

Hence, from (5.6.9),

$$\begin{aligned} \hat{\pi} - \hat{\hat{\pi}} &\sim \mathbf{P}[\mathbf{X}_1(\hat{\theta}_1 - \theta_1) + \mathbf{X}_2\hat{\theta}_2 - \mathbf{X}_1(\hat{\theta}_1 - \theta_1) - \mathbf{X}_1'(\mathbf{X}_1\mathbf{P}\mathbf{X}_1)^{-1}(\mathbf{X}_1'\mathbf{P}\mathbf{X}_2)\hat{\theta}_2] \\ &= \mathbf{P}[\mathbf{X}_2 - \mathbf{X}_1(\mathbf{X}_1'\mathbf{P}\mathbf{X}_1)^{-1}(\mathbf{X}_1'\mathbf{P}\mathbf{X}_2)]\hat{\theta}_2 \\ &= \mathbf{P}\tilde{\mathbf{X}}_2\hat{\theta}_2. \end{aligned} \tag{5.6.13}$$

Therefore,

$$X_P^2(2|1) = n(\hat{\pi} - \hat{\hat{\pi}})'\mathbf{D}_{\hat{\pi}}^{-1}(\hat{\pi} - \hat{\hat{\pi}}) \sim n\hat{\theta}_2'(\tilde{\mathbf{X}}_2'\mathbf{P}\tilde{\mathbf{X}}_2)\hat{\theta}_2, \tag{5.6.14}$$

since $\mathbf{P}\mathbf{D}_{\pi}^{-1}\mathbf{P} = \mathbf{P}$ and $\mathbf{P}\mathbf{D}_{\hat{\pi}}^{-1}\mathbf{P} \approx \mathbf{P}\mathbf{D}_{\pi}^{-1}\mathbf{P}$.

Also, it follows from (5.4.4) and the formula for the inverse of a partitioned matrix that the covariance matrix of $\hat{\theta}_2$ is

$$\mathbf{D}(\hat{\theta}_2) = n^{-1}(\tilde{\mathbf{X}}_2'\mathbf{P}\tilde{\mathbf{X}}_2)^{-1}(\tilde{\mathbf{X}}_2'\mathbf{V}\tilde{\mathbf{X}}_2)(\tilde{\mathbf{X}}_2'\mathbf{P}\tilde{\mathbf{X}}_2)^{-1}. \tag{5.6.15}$$

Therefore, under $H_{2|1}$, $\hat{\theta}_2 \sim N_u(\mathbf{0}, \mathbf{D}(\hat{\theta}_2))$. Normality and zero mean of $\hat{\theta}_2$ follow from Eq. (5.4.2) and Theorem A.4.2 in the appendix. Using Theorem 4.2.1, we see that $X_P^2(2|1) \sim \sum_{i=1}^{u} \delta(2|1)_i\psi_i$ where the $\delta(2|1)_i$'s are the eigenvalues of

$$n\mathbf{D}(\hat{\theta}_2)(\tilde{\mathbf{X}}_2'\mathbf{P}\tilde{\mathbf{X}}_2) = (\tilde{\mathbf{X}}_2'\mathbf{P}\tilde{\mathbf{X}}_2)^{-1}(\tilde{\mathbf{X}}_2'\mathbf{V}\tilde{\mathbf{X}}_2).$$

□

The $\delta(2|1)_i$'s may be interpreted as the generalized deff's, with $\delta(2|1)_1$ as the largest possible deff taken over all linear combinations of the elements of $\tilde{\mathbf{X}}_2'\hat{\mathbf{p}}$.

Corollary 5.6.1.1 *For multinomial sampling,* $\mathbf{V} = \mathbf{P}$ *and hence* $\delta(2|1)_i = 1\ \forall\, i = 1, \ldots, u$ *and we get the standard result* $X_P^2(2|1) \sim \chi^2_{(u)}$ *under* $H_{2|1}$. □

Note that the asymptotic distribution of $X_P^2(2|1)$ or $G^2(2|1)$ depends on $\mathbf{V}$ through the eigenvalues δ_i's.

5.6.2 A Wald Statistic

If a consistent estimator $\hat{\mathbf{V}}/n$ of $\mathbf{V}/n$, the covariance matrix of $\hat{\mathbf{p}}$ is available, we can construct Wald statistic, X_W^2 for testing $H_{2|1} : \theta_2 = \mathbf{0}$.

Let $\mathbf{C}$ be a $T \times u$ matrix of rank u, such that $\mathbf{C}'\mathbf{X}_1 = \mathbf{0}$, $\mathbf{C}'\mathbf{1}_T = \mathbf{0}$ and $\mathbf{C}'\mathbf{X}_2$ is nonsingular. If $\mathbf{X}_1'\mathbf{X}_2 = \mathbf{0}$, a convenient choice for $\mathbf{C}$ is $\mathbf{X}_2$.

Now from model (5.2.4),

$$\mu = \mathbf{X}\theta + \text{ constant} = \mathbf{X}_1\theta_1 + \mathbf{X}_2\theta_2 + \text{ constant}.$$

Hence, under $H_{2|1}$,

$$\mathbf{C}'\mu = \phi \text{ (say)} = \mathbf{C}'\mathbf{X}_1\theta_1 + \mathbf{C}'\mathbf{X}_2\theta_2 = \mathbf{0}.$$

Therefore, the hypothesis $H_{2|1}$ is equivalent to the hypothesis $H_0 : \phi = \mathbf{0}$. Hence, a Wald statistic for testing H_0 is

$$X_W^2(2|1) = \hat{\phi}[\hat{\mathbf{D}}(\hat{\phi})]^{-1}\hat{\phi} \tag{5.6.16}$$

where $\hat{\phi} = \mathbf{C}'\hat{\mu}$ and $\hat{\mathbf{D}}(\hat{\phi})$ is the estimated covariance matrix of $\hat{\phi}$.

Now,

$$\begin{aligned} \hat{\mu}_i - \mu_i &= \ln \hat{\pi}_i - \ln \pi_i \approx (\hat{\pi}_i - \pi_i) \left.\frac{\partial \ln(\pi_i)}{\partial \pi_i}\right]_{\pi_i} \\ &= (\hat{\pi}_i - \pi_i)\frac{1}{\pi_i}, \end{aligned} \tag{5.6.17}$$

by delta method. Hence,

$$\hat{\mu} - \mu \approx \mathbf{D}_{\pi}^{-1}(\hat{\pi} - \pi) \tag{5.6.18}$$

where $\mathbf{D}_{\pi} =$ Diag.$(\pi_1, \ldots, \pi_T)$. Therefore,

$$\mathbf{D}(\hat{\phi}) = \mathbf{C}'\mathbf{D}(\hat{\mu})\mathbf{C} = \mathbf{C}'\mathbf{D}_{\pi}^{-1}\mathbf{D}(\hat{\pi})\mathbf{D}_{\pi}^{-1}\mathbf{C} = \Sigma_{\phi} \text{ (say)}, \tag{5.6.19}$$

using (5.6.17), where $\mathbf{D}(\hat{\pi})$ is the asymptotic covariance matrix of $\hat{\pi}$. The expression for $\mathbf{D}(\hat{\phi})$ is obtained from (5.4.4) and (5.4.5). The estimator $\hat{\mathbf{D}}(\hat{\pi})$ is obtained by replacing π by $\hat{\pi}$ and $\mathbf{V}$ by $\hat{\mathbf{V}}$. The Wald statistic (5.6.16) is independent of the choice of $\mathbf{C}$.

If a consistent estimator $\hat{\mathbf{V}}$ of $\mathbf{V}$ is not available, the effect of survey design is sometimes ignored. In this case $\mathbf{V}/n$ is replaced by the multinomial covariance matrix $\mathbf{P}/n$ in $\mathbf{D}(\hat{\phi})$ when $\mathbf{D}(\hat{\phi})$ reduces to

$$\Sigma_0 = n^{-1}\mathbf{C}'\mathbf{X}(\mathbf{X}'\mathbf{P}\mathbf{X})^{-1}\mathbf{X}'\mathbf{C} \tag{5.6.20}$$

since $\mathbf{PD}_\pi^{-1}\mathbf{C} = (\mathbf{D}_\pi - \pi\pi')\mathbf{D}_\pi^{-1}\mathbf{C} = \mathbf{C}$, since $\mathbf{C}'\mathbf{1} = \mathbf{0}$. This gives a test statistic alternative to $X_P^2(2|1)$ or $G^2(2|1)$:

$$\tilde{X}_W^2(2|1) = n\hat{\phi}'[\mathbf{C}'\mathbf{X}(\mathbf{X}'\hat{\mathbf{P}}\mathbf{X})^{-1}\mathbf{X}'\mathbf{C}]^{-1}\hat{\phi}. \tag{5.6.21}$$

As in the case of $X_P^2(2|1)$ or $G^2(2|1)$, the true asymptotic null distribution of $X_W^2(2|1)$ is a weighted sum of independent $\chi^2_{(1)}$ variables, $\sum_{i=1}^u \gamma(2|1)_i\psi_i$, where $\gamma(2|1)_1, \ldots, \gamma(2|1)_u$ are eigenvalues of $\Sigma_0^{-1}\Sigma_\phi$.

Note 5.6.1 In case of saturated model $s + u = T - 1$. Here $\hat{\pi} = \hat{\mathbf{p}} = \mathbf{n}/n$ and hence $\mathbf{D}(\hat{\mathbf{p}}) = \mathbf{P}/n$. Therefore, $\Sigma_0 = n^{-1}\mathbf{C}'\mathbf{D}_\pi^{-1}\mathbf{P}\mathbf{D}_\pi^{-1}\mathbf{C} = n^{-1}\mathbf{C}'\mathbf{D}_\pi^{-1}\mathbf{C}$. Therefore, $\delta(2|1)_i$'s are eigenvalues of $\Sigma_0^{-1}\Sigma_\phi = (\mathbf{C}'\mathbf{D}_\pi^{-1}\mathbf{C})^{-1}(\mathbf{C}'\mathbf{D}_\pi^{-1}\mathbf{V}\mathbf{D}_\pi^{-1}\mathbf{C})$.

Theorem 5.6.2 below shows that $X_W^2(2|1)$ is, in fact, asymptotically equivalent to $X_P^2(2|1)$ under $H_{2|1}$, which implies that $X_W^2(2|1) \sim \sum_{i=1}^u \delta_i\psi_i$ asymptotically and $[\delta(2|1)_1, \ldots, \delta(2|1)_u]$ is asymptotically identical to $[\gamma(2|1)_1, \ldots, \gamma(2|1)_u]$.

Theorem 5.6.2 *Under* $H_{2|1} : \theta_2 = \mathbf{0}$, $X_W^2(2|1) \sim X_P^2(2|1)$.

Proof Under $H_{2|1}$, we have $\hat{\phi} = \hat{\phi} - \phi = \mathbf{C}'(\hat{\mu} - \mu) \sim \mathbf{C}'\mathbf{D}_\pi^{-1}(\hat{\pi} - \pi)$ (by (5.6.18)) $\sim \mathbf{C}'\mathbf{D}_\pi^{-1}\mathbf{PX}(\hat{\theta} - \theta)$ (by (5.4.1)). Now, $\mathbf{C}'\mathbf{D}_\pi^{-1}\mathbf{PX} = \mathbf{C}'\mathbf{X} = (\mathbf{C}'\mathbf{X}_1, \mathbf{C}'\mathbf{X}_2) = (\mathbf{0}, \mathbf{C}'\mathbf{X}_2)$. Hence,

$$\begin{aligned}\hat{\phi} &= (\mathbf{0}(u \times s, \mathbf{C}'\mathbf{X}_2(u \times u))(\hat{\theta} - \theta) \\ &= \mathbf{C}'\mathbf{X}_2\hat{\theta}_2\end{aligned} \tag{5.6.22}$$

under $H_{2|1}$. Also,

$$\mathbf{C}'\mathbf{X}(\mathbf{X}'\mathbf{PX})^{-1}\mathbf{X}'\mathbf{C} = \mathbf{C}'\mathbf{X}_2(\tilde{\mathbf{X}}_2'\mathbf{P}\tilde{\mathbf{X}}_2)^{-1}\mathbf{X}_2'\mathbf{C} \tag{5.6.23}$$

using the formula for the inverse of the partitioned matrix $\mathbf{X}'\mathbf{PX}$. Hence, from (5.6.21),

$$\begin{aligned}X_W^2(2|1) &= n\hat{\theta}_2'\mathbf{X}_2'\mathbf{C}[\mathbf{C}'\mathbf{X}_2(\tilde{\mathbf{X}}_2'\mathbf{P}\tilde{\mathbf{X}}_2)^{-1}\mathbf{X}_2'\mathbf{C}]^{-1}\mathbf{C}'\mathbf{X}_2\theta_2 \\ &= n\theta_2(\tilde{\mathbf{X}}_2'\mathbf{P}\tilde{\mathbf{X}}_2)^{-1}\hat{\theta}_2 \\ &= X_P^2(2|1),\end{aligned} \tag{5.6.24}$$

since $\mathbf{C}'\mathbf{X}_2(u \times u)$ is nonsingular.

5.6.3 Modifications to Test Statistics

A primary modification to $X_P^2(2|1)$ (or $G^2(2|1)$) is to treat $X_P^2(2|1)/\hat{\delta}_0$ or $G^2(2|1)/\hat{\delta}_0$ under H_0, where δ_0 may be written as

$$
\begin{aligned}
u\delta(2|1)_0 &= \text{tr}[(\tilde{\mathbf{X}}_2'\mathbf{P}\tilde{\mathbf{X}}_2)^{-1}(\tilde{\mathbf{X}}_2'\mathbf{V}\tilde{\mathbf{X}}_2)] \\
&= \text{tr}[(\mathbf{X}'\mathbf{P}\mathbf{X})^{-1}(\mathbf{X}'\mathbf{V}\mathbf{X})] - \text{tr.}[(\mathbf{X}_1'\mathbf{P}\mathbf{X}_1)^{-1}(\mathbf{X}_1'\mathbf{V}\mathbf{X}_1)] \\
&= (s+u)\lambda_0 - s\lambda_{10} \text{ (say).}
\end{aligned} \tag{5.6.25}
$$

We note that $(\mathbf{X}'\mathbf{P}\mathbf{X})^{-1}(\mathbf{X}''\mathbf{V}\mathbf{X})$ is the design effect matrix for the contrast vector $\mathbf{X}'\hat{\mathbf{p}}$ so that λ_0 is the average generalized deff of $\mathbf{X}'\hat{\mathbf{p}}$. Similarly, $(\mathbf{X}_1'\mathbf{P}\mathbf{X}_1)^{-1}(\mathbf{X}_1'\mathbf{V}\mathbf{X}_1)$ is the deff matrix for the contrast vector $\mathbf{X}_1'\hat{\mathbf{p}}$ and λ_{10} is the average generalized deff of $\mathbf{X}_1'\hat{\mathbf{p}}$.

In the special case of saturated model, $s + u = T - 1$ and $u\delta(2|1)_0$ reduces to

$$
(T - s - 1)\delta(2|1)_0 = (T - 1)\lambda_0 - s\lambda_{10} \tag{5.6.26}
$$

where

$$
(T - 1)\lambda_0 = \sum_{i=1}^{u}(1 - \pi)d_i \text{ (vide (4.2.29))} \tag{5.6.27}
$$

and

$$
d_i = \frac{v_{ii}}{\pi_i(1 - \pi_i)} \tag{5.6.28}
$$

is the (cell) deff of $\hat{p}_i$ and $\mathbf{V} = ((v_{ij}))$.

If T is much larger than s, then $\delta(2|1)_0 \approx \lambda_0$ and we might expect $X_P^2(2|1)/\hat{\lambda}_0$ to perform well in large tables if s is fairly small. Note that unlike $\delta(2|1)_0$, λ_0 is independent of $H_{2|1}$.

Another approximation $X_P^2(2|1)/\hat{d}_0$ (Fellegi 1980), where $\hat{d}_0 = \sum_{i=1}^{T} \hat{d}_i/T$ is the average estimated cell deff., is also independent of $H_{2|1}$.

Empirical result reported in Holt et al. (1980) and Hidiroglou and Rao (1987) for testing independence in a two-way table indicates that both $X_P^2(2|1)/\hat{\lambda}_0$ and $X_P^2(2|1)/\hat{d}_0$ tend to be conservative, that is, their asymptotic significance level is less than α and sometimes very conservative, whereas if the coefficient of variations of $\hat{\delta}(2|1)_i$'s is small, $X_P^2(2|1)/\hat{\delta}(2|1)_0$ works fairly well. The Pearson statistic $X_P^2(2|1)$ often leads to unacceptably high values of significance level of χ^2.

Fay's (1979, 1984, 1985) jackknifed methods can also be applied to the $X_P^2(2|1)$ and $G^2(2|1)$ statistics. The jackknifed version $X_{PJ}(2|1)$ for testing the nested hypothesis $H_{2|1}$ is obtained by replacing X_P^2 in Eqs. 5.3.21 through 5.3.24 by $X_P^2(2|1)$. The jackknifed version $G_J(2|1)$ can be obtained in an analogous manner.

5.6.4 *Effects of Survey Design on* $X_P^2(2|1)$

The asymptotic null distribution of $X_P^2(2|1)$ may be approximated to a χ^2 variable, following Satterthwaite (1946):

$$X_S^2(2|1) = \frac{X_P^2(2|1)}{(1 + a(2|1)^2)\delta(2|1)_0} \quad (5.6.29)$$

is treated as a $\chi^2_{(\nu)}$ variable, where $\nu = u/(1 + a(2|1)^2)$, $u\delta(2|1)_0 = \sum_{i=1}^u \delta(2|1)_i$ and $a(2|1)^2 = \sum_{i=1}^u (\delta(2|1)_i - \delta(2|1)_0)^2/[u\delta(2|1)_0^2]$ is the coefficient of variations of $\delta(2|1)_i$'s and $\delta(2|1)_i$'s are the eigenvalues of Theorem 5.6.1

The effect of survey design may be studied by computing the asymptotic significance level (SL) of $X_P^2(2|1)$ for a desired nominal level α, that is,

$$SL[X_P^2(2|1)] = P\left[X_P^2(2|1) \geq \chi^2_{(u)}(\alpha)] \approx P[\chi^2_{(\nu} \geq \frac{\chi^2_{(u)}(\alpha)}{(1 + a(2|1)^2)\delta_0}\right] \quad (5.6.30)$$

is compared with α. In practice, SL$[X_P^2(2|1)]$ is estimated using $\hat{\mathbf{V}}$ for $\mathbf{V}$ and $\hat{\pi}$ (or $\tilde{\pi}$) for π.

Further references in this area are due to Holt et al. (1980) and Hidiroglou and Rao (1987). Rao and Thomas (1988) illustrated all the results in this chapter and previous chapter using survey data from Canadian Class Structure (1983) and Canadian Health Survey (1978–79).

Ballin et al. (2010) considered applications of a set of graphical models known as Probabilistic Expert Systems (PES) to define two classes of estimators of a multiway contingency table where a sample is drawn according to a stratified sampling design. Two classes are characterized by the different roles of the sampling design. In the first, the sampling design is treated as an additional variable and in the second it is used only for estimation purposes by means of sampling weights.

Chapter 6
Analysis of Categorical Data Under Logistic Regression Model

Abstract This chapter considers analysis of categorical data under logistic regression models when the data are generated from complex surveys. Section 6.2 addresses binary logistic regression model due to Roberts et al. (Biometrika 74:1–12, 1987), and finds the pseudo ML estimators of the population parameter along with its asymptotic covariance matrix. The goodness-of-fit statistics X_P^2 and G^2, and a Wald statistic have been considered and their asymptotic distributions derived. The modifications of these statistics using Rao-Scott corrections and F ratio have been examined. All the above problems have been considered in the light of nested models. We also considered problem of choosing appropriate cell-sample-sizes for running logistic regression program in a standard computer package. Following Morel (Surv Methodol 15:203–223, 1989) polytomous logistic regression has been considered in Sect. 6.5. Finally, using empirical logits the model has been converted into general linear model which uses generalized least square procedures for estimation. The model has been extended to accommodate cluster effects and procedures for testing of hypotheses under the extended model investigated.

Keywords Pseudo-likelihood · Empirical logit · Binary logistic regression · Polytomous logistic regression · Generalized least square estimator · Nested models · Cluster effects

6.1 Introduction

In Sect. 3.7, we introduced briefly Logistic regression models, for both binary and polytomous data and hinted on their method of analysis under IID assumptions. Some modifications in the classical methods of analyzing data under binary logistic regression model when such data are generated from complex sample surveys were also discussed. Here, we consider the analysis under these models when the data are obtained from complex surveys and hence do not generally respect IID assumptions.

In Sect. 6.2, we consider binary logistic regression model due to Roberts et al. (1987), and find the pseudo ML estimators of the population parameter along with its asymptotic covariance matrix. Standard Pearson X_P^2, likelihood ratio G^2 test

P. Mukhopadhyay, *Complex Surveys*, DOI 10.1007/978-981-10-0871-9_6

statistics and a Wald statistic have been considered and their asymptotic distributions derived. These statistics have been adjusted on the basis of certain generalized design effects. All the above problems have also been considered in the light of nested models. In Sect. 6.4, we considered problem of choosing appropriate cell-sample-sizes for running logistic regression program in a standard computer package. In the next section following Morel (1989) analysis of complex survey data under polytomous logistic regression has been addressed. Finally, using empirical logits the model has been converted into general linear models which use generalized least square procedures for estimation. The model has been extended to accommodate cluster effects and procedures for testing of hypotheses under the extended model investigated.

6.2 Binary Logistic Regression

Roberts et al. (1987) considered the following binary logistic regression model. The population is divided into I cells or domains or strata. A binary (0, 1)-response variable Y is defined on each of the N_i population units in the ith cell such that its value $Y_i = 1$ if the unit possesses a specific characteristic, 0 otherwise ($i = 1, \ldots, I$). Let $\hat{N}_i$ denote the survey estimate of N_i and $\hat{N}_{i1}$ the survey estimate of N_{i1}, the ith domain total of the binary response variable. Then, $p_i = \frac{\hat{N}_{i1}}{\hat{N}_i}$ is a design-based estimate of the population proportion $\pi_i = \frac{N_{i1}}{N_i}$ of units in the ith domain.

Binomial samples are drawn independently from each stratum, sample size from the ith domain being n_i and the variable total n_{i1}. The n_i's may be fixed or random depending on the sampling design adopted.

Suppose we have access to the values of p auxiliary variables $\mathbf{x} = (x_1, \ldots, x_p)'$ for each domain, $\mathbf{x}_i = (x_{i1}, \ldots, x_{ip})'$ being the value of $\mathbf{x}$ on the ith domain. Note that the value of $\mathbf{x}$ remains the same, namely, $\mathbf{x}_i$, for all the units in the ith domain ($i = 1, \ldots, I$).

The logistic regression model for π_i, the population proportion in the ith domain, is

$$\pi_i = \frac{\exp(\mathbf{x}_i'\beta)}{1 + \exp(\mathbf{x}_i'\beta)} \tag{6.2.1}$$

where $\beta = (\beta_1, \ldots, \beta_p)'$ is a vector of unknown regression coefficients. Hence,

$$\frac{\pi_i(\mathbf{x}_i)}{1 - \pi_i(\mathbf{x}_i)} = \exp(\mathbf{x}_i'\beta)$$

or

$$\log\left\{\frac{\pi_i(\mathbf{x}_i)}{1 - \pi_i(\mathbf{x}_i)}\right\} = \text{logit } (\pi_i(\mathbf{x}_i)) = \mathbf{x}_i'\boldsymbol{\beta}. \tag{6.2.2}$$

Letting $\pi_i(\mathbf{x}_i) = \pi_i = f_i(\beta)$, $\pi = (\pi_1, \ldots, \pi_I)'$, $\mathbf{f}(\beta) = (f_1(\beta), \ldots, f_I(\beta))'$, (6.2.2) can be written as

$$\nu_i = \text{ logit } (\pi_i) = \log\left\{\frac{f_i(\beta)}{1 - f_i(\beta)}\right\} = \mathbf{x}_i'\beta,\ i = 1, \ldots I,$$

or

$$\boldsymbol{\nu} = \mathbf{X}\boldsymbol{\beta} \tag{6.2.3}$$

where $\mathbf{X} = (\mathbf{x}_1, \mathbf{x}_2, \ldots, \mathbf{x}_I)'$ is the $I \times p$ matrix of observations x_{ij}, $i = 1, \ldots, I; j = 1, \ldots, p$ and is of full rank p and $\boldsymbol{\nu} = (\nu_1, \ldots, \nu_I)'$.

Note that the I parameters $\pi_1, \ldots, \pi_I$, here, depend on the p parameters in β and therefore, the theory in Sect. A.4 has to be applied for estimation.

6.2.1 Pseudo-MLE of π

Now, the joint distribution of $\mathbf{n}_1 = (n_{11}, \ldots, n_{I1})'$, $g(\mathbf{n}_1)$ is proportional to the product of I binomial functions

$$\begin{aligned} g(\mathbf{n}_1|\pi_1, \ldots, \pi_I, \mathbf{X}) &\propto \Pi_{i=1}^I \pi_i(\mathbf{x}_i)^{n_{i1}}[1 - \pi_i(\mathbf{x}_i)]^{n_i - n_{i1}} \\ &= \left[\prod_{i=1}^I \left\{\frac{\pi_i(\mathbf{x}_i)}{1-\pi_i(\mathbf{x}_i)}\right\}^{n_{i1}}\right]\left[\prod_{i=1}^I \{1 - \pi_i(\mathbf{x}_i)\}^{n_i}\right] \\ &= \left[\exp\left\{\sum_{i=1}^I n_{i1} \log \frac{\pi_i(\mathbf{x}_i)}{1-\pi_i(\mathbf{x}_i)}\right\}\right]\left[\prod_{i=1}^I \{1 - \pi_i(\mathbf{x}_i)\}^{n_i}\right]. \end{aligned} \tag{6.2.4}$$

Again,

$$\log \frac{\pi_i}{1 - \pi_i} = \mathbf{x}_i'\beta = \sum_{j=1}^p \beta_j x_{ij},$$

and

$$1 - \pi_i = \left[1 + \exp\left(\sum_{j=1}^p \beta_j x_{ij}\right)\right]^{-1}.$$

Hence, from (6.2.4), the log-likelihood of β is

$$\mathcal{L}(\beta|\mathbf{n}_1, \mathbf{X}) = \log g(\mathbf{n}_1|\pi_1, \ldots, \pi_I, \mathbf{X}) = \text{ Constant } + \sum_{j=1}^{p} \beta_j \left(\sum_i n_{i1} x_{ij} \right)$$
$$- \sum_{i=1}^{I} n_i \log \left[1 + \exp \left(\sum_{j=1}^{p} \beta_j x_{ij} \right) \right] \tag{6.2.5}$$

which depends on the binomial counts $n_{i1} (i = 1, \ldots, I)$ only through $\sum_i n_{i1} x_{ij}$, $j = 1, \ldots, p$. Now,

$$\frac{\partial \mathcal{L}(\beta)}{\partial \beta_j} = \sum_{i=1}^{I} n_{i1} x_{ij} - \sum_{i=1}^{I} n_i \frac{x_{ij} \exp\left(\sum_k \beta_k x_{ik}\right)}{1 + \exp\left(\sum_k \beta_k x_{ik}\right)}, \; j = 1, \ldots, p.$$

Hence, the likelihood equations for finding the MLE $\hat{\beta}$ of β are

$$\sum_{i=1}^{I} n_{i1} x_{ij} - \sum_i n_i x_{ij} \hat{f}_i = 0, \; j = 1, \ldots, p. \tag{6.2.6}$$

Let $\mathbf{D}(\mathbf{n}) =$ Diag. $(n_1, \ldots, n_I)$, $\mathbf{q} = (q_1, \ldots, q_I)'$ where $q_i = \frac{n_{i1}}{n_i}$. Then, Eq. (6.2.6) can be written as

$$\mathbf{X}'\mathbf{D}(\mathbf{n})\hat{\mathbf{f}} = \mathbf{X}'\mathbf{D}(\mathbf{n})\mathbf{q}. \tag{6.2.7}$$

For general sampling designs, appropriate likelihood functions are difficult to obtain and maximum likelihood estimates are hard to get by. Hence, it is a common practice to obtain pseudo maximum likelihood equations by replacing in (6.2.7), n_i/n by w_i, an estimate of the domain relative size $W_i = N_i/N$ and $q_i = n_{i1}/n_i$ by p_i, a design-based estimate of $\pi_i = N_{i1}/N_i$, $(i = 1, \ldots, I)$. The pseudo-MLE's $\tilde{\beta}$ of β is then obtained by solving the equations

$$\mathbf{X}'\mathbf{D}(\mathbf{w})\tilde{\mathbf{f}} = \mathbf{X}'\mathbf{D}(\mathbf{w})\mathbf{p} \tag{6.2.8}$$

where $\mathbf{D}(\mathbf{w}) =$ Diag. $(w_1, \ldots, w_I)$ and $\mathbf{p} = (p_1, \ldots, p_I)'$ and $\tilde{\mathbf{f}}$ is the pseudo-MLE of $\mathbf{f}$.

6.2.2 *Asymptotic Covariance Matrix of the Estimators*

Instead of considering $\mathbf{D}(\mathbf{w})\mathbf{f}$ we shall consider $\mathbf{D}(\mathbf{W})\mathbf{f}$ as a function of $\boldsymbol{\beta}$, where $\mathbf{D}(\mathbf{W}) =$ Diag. $(W_1, \ldots, W_I)$, since $w_i - W_i = o_p(1), i = 1, \ldots, I$. We will now use Birch's result (A.4.10). Let us write

$$\mathbf{b} = \mathbf{b}(\beta) = \mathbf{D}(\mathbf{W})\mathbf{f} = (W_1 f_1, \ldots, W_I f_I)',$$

$$\mathbf{B} = \mathbf{D}(\mathbf{b})^{-1/2} \left(\frac{\partial \mathbf{b}(\beta)}{\partial \beta} \right)$$

where $\mathbf{D}(\mathbf{b}) =$ Diag. $(b_1, \ldots, b_I)$. Now,

$$\frac{\partial b_i(\beta)}{\partial \beta_j} = W_i \frac{\partial f_i(\beta)}{\partial \beta_j} = W_i f_i (1 - f_i) x_{ij}.$$

Therefore,

$$\frac{\partial \mathbf{b}(\beta)}{\partial \beta} = \mathbf{\Delta X}$$

where

$$\mathbf{\Delta} = \text{Diag. } (W_i f_i (1 - f_i), i = 1, \ldots, I). \tag{6.2.9}$$

Hence,

$$\mathbf{B} = \mathbf{D}(\mathbf{b})^{-1/2} \mathbf{\Delta X}. \tag{6.2.10}$$

Therefore, from (A.4.10),

$$\tilde{\beta} - \beta \sim (\mathbf{B}'\mathbf{B})^{-1}\mathbf{B}'\mathbf{D}(\mathbf{b})^{-1/2}(\mathbf{a} - \mathbf{b}(\beta)) \tag{6.2.11}$$

where $\mathbf{a} = (a_1, \ldots, a_I)'$, $a_i = W_i p_i$. Hence, (6.2.11) reduces to

$$\begin{aligned} \sqrt{n}(\tilde{\beta} - \beta) &\sim \sqrt{n}(\mathbf{B}'\mathbf{B})^{-1}\mathbf{X}'\mathbf{\Delta D}(\mathbf{b})^{-1}\mathbf{D}(\mathbf{W})(\mathbf{p} - \mathbf{f}) \\ &= \sqrt{n}(\mathbf{X}'\mathbf{\Delta D}(\mathbf{b})^{-1}\mathbf{\Delta X})^{-1}\mathbf{X}'\mathbf{\Delta D}(\mathbf{b})^{-1}\mathbf{D}(\mathbf{W})(\mathbf{p} - \mathbf{f}) \\ &= \sqrt{n}(\mathbf{X}'\mathbf{\Delta X})^{-1}\mathbf{X}'\mathbf{D}(\mathbf{W})(\mathbf{p} - \mathbf{f}). \end{aligned} \tag{6.2.12}$$

Assuming

$$\sqrt{n}(\mathbf{p} - \mathbf{f}) \to^{L} (\mathbf{0}, \mathbf{V}), \tag{6.2.12'}$$

the asymptotic covariance matrix of $\tilde{\beta}$ is

$$\mathbf{V}_{\tilde{\beta}} = n^{-1}(\mathbf{X}'\mathbf{\Delta X})^{-1}\{\mathbf{X}'\mathbf{D}(\mathbf{W})\mathbf{V}\mathbf{D}(\mathbf{W})\mathbf{X}\}(\mathbf{X}'\mathbf{\Delta X})^{-1}. \tag{6.2.13}$$

Replacing the parameters in (6.2.13) by their estimates, we get the estimated asymptotic covariance matrix

$$\hat{\mathbf{V}}_{\tilde{\beta}} = n^{-1}(\mathbf{X}'\hat{\boldsymbol{\Delta}}\mathbf{X})^{-1}\{\mathbf{X}'\mathbf{D}(\mathbf{w})\hat{\mathbf{V}}\mathbf{D}(\mathbf{w})\mathbf{X}\}(\mathbf{X}'\boldsymbol{\Delta}\mathbf{X})^{-1} \quad (6.2.14)$$

where $\hat{\boldsymbol{\Delta}} =$ Diag. $\{w_i\hat{f}_i(1-\hat{f}_i)\}$.

Again, by (A.4.14),

$$\begin{aligned}\sqrt{n}\{\mathbf{b}(\tilde{\beta}) - \mathbf{b}(\beta)\} &\sim \sqrt{n}(\tfrac{\partial \mathbf{b}(\beta)}{\partial \beta})(\tilde{\beta} - \beta) \\ &= \sqrt{n}\boldsymbol{\Delta}\mathbf{X}(\tilde{\beta} - \beta).\end{aligned} \quad (6.2.15)$$

Therefore, by (6.2.12) and (6.2.14), the asymptotic covariance matrix of the fitted cell-frequencies $\tilde{\mathbf{f}}$ is

$$\mathbf{V}_{\tilde{\mathbf{f}}} = \mathbf{D}(\mathbf{W})^{-1}\boldsymbol{\Delta}\mathbf{X}\mathbf{V}_{\tilde{\beta}}\mathbf{X}'\boldsymbol{\Delta}\mathbf{D}(\mathbf{W})^{-1}. \quad (6.2.16)$$

Hence, the estimated covariance matrix is

$$\hat{\mathbf{V}}_{\tilde{\mathbf{f}}} = \mathbf{D}(\mathbf{w})^{-1}\hat{\boldsymbol{\Delta}}\mathbf{X}\hat{\mathbf{V}}_{\tilde{\beta}}\mathbf{X}'\hat{\boldsymbol{\Delta}}\mathbf{D}(\mathbf{w})^{-1}. \quad (6.2.17)$$

The residual error vector

$$\mathbf{r} = \mathbf{p} - \tilde{\mathbf{f}} = (\mathbf{p} - \mathbf{f}) - (\tilde{\mathbf{f}} - \mathbf{f}) \quad (6.2.18)$$

Again,

$$\tilde{\mathbf{f}} - \mathbf{f} = \mathbf{D}(\mathbf{W})^{-1}\boldsymbol{\Delta}\mathbf{X}(\mathbf{X}'\boldsymbol{\Delta}\mathbf{X})^{-1}\mathbf{X}'\mathbf{D}(\mathbf{W})(\mathbf{p} - \mathbf{f}) \quad (6.2.19)$$

by (6.2.15) and (6.2.12). Therefore, from (6.2.18),

$$\sqrt{n}\mathbf{r} = \sqrt{n}(\mathbf{p} - \tilde{\mathbf{f}}) \sim [\mathbf{I} - \mathbf{D}(\mathbf{W})^{-1}\boldsymbol{\Delta}\mathbf{X}(\mathbf{X}'\boldsymbol{\Delta}\mathbf{X})^{-1}\mathbf{X}'\mathbf{D}(\mathbf{W})]\sqrt{n}(\mathbf{p} - \mathbf{f}). \quad (6.2.20)$$

Its estimated asymptotic covariance matrix is

$$\hat{\mathbf{V}}_{\mathbf{r}} = n^{-1}\mathbf{A}\hat{\mathbf{V}}\mathbf{A}' \quad (6.2.21)$$

where

$$\mathbf{A} = \mathbf{I} - \mathbf{D}(\mathbf{w})^{-1}\hat{\boldsymbol{\Delta}}\mathbf{X}(\mathbf{X}'\hat{\boldsymbol{\Delta}}\mathbf{X})^{-1}\mathbf{X}'\mathbf{D}(\mathbf{w}). \quad (6.2.22)$$

6.2.3 Goodness-of-Fit Tests

The usual goodness-of-fit tests are:

$$X_P^2 = n\sum_{i=1}^{I} w_i \frac{(p_i - \hat{f}_i)^2}{\hat{f}_i(1-\hat{f}_i)}; \tag{6.2.23}$$

and

$$\begin{aligned} G^2 &= 2n\sum_{i=1}^{I} w_i \left[p_i \log\left(\tfrac{p_i}{\hat{f}_i}\right) + (1-p_i)\log\left(\tfrac{1-p_i}{1-\hat{f}_i}\right)\right] \\ &= \sum_{i=1}^{I} G_i^2 \ \text{(say)}, \end{aligned} \tag{6.2.24}$$

where $\hat{f}_i$ is any estimator of f_i including the pseudo-MLE.

If $p_i = 0$, $G_i^2 = -2nw_i\log(1-\hat{f}_i)$; if $p_i = 1$, $G_i^2 = -2nw_i\log\hat{f}_i$.

Under independent binomial sampling, both X_P^2 and G^2 are asymptotically distributed as a $\chi^2_{(I-p)}$ variable, when the model (6.2.3) holds. However, this result is not valid for general sampling design.

For a general sampling design, X_P^2 and G^2 are asymptotically distributed as a weighted sum $\sum_{i=1}^{I-p}\delta_i\psi_i$ of independent $\chi^2_{(1)}$ variables ψ_i. Here, the weights δ_i's are the eigenvalues of the matrix

$$\mathbf{\Sigma}_{0\phi}^{-1}\mathbf{\Sigma}_{\phi} \tag{6.2.25}$$

where

$$\mathbf{\Sigma}_{\phi} = n^{-1}\mathbf{C}'\mathbf{\Delta}^{-1}\mathbf{D}(\mathbf{W})\mathbf{V}\mathbf{D}(\mathbf{W})\mathbf{\Delta}^{-1}\mathbf{C} \tag{6.2.26}$$

and

$$\mathbf{\Sigma}_{0\phi} = n^{-1}\mathbf{C}'\mathbf{\Delta}^{-1}\mathbf{C} \tag{6.2.27}$$

and $\mathbf{C}$ is any $I\times(I-p)$ matrix of rank $(I-p)$ such that $\mathbf{C}'\mathbf{X} = \mathbf{0}$. For proof, see Corollary 6.3.1.2.

The eigenvalues are independent of the choice of $\mathbf{C}$. The matrix $\mathbf{\Sigma}_{0\phi}$ and δ_i's are, as before, termed a 'generalized design effect matrix' and 'generalized design effects', respectively.

A Wald statistic:
Testing fit of the model (6.2.3) is equivalent to testing $\mathbf{C}'\boldsymbol{\nu} = \boldsymbol{\phi}$ (say) $= \mathbf{C}'\mathbf{X}\beta = \mathbf{0}$ and hence, as shown by Roberts et al. (1987), a Wald statistic is

$$X_W^2 = \hat{\boldsymbol{\phi}}'\hat{\mathbf{V}}_{\hat{\phi}}^{-1}\hat{\boldsymbol{\phi}} = \hat{\boldsymbol{\nu}}'\mathbf{C}\hat{\mathbf{V}}_{\hat{\phi}}^{-1}\mathbf{C}'\hat{\boldsymbol{\nu}} \tag{6.2.28}$$

where $\hat{\nu}$ is the vector of estimated logits $\hat{\nu}_i = \{p_i/(1-p_i)\}$ and $\hat{\mathbf{V}}_{\hat{\phi}}$ is the estimated true-design-covariance matrix of $\hat{\phi}$. The statistic X^2_W is invariant to the choice of $\mathbf{C}$ and is asymptotically distributed as $\chi^2_{(I-p)}$ when the model (6.2.3) holds. The statistic X^2_W, however, is not defined if $p_i = 0$ or 1 for some i and may be very unstable if any p_i is close to 1 or when the number of degrees of freedom for estimating $\hat{\mathbf{V}}_{\hat{\phi}}$ is not large in comparison to $I-p$ (Fay 1985).

6.2.4 Modifications of Tests

As in (4.2.30), Rao-Scott adjustment to X^2_P or G^2 is obtained by treating $X^2_{P(c)} = X^2_P/\hat{\delta}_0$ or $G^2_c = G^2/\hat{\delta}_0$ as a $\chi^2_{(I-p)}$ random variable, where

$$(I-p)\hat{\delta}_0 = \sum_i \hat{\delta}_i = n\sum_{i=1}^{I} \frac{\hat{V}_{ii,r}w_i}{\hat{f}_i(1-\hat{f}_i)} \tag{6.2.29}$$

where $\hat{V}_{ii,r}$ is the ith diagonal element of the estimated covariance matrix of the residual $\mathbf{r} = \mathbf{p} - \hat{\mathbf{f}}$ given in (6.2.20). This adjustment is satisfactory if the c.v. among δ_i's is small.

A better approximation, based on Satterthwaite approximation, is given by $X^2_{P(cc)} = X^2_{P(c)}/(1+\hat{a}^2)$ or $G^2_{cc} = G^2_c/(1+\hat{a}^2)$ where

$$\hat{a}^2 = \sum_{i=1}^{I-p}(\hat{\delta}_i^2 - \hat{\delta}_0)^2/\{(I-p)\hat{\delta}_0^2\} \tag{6.2.30}$$

and $\sum_i \hat{\delta}_i^2$ is given by

$$\sum_{i=1}^{I-p}\hat{\delta}_i^2 = \sum_{i=1}^{I}\sum_{j=1}^{I}\frac{\hat{V}_{ij,r}(nw_i)(nw_j)}{\{\hat{f}_i\hat{f}_j(1-\hat{f}_i)(1-\hat{f}_j)\}} \tag{6.2.31}$$

where $\hat{V}_{ij,r}$ is the (i,j)th element of $\hat{\mathbf{V}}_r$.

6.3 Nested Model

As in Sect. 5.6, we call the full model (6.2.3) as model M_1. Let $\mathbf{X} = (\mathbf{X}_1(I\times r), \mathbf{X}_2(I\times u))$ where $r+u=p$. The model (6.2.3) can then be written as

$$\nu = \mathbf{X}\beta = \mathbf{X}_1\beta^{(1)} + \mathbf{X}_2\beta^{(2)} \tag{6.3.1}$$

where $\beta = (\beta^{(1)'}(1 \times r), \beta^{(2)'}(1 \times u))'$. We are interested in testing the null hypothesis $H_{2|1} : \beta^{(2)} = \mathbf{0}$, given that the unconstrained model is M_1, so that under $H_{2|1}$, we get the reduced model M_2 as

$$\nu = \mathbf{X}_1 \beta^{(1)}. \tag{6.3.2}$$

The pseudo-MLE $\tilde{\tilde{\beta}}^{(1)}$ of $\beta^{(1)}$ under M_2 can be obtained from the equations

$$\mathbf{X}_1' \mathbf{D}(\mathbf{w}) \tilde{\tilde{\mathbf{f}}} = \mathbf{X}_1' \mathbf{D}(\mathbf{w}) \mathbf{p} \tag{6.3.3}$$

by iterative calculations, where $\tilde{\tilde{\mathbf{f}}} = \mathbf{f}(\tilde{\tilde{\beta}}^{(1)})$. The standard Pearson and likelihood ratio tests for $H_{2|1}$ are then given by

$$X_P^2(2|1) = n \sum_{i=1}^{I} w_i \frac{(\tilde{f}_i - \tilde{\tilde{f}}_i)^2}{\tilde{\tilde{f}}_i (1 - \tilde{\tilde{f}}_i)} \tag{6.3.4}$$

and

$$G^2(2|1) = 2n \sum_{i=1}^{I} \left[\tilde{f}_i \log \frac{\tilde{f}_i}{\tilde{\tilde{f}}_i} + (1 - \tilde{f}_i) \log \frac{1 - \tilde{f}_i}{1 - \tilde{\tilde{f}}_i} \right]. \tag{6.3.5}$$

We shall now derive the asymptotic null distribution of $X_P^2(2|1)$ or $G^2(2|1)$ for any survey design $p(s)$.

Theorem 6.3.1 *Under $H_{2|1}$, asymptotic distribution of $X_P^2(2|1)$ given in (6.3.4) is given by*

$$X_P^2(2|1) \sim n \tilde{\beta}^{(2)'} (\tilde{\mathbf{X}}_2' \mathbf{\Delta} \tilde{\mathbf{X}}_2) \tilde{\beta}^{(2)} \tag{6.3.6}$$

where

$$\tilde{\mathbf{X}}_2 = \mathbf{X}_2 - \mathbf{X}_1 (\mathbf{X}_1' \mathbf{\Delta} \mathbf{X}_1)^{-1} (\mathbf{X}_1' \mathbf{\Delta} \mathbf{X}_2) \tag{6.3.7}$$

and $\mathbf{\Delta}$ is given in (6.2.9). Moreover, under $H_{2|1}$,

$$X_P^2(2|1) = \sum_{i=1}^{u} \hat{\delta}_i(2|1) \psi_i \tag{6.3.8}$$

where ψ_i's are independent $\chi^2_{(1)}$ variables and $\hat{\delta}_i(2|1)$'s (all >0) are the eigenvalues of the matrix

$$\hat{\mathbf{\Delta}}(2|1) = (\tilde{\mathbf{X}}_2' \hat{\mathbf{\Delta}} \tilde{\mathbf{X}}_2)^{-1} \{\tilde{\mathbf{X}}_2' \mathbf{D}(\mathbf{w}) \hat{\mathbf{V}} \mathbf{D}(\mathbf{w}) \tilde{\mathbf{X}}_2\}. \tag{6.3.9}$$

Proof From (6.3.4),

$$X_P^2(2|1) \sim n(\tilde{\mathbf{f}} - \tilde{\tilde{\mathbf{f}}})'\mathbf{D}(\mathbf{W})\tilde{\mathbf{\Delta}}^{-1}\mathbf{D}(\mathbf{W})(\tilde{\mathbf{f}} - \tilde{\tilde{\mathbf{f}}}) \tag{6.3.10}$$

where

$$\tilde{\mathbf{\Delta}} = \text{ Diag. } (W_i\tilde{f}_i(1 - \tilde{f}_i), i = 1, \ldots, I). \tag{6.3.11}$$

Following (6.2.12) and (6.2.19)

$$\sqrt{n}(\tilde{\tilde{\mathbf{f}}} - \mathbf{f}) = \mathbf{D}(\mathbf{W})^{-1}\mathbf{\Delta}\mathbf{X}_1\{\sqrt{n}(\tilde{\tilde{\beta}}^{(1)} - \beta^{(1)})\}, \tag{6.3.12}$$

where

$$\sqrt{n}(\tilde{\tilde{\beta}}^{(1)} - \beta^{(1)}) \sim (\mathbf{X}_1'\mathbf{\Delta}X_1)^{-1}\mathbf{X}_1\mathbf{D}(\mathbf{W})\{\sqrt{n}(\mathbf{p} - \mathbf{f})\}, \tag{6.3.13}$$

as in (6.2.12). Hence, from (6.2.15) and (6.3.11),

$$\sqrt{n}(\tilde{\mathbf{f}} - \tilde{\tilde{\mathbf{f}}}) \sim \mathbf{D}(\mathbf{W})^{-1}\mathbf{\Delta}\sqrt{n}\{\mathbf{X}_1(\tilde{\beta}^{(1)} - \beta^{(1)}) + \mathbf{X}_2\tilde{\beta}^{(2)} - \mathbf{X}_1(\tilde{\tilde{\beta}}^{(1)} - \beta^{(1)})\} \tag{6.3.14}$$

under H_0.

Again $\mathbf{X}'\mathbf{\Delta}\mathbf{X}$ can be expressed as the partitioned matrix

$$\mathbf{X}'\mathbf{\Delta}\mathbf{X} = \begin{bmatrix} \mathbf{X}_1'\mathbf{\Delta}\mathbf{X}_1 & \mathbf{X}_1'\mathbf{\Delta}\mathbf{X}_2 \\ \mathbf{X}_2'\mathbf{\Delta}\mathbf{X}_1 & \mathbf{X}_2'\mathbf{\Delta}\mathbf{X}_2 \end{bmatrix}.$$

As in (5.6.12), using the standard formula for the inverse of a partitioned matrix and thereby calculating $(\tilde{\tilde{\beta}}^{(1)} - \beta^{(1)})$, it can be shown from (6.2.12) and (6.3.12) that

$$\sqrt{n}(\tilde{\tilde{\beta}}^{(1)} - \beta^{(1)}) \sim \sqrt{n}(\tilde{\beta}^{(1)} - \beta^{(1)}) + (\mathbf{X}_1'\mathbf{\Delta}\mathbf{X}_1)^{-1}(\mathbf{X}_1'\mathbf{\Delta}\mathbf{X}_1)\sqrt{n}\tilde{\beta}^{(2)}. \tag{6.3.15}$$

Substitution of (6.3.14) into (6.3.13) gives

$$\sqrt{n}(\tilde{\mathbf{f}} - \tilde{\tilde{\mathbf{f}}}) \sim \sqrt{n}\mathbf{D}(\mathbf{W})^{-1}\mathbf{\Delta}\tilde{\mathbf{X}}_2\tilde{\beta}^{(2)}. \tag{6.3.16}$$

Hence, from (6.3.6) and (6.3.13),

$$X_P^2(2|1) \sim n\tilde{\beta}^{(2)'}(\tilde{\mathbf{X}}_2'\mathbf{\Delta}\tilde{\mathbf{X}}_2)\tilde{\beta}^{(2)}, \tag{6.3.17}$$

assuming $\tilde{\mathbf{\Delta}} \approx \mathbf{\Delta}$.

It follows from the formula for $\mathbf{V}_{\tilde{\beta}}$ given in (6.2.13) and the formula for the inverse of a partitioned matrix that the estimated asymptotic covariance matrix of $\tilde{\beta}^{(2)}$ is

$$\hat{\mathbf{V}}_{\tilde{\beta}^{(2)}} = n^{-1}(\tilde{\mathbf{X}}_2\hat{\mathbf{\Delta}}\tilde{\mathbf{X}}_2)^{-1}\{\tilde{\mathbf{X}}_2'\mathbf{D}(\mathbf{w})\hat{\mathbf{V}}\mathbf{D}(\mathbf{w})\tilde{\mathbf{X}}_2\}(\tilde{\mathbf{X}}_2'\hat{\mathbf{\Delta}}\tilde{\mathbf{X}}_2)^{-1} \tag{6.3.18}$$

so that $\tilde{\beta}^{(2)}$ is approximately $N_u(\mathbf{0}, \mathbf{V}_{\tilde{\beta}^{(2)}})$ under $H_{2|1}$.

Hence, by Theorem 4.2.1, $X_P^2(2|1)$ is asymptotically distributed as $\sum_{i=1}^{u}\hat{\delta}_i(2|1)\psi_i$ where ψ_i's are independent $\chi^2_{(1)}$ variables and $\hat{\delta}_i(2|1)$'s are the eigenvalues of the matrix $\hat{\mathbf{\Delta}}(2|1)$ given in (6.3.8). As stated before (see Eq. (6.2.12′), $\hat{\mathbf{V}}$ is the asymptotic estimated design covariance matrix of $\sqrt{n}\mathbf{p}$.

The $\hat{\delta}_i(2|1)$'s may be interpreted as the generalized deff's, with $\hat{\delta}_1(2|1)$ as the largest possible deff taken over all linear combinations of the elements of $\tilde{\mathbf{X}}_2'\mathbf{p}$.

Corollary 6.3.1.1 *For independent binomial sampling in each domain, $\hat{\mathbf{V}}$ reduces to $\mathbf{D}(\mathbf{w})^{-1}\hat{\mathbf{\Delta}}\mathbf{D}(\mathbf{w})^{-1}$ and hence $\hat{\delta}_i(2|1) = 1\ \forall\ i = 1, \ldots, u$ and we get the result $X_P^2(2|1) \sim \chi^2_{(u)}$ under $H_{2||1}$. It is well-known that $G^2(2|1)$ and $X_P^2(2|1)$ have the same asymptotic distribution.*

Note that the asymptotic distribution of $X_P^2(2|1)$ or $G^2(2|1)$ depend on $\mathbf{V}$ only through the eigenvalues $\delta_i(2|1)$'s.

Corollary 6.3.1.2 *The asymptotic distribution of X_P^2 and G^2 under the full model M_1, i.e., under model (6.2.3) can be obtained as a special case of the above result. For this, let $\mathbf{X}_1 = \mathbf{X}$ and $\mathbf{X}_2$ be any $I\times(I-p)$ matrix of rank $I-p$ so that $(\mathbf{X}_1, \mathbf{X}_2)(I\times I)$ is of rank I. Let $\mathbf{C} = \mathbf{\Delta}\tilde{\mathbf{X}}_2$ so that rank of $\mathbf{C}$ = rank $\tilde{\mathbf{X}}_2 = I-p$ and $\mathbf{C}'\mathbf{X} = \tilde{\mathbf{X}}_2'\mathbf{\Delta}\mathbf{X} = \mathbf{0}$. Hence,*

$$(\tilde{\mathbf{X}}_2'\hat{\mathbf{\Delta}}\tilde{\mathbf{X}}_2)^{-1}\{\tilde{\mathbf{X}}_2'\mathbf{D}(\mathbf{w})\hat{\mathbf{V}}\mathbf{D}(\mathbf{w})\tilde{\mathbf{X}}_2\} = (\mathbf{C}'\hat{\mathbf{\Delta}}^{-1}\mathbf{C})^{-1}(\mathbf{C}'\hat{\mathbf{\Delta}}^{-1}\mathbf{D}(\mathbf{w})\hat{\mathbf{V}}\mathbf{D}(\mathbf{w})\hat{\mathbf{\Delta}}\mathbf{C}\}.$$

Therefore, in the saturated case, X_P^2 or G^2 is asymptotically distributed as $\sum_i \delta_i\psi_i$ where the weights δ_i's are the eigenvalues of

$$(\mathbf{C}'\mathbf{\Delta}^{-1}\mathbf{C})^{-1}(\mathbf{C}'\mathbf{\Delta}^{-1}\mathbf{D}(\mathbf{W})\mathbf{V}\mathbf{D}(\mathbf{W})\mathbf{\Delta}^{-1}\mathbf{C}). \tag{6.3.19}$$

It can be easily seen that δ_i's are invariant to the choice of $\mathbf{C}$.

6.3.1 A Wald Statistic

Let $\mathbf{C}$ be a $I\times u$ matrix of rank u, such that $\mathbf{C}'\mathbf{X}_1 = \mathbf{0}$, $\mathbf{C}'\mathbf{1}_I = \mathbf{0}$, $\mathbf{C}'\mathbf{X}_2$ is non-singular, where $\mathbf{X}_1, \mathbf{X}_2$ have been defined in (6.3.1). From model (6.2.3),

$$\mathbf{C}'\nu = \phi \text{ (say) } = \mathbf{C}'\mathbf{X}_1\beta^{(1)} + \mathbf{C}'\mathbf{X}_2\beta^{(2)} = \mathbf{0}$$

under $H_{2|1} : \beta^{(2)} = \mathbf{0}$. Hence, $H_{2|1}$ is equivalent to the hypothesis $H_0 : \phi = \mathbf{0}$. Therefore, an alternative test statistic is Wald statistic,

$$\begin{aligned} X^2_W(2|1) &= \hat{\phi}'\hat{\mathbf{V}}^{-1}_{\hat{\phi}}\hat{\phi} \\ &= \hat{\nu}'\mathbf{C}(\mathbf{C}'\hat{\mathbf{V}}_{\hat{\nu}}\mathbf{C})^{-1}\mathbf{C}'\hat{\nu} \end{aligned} \tag{6.3.20}$$

where $\hat{\mathbf{V}}_{\hat{\nu}}$ is the estimated covariance matrix of $\hat{\nu}$ under the true sampling design $p(s)$.

If a consistent estimator $\hat{\mathbf{V}}_{\hat{\nu}}$ is not available, the effect of survey design is sometimes ignored. In this case, $\hat{\mathbf{V}}_{\hat{\nu}}$ is replaced by the covariance matrix under independent binomial sampling, which is asymptotically,

$$n^{-1}\hat{\mathbf{\Delta}}^{-1} = \hat{\mathbf{V}}_{\hat{\nu}0} \text{ (say)}. \tag{6.3.21}$$

As in the case of $X^2_P(2|1)$ or $G^2(2|1)$, the true asymptotic null distribution of $X^2_W(2|1)$ is a weighted sum of independent $\chi^2_{(1)}$ variables $\sum_{i=1}^{u} \gamma_i(2|1)\psi_i$, where $\gamma_i(2|1)$'s are the eigenvalues of $\hat{\mathbf{V}}^{-1}_{\hat{\phi}0}\hat{\mathbf{V}}_{\hat{\phi}}$, $\hat{\mathbf{V}}_{\hat{\phi}0}$ being defined similarly. It can shown that $X^2_W(2|1)$ is invariant to the choice of $\mathbf{C}$ subject to the above conditions. Under $H_{2|1}$, the statistic $X^2_W(2|1)$ is asymptotically distributed as a $\chi^2_{(u)}$ random variable. Also, $X^2_W(2|1)$ is well-defined even if p_i is 0 or 1 for some i, unlike X^2_W in (6.2.26).

6.3.2 Modifications to Tests

An adjustment to $X^2_P(2|1)$ or $G^2(2|1)$ is obtained by treating $X^2_P(2|1)/\hat{\delta}_0(2|1)$ or $G^2(2|1)/\hat{\delta}_0(2|1)$ as a $\chi^2_{(u)}$ random variable under $H_{2|1}$, where

$$u\hat{\delta}_0(2|1) = \sum_{i=1}^{u} \hat{\delta}_i(2|1) = n\sum_{i=1}^{I} \frac{\tilde{\tilde{V}}_{ii,r}w_i}{\{\tilde{\tilde{f}}_i(1-\tilde{\tilde{f}}_i)\}} \tag{6.3.22}$$

where $\tilde{\tilde{V}}_{ii,r}$ is the ith diagonal element of estimated covariance matrix of the residual $\tilde{\tilde{\mathbf{r}}} = \tilde{\mathbf{f}} - \tilde{\tilde{\mathbf{f}}}$ obtainable from (6.3.15) and (6.3.17) and is given by

$$\tilde{\tilde{\mathbf{V}}}_{\tilde{\tilde{\mathbf{r}}}} = n^{-1}\mathbf{D}(\mathbf{w})^{-1}\hat{\mathbf{\Delta}}\tilde{\mathbf{X}}_2\tilde{\tilde{\mathbf{A}}}\tilde{\mathbf{X}}_2'\hat{\mathbf{\Delta}}\mathbf{D}(\mathbf{w})^{-1}, \tag{6.3.23}$$

where

$$\tilde{\tilde{\mathbf{A}}} = (\tilde{\mathbf{X}}_2'\hat{\mathbf{\Delta}}\tilde{\mathbf{X}}_2)^{-1}\{\tilde{\mathbf{X}}_2'\mathbf{D}(\mathbf{w})\hat{\mathbf{V}}\mathbf{D}(\mathbf{w})\tilde{\mathbf{X}}_2\}(\tilde{\mathbf{X}}_2'\hat{\mathbf{\Delta}}\tilde{\mathbf{X}}_2)^{-1}. \tag{6.3.24}$$

As in (6.2.30), a better adjustment based on Sarrerthwaite approximation can be obtained, using the elements of $\tilde{\tilde{V}}_{\tilde{\tilde{\mathbf{r}}}}$.

Another adjustment is to treat

$$F = \frac{G^2(2|1)/u}{G^2/(I-p)} \tag{6.3.25}$$

as a $F_{u,(I-p)}$ variable under $H_{2|1}$. Rao and Scott (1987) have studied the behavior of this test. This test does not require the knowledge of any design effect.

6.4 Choosing Appropriate Cell-Sample Sizes for Running Logistic Regression Program in a Standard Computer Package

If we have an estimate $\hat{\mathbf{V}}$ of $\mathbf{V}$, covariance matrix of $\mathbf{p}$, a design-based estimate of the population cell-proportions $\boldsymbol{\pi}$, we can obtain a generalized least square estimate of $\boldsymbol{\beta}$ based on the empirical logits, $\hat{\nu}_i = \log[\hat{f}_i/(1-\hat{f}_i)]$ and the model (6.2.3). It follows from the assumption (6.2.12′) and the standard asymptotic theory that $\sqrt{n}(\hat{\boldsymbol{\nu}} - \boldsymbol{\nu}) \to^L N_I(\mathbf{0}, \mathbf{V}_{\hat{\nu}})$ where

$$\mathbf{V}_{\hat{\nu}} = \mathbf{F}^{-1}\mathbf{V}\mathbf{F}^{-1} \tag{6.4.1}$$

with $\mathbf{F} =$ Diag. $(f_i(1-f_i))$. The generalized least square estimate of $\boldsymbol{\beta}$ is

$$\hat{\boldsymbol{\beta}}_G = (\mathbf{X}'\hat{\mathbf{V}}_{\hat{\nu}}^{-1}\mathbf{X})^{-1}\mathbf{X}'\hat{\mathbf{V}}_{\hat{\nu}}^{-1}\hat{\boldsymbol{\nu}} \tag{6.4.2}$$

with estimated covariance matrix $(\mathbf{X}'\hat{\mathbf{V}}_{\hat{\nu}}^{-1}\mathbf{X})^{-1}$. Asymptotic tests for linear hypotheses about $\boldsymbol{\beta}$ can then be produced immediately (Koch et al. 1975).

All these results hold if a good estimate of $\mathbf{V}$, the covariance matrix of $\mathbf{p}$ is available. Such estimates are, however, rarely available. Even if an estimate $\hat{\mathbf{V}}$ is available, it will usually be available using a random group method or a sampling design with a small number of primary stage units per stratum. In any case, the degrees of freedom of the estimate will be relatively low and $\mathbf{V}_{\hat{\nu}}^{-1}$ will be rather unstable. For these reasons, investigators often simply run their data through the logistic regression program in a standard computer package. Typically these packages produce the MLE of $\boldsymbol{\beta}$ along with its estimated covariance matrix and the likelihood ratio test statistic for the hypothesis $H_{2|1} : \boldsymbol{\beta}^{(2)} = \mathbf{0}$ in the model (6.3.1) under the assumption of an independent binomial sample of n_i observations in the ith cell ($i = 1, \ldots, I$).

Scott (1989) examined the consequences of using pseudo-cell-sample sizes in a standard logistic regression computer program. Let $\tilde{\boldsymbol{\beta}}$ be the pseudo *mle* of $\boldsymbol{\beta}$ obtained by running the observed vector of proportions, together with a vector of pseudo-sample sizes $\tilde{\mathbf{n}} = (\tilde{n}_1, \ldots, \tilde{n}_I)'$ through a standard package. As depicted above asymptotic properties of $\tilde{\beta}$ with $\tilde{n}_i = n_i = nw_i$ where $w_i = n_i/n$, an estimator of the cell relative size $W_i = N_i/N$ have been developed by Roberts et al. (1987) using the methods in Rao and Scott (1984) for general log-linear models. As stated by Scott (1989), the same method holds for more general choices of $\tilde{n}_i$ so long as $\tilde{n}_i/n \to \tilde{w}_i$ with $0 < \tilde{w}_i < 1$ as $n \to \infty$. Writing

$$\begin{aligned} D(\tilde{\mathbf{w}}) &= \text{Diag. } (\tilde{w}_1, \ldots, \tilde{w}_I), \\ \tilde{\Delta} &= \text{Diag. } (\tilde{w}_i f_i(1 - f_i)) \end{aligned} \tag{6.4.3}$$

the asymptotic covariance matrix of $\tilde{\beta}$ is now,

$$D(\tilde{\beta}) = (\mathbf{X}'\tilde{\Delta}\mathbf{X})^{-1}\{\mathbf{X}'\mathbf{D}(\tilde{\mathbf{w}})\mathbf{V}\mathbf{D}(\tilde{\mathbf{w}})\mathbf{X}\}(\mathbf{X}'\tilde{\boldsymbol{\Delta}}\mathbf{X})^{-1}/n. \tag{6.4.4}$$

The last factor in (6.4.4) is the asymptotic covariance matrix of $\tilde{\beta}$ under the standard assumptions with $\tilde{n}_i = n\tilde{w}_i$. Hence the product of the first two factors is the adjustment needed to be applied to the output from a standard package to make room for the complexity of the sampling design.

The choice of $\tilde{n}_i$ may considerably affect the properties of the resulting estimator. Common choices are actual sample sizes n_i or if these are not known, $n\hat{\pi}_i$ where $\hat{\pi}_i$ is some consistent estimator of π_i. If the covariance structure of $\mathbf{p}$ is known it is possible to improve the performance. For example if the estimated cell variances v_{ii} are known one can take $\tilde{n}_i = f_i(1 - f_i)/v_{ii}$ in which case diagonal elements of $\mathbf{D}(\tilde{\mathbf{w}})\mathbf{V}\mathbf{D}(\tilde{\mathbf{w}})$ will be identical to those of $\tilde{\boldsymbol{\Delta}}$. Scott (1989) reported that this choice of $\tilde{n}_i$ have worked well in many situations making modifications in standard computer output almost unnecessary.

Similar results also hold for testing the hypothesis $H_{2|1} : \beta^{(2)} = \mathbf{0}$ under the model M_2 nested within model M_1 as depicted in Eqs. (6.3.1) and (6.3.2).

6.5 Model in the Polytomous Case

Morel (1989) considered the case where the dependent variable has more than two categories.

Consider first-stage cluster sampling where n clusters or first stage units are selected with known probabilities with replacement from a finite population or without replacement from a very large population.

Let m_i be the number of units sampled from the ith cluster and $\mathbf{y}^*_{ik}(k = 1, \ldots, m_i)$, the $(d+1)$-dimensional classification vector. The vector $\mathbf{y}^*_{ik}$ consists entirely of zeros except for position r which will contain a 1 if the kth unit from the ith cluster falls in the rth category, i.e., $y_{ikt} = 1$ if $t = r$; $= 0$ otherwise $(r = 1, 2, \ldots, d + 1)$. Let also $\mathbf{x}_{ik} = (x_{ik1}, \ldots, x_{ikp})'$ be a p-dimensional vector of values of explanatory variables $\mathbf{x} = (x_1, \ldots, x_p)'$ associated with the kth unit in the ith cluster. Note that unlike the model (6.2.3), here different units in the ith cluster may have different values of $\mathbf{x}$.

Consider the following logistic regression model.

$$\pi_{ikt} = \pi_{ikt}(\mathbf{x}_{ik}) = P\{y_{ikt} = 1|\mathbf{x}_{ik}\}$$
$$= \begin{cases} \frac{\exp(\mathbf{x}'_{ik}\beta^0_t)}{1+\sum_{s=1}^d \exp(\mathbf{x}'_{ik}\beta^0_s)}, & t = 1, \ldots, d \\ 1 - \sum_{s=1}^d \pi_{iks}, & t = d+1. \end{cases} \tag{6.5.1}$$

Note that for each category t, there is associated a regression vector $\beta^0_t (t = 1, \ldots, d)$. Let

$$\beta^0 = (\beta^{0'}_1, \ldots, \beta^{0'}_d)'_{pd\times 1} \tag{6.5.2}$$

Since, $y_{ikd+1} = 1 - \sum_{s=1}^d y_{iks}$, the contribution to the log-likelihood, $\log \tilde{L}_n(\beta^0)$ by subject in the (i, k) is

$$\log[\Pi_{t=1}^{d+1} \pi_{ikt}(\mathbf{x}_{ik})^{y_{ikt}}]$$

$$\sum_{t=1}^{d} y_{ikt} \log \pi_{ikt}(\mathbf{x}_{ik}) + \left(1 - \sum_{t=1}^{d} y_{ikt}\right) \log\left(1 - \sum_{t=1}^{d} \pi_{ikt}(\mathbf{x}_{ik})\right)$$

$$= \sum_{t=1}^{d} y_{ikt} \log \frac{\pi_{ikt}(\mathbf{x}_{ik})}{1 - \sum_{s=1}^d \pi_{iks}(\mathbf{x}_{ik})} + \log\left[1 - \sum_{t=1}^{d} \pi_{ikt}(\mathbf{x}_{ik})\right]. \tag{6.5.3}$$

Now,

$$1 - \sum_{t=1}^{d} \pi_{ikt}(\mathbf{x}_{ik}) = \left[1 + \sum_{t=1}^{d} \exp(\mathbf{x}'_{ik}\beta^0_t)\right]^{-1}. \tag{6.5.4}$$

Hence,

$$\log \frac{\pi_{ikt}(\mathbf{x}_{ik})}{1 - \sum_{s=1}^d \pi_{iks}(\mathbf{x}_{ik})} = \mathbf{x}'_{ik}\beta^0_t, \quad t = 1, \ldots, d. \tag{6.5.5}$$

Substituting from (6.5.4) and (6.5.5), (6.5.3) reduces to

$$\sum_{t=1}^{d} y_{ikt}(\mathbf{x}'_{ik}\beta^0_t) - \log\left\{1 + \sum_{t=1}^{d} \exp(\mathbf{x}'_{ik}\beta^0_t)\right\}. \tag{6.5.6}$$

Writing $\beta_t^0 = (\beta_{t1}, \ldots, \beta_{tp})'$, $\mathbf{x}'_{ik}\beta_t^0 = \sum_{j=1}^{p} x_{ikj}\beta_{tj}$. Log-likelihood is therefore,

$$\begin{aligned}\log \tilde{L}_n(\beta^0) &= \sum_{i=1}^{n}\sum_{k=1}^{m_i}\left[\sum_{t=1}^{d} y_{ikt}\sum_{j=1}^{p} x_{ikj}\beta_{tj}\right. \\ &\quad \left. - \log\left\{1+\sum_{t=1}^{d}\exp\left(\sum_{j=1}^{p} x_{ikj}\beta_{tj}\right)\right\}\right] \\ &= \sum_{t=1}^{d}\left[\sum_{j=1}^{p}\beta_{tj}\left(\sum_{i=1}^{n}\sum_{k=1}^{m_i} x_{ikj}y_{ikt}\right)\right. \\ &\quad \left. - \sum_{i=1}^{n}\sum_{k=1}^{m_i}\log\left\{1+\sum_{t=1}^{d}\exp\left(\sum_{j=1}^{p} x_{ikj}\beta_{tj}\right)\right\}\right].\end{aligned} \tag{6.5.7}$$

The author considered the pseudo-likelihood as

$$\propto \Pi_{i=1}^{n}\Pi_{k=1}^{m_i}\{\Pi_{t=1}^{d+1}\pi_{ikt}(\mathbf{x}_{ik})^{y_{ikt}w_i}\} \tag{6.5.8}$$

where w_i is the sample weight attached to the ith cluster. Therefore pseudo-log-pseudo likelihood

$$\log L_n(\beta^0) \propto \sum_{i=1}^{n}\sum_{k=1}^{m_i}\left\{\sum_{t=1}^{d+1} w_i y_{ikt}\log(\pi_{ikt})\right\}. \tag{6.5.9}$$

Writing $\log(\pi_{ik}^*) = (\log\pi_{ik1}, \ldots, \log\pi_{ikd+1})'$, (6.5.9) reduces to

$$\sum_{i=1}^{n}\sum_{k=1}^{m_i} w_i(\log\pi_{ik}^*)'\mathbf{y}_{ik}^*. \tag{6.5.10}$$

The function can be viewed as the weighted likelihood function, where the $\mathbf{y}_{ik}^*$ are distributed as multinomial random variables. If the sampling weights w_j are all one, then (6.5.10) becomes the likelihood function under the assumption that $\mathbf{y}_{ik}^*$ are independently multinomially distributed. For further details, the reader may refer to Morel (1989).

6.6 Analysis Under Generalized Least Square Approach

Consider first the model (6.2.3). Suppose, therefore, there are I domains, the ith domain being characterized by the value $\mathbf{x}_i = (x_{i1}, \ldots, x_{ip})'$ of a auxiliary random vector $\mathbf{x} = (x_1, \ldots, x_p)'$.

Suppose first that the dependent variable Y is dichotomous, taking values 0 or 1. Hence, there are $2I$ cells. We assume

$$
\begin{aligned}
P\{Y = 1|\mathbf{x} = \mathbf{x}_i\} &= \pi_i \\
P\{Y = 0|\mathbf{x} = \mathbf{x}_i\} &= 1 - \pi_i.
\end{aligned}
\tag{6.6.1}
$$

We have already defined the logistic regression model,

$$
\nu_i = \log \frac{\pi_i}{1 - \pi_i} = \mathbf{x}_i'\beta, \quad i = 1, \ldots, I.
$$

Hence, the model is

$$
\boldsymbol{\nu} = \mathbf{X}\beta \tag{6.6.2}
$$

where $\mathbf{X}$ is a $I \times p$ matrix of values of $\mathbf{x}$ and $\beta = (\beta_1, \ldots, \beta_p)'$ is the vector of logistic regression parameters.

Now, $\boldsymbol{\nu}$ cannot be observed and we replace it by a vector of empirical logits $\mathbf{L} = (l_1, \ldots, l_I)'$ where $l_i = \log\{\hat{\pi}_i/(1 - \hat{\pi}_i)\}$. If there are n_i observations in the domain i, of which m_i observations belong to category 1 of Y, then $l_i = \log\{m_i/(n_i - m_i)\}$. Hence, corresponding to (6.6.2), we have the empirical logistic regression model

$$
\mathbf{L} = \mathbf{X}\boldsymbol{\beta} + \mathbf{u} \tag{6.6.3}
$$

where $\mathbf{u}$ is a random vector with mean $\mathbf{0}$. To find its variance, we first find $Var(l_i)$.

Setting $m_i/n_i = r_i$, we have $E(r_i) = \pi_i$. Again, $l_i = h(r_i)$ where $h(.)$ denotes a function of (.). Provided that the variations in r_i is relatively small, we can write

$$
h(r_i) \approx h(\pi_i) + (r_i - \pi_i)h'(r_i) \tag{6.6.4}
$$

from which it can be shown that $h(r_i)$ is approximately normally distributed with mean $h(\pi_i)$ and variance $\{h'(r_i)\}^2\big]_{r_i=\pi_i} Var(r_i)$. For the present case, $Var(l_i) = (n_i\pi_i(1 - \pi_i))^{-1}$. Also, sampling being independent from domain to domain, Cov $(l_i, l_j) = 0, i \neq j$. Therefore

$$
Var(\mathbf{u}) = \text{ Diag. } \left[\frac{1}{n_i}\left(\frac{1}{\pi_i} + \frac{1}{1 - \pi_i}\right), i = 1, \ldots, I\right] = \mathbf{\Sigma}_{I\times I} \text{ (say)}. \tag{6.6.5}
$$

Suppose now the dependent variable Y is polytomous with $d + 1$ categories $\{0, 1, \ldots, d\}$ and let l_{ik} be empirical logit corresponding to the baseline logit

$$
\nu_{ik} = \log \frac{\pi_{ik}}{\pi_{i0}} \tag{6.6.6}
$$

where

$$
\pi_{ik} = P\{Y \text{ belongs to category k } |\mathbf{x} = \mathbf{x}_i\}, i = 1, \ldots, I; k = 1, \ldots, d \tag{6.6.7}
$$

and $\pi_{i0} = 1 - \sum_{s=1}^{d} \pi_{is}$ being defined similarly. The model is therefore,

$$l_{ik} = \mathbf{x}_i'\beta^{(k)} + u_{ik}, i = 1, \ldots, I; k = 1, \ldots, d. \tag{6.6.8}$$

Writing

$$\mathbf{L} = (l_{11}, \ldots, l_{1d}, l_{21}, \ldots, l_{2d}, \ldots, l_{I1}, \ldots, l_{Id})'_{Id \times 1}$$

$$\mathbf{B} = (\beta^{(1)'}, \beta^{(2)'}, \ldots, \beta^{(d)'})'_{pd \times 1}$$

$$\mathbf{U} = (u_{11}, \ldots, u_{1d}, u_{21}, \ldots, u_{2d}, \ldots, u_{I1}, \ldots, u_{Id})' \tag{6.6.9}$$

$$\mathbf{X} = \begin{bmatrix} \mathbf{1}_d \otimes \mathbf{x}_1' \\ \mathbf{1}_d \otimes \mathbf{x}_2' \\ \cdot \\ \mathbf{1}_d \otimes \mathbf{x}_I' \end{bmatrix}_{(dI) \times (dp)},$$

where $\mathbf{A} \otimes \mathbf{B}$ denotes direct product between matrices $\mathbf{A}$ and $\mathbf{B}$, we have the model

$$\mathbf{L} = \mathbf{XB} + \mathbf{U} \tag{6.6.10}$$

with

$$\text{Cov } (\mathbf{U}) = \oplus_{i=1}^{I} \mathbf{\Sigma}_i = \mathbf{\Sigma} \quad \text{(say)}, \tag{6.6.11}$$

$\mathbf{\Sigma}_i$ being the Cov $(\mathbf{l}_i)$, $\mathbf{l}_i = (l_{i1}, \ldots, l_{id})'$. Here, $\oplus_{i=1}^{a} \mathbf{A}_i$ denotes the block-diagonal matrix whose diagonal-elements are $\mathbf{A}_i$. The model (6.6.10) is in the general linear form and $\mathbf{\Sigma}$ is a function of the unknown cell-probabilities $\pi_{ik}, i = 1, \ldots, I; k = 1, \ldots, d$ only. For estimation purposes, an iterative form of the generalized least squares is, therefore, needed. The usual procedure is to choose an initial set of values of the elements of the covariance matrix, estimate the logistic regression parameters, subsequently refine the variance matrix and iteratively estimate the β's.

We now assume that the above-mentioned sampling is done for m different clusters, so that the empirical logits may now accommodate a cluster effect. Suppose, therefore, that the observations from the cth cluster contains a random cluster effect term η_c with $E(\eta_c) = 0$, $Var(\eta_c) = w^2$ and $\text{Cov}(\eta_c, \eta_{c'}) = 0, c \neq c' = 1, \ldots, m$.

If the empirical logits are calculated for each cluster separately, then for the dichotomous case, the model is

$$\mathbf{L}_{Im \times 1} = \mathbf{X}_{Im \times p}\beta_{p \times 1} + \mathbf{u}_{Im \times 1} + \boldsymbol{\eta}_{Im \times 1}. \tag{6.6.12}$$

Here **X** is made up of m identical $(I \times p)$ subvectors described in Eq. (6.6.2), **L** and **u** defined suitably and

$$\boldsymbol{\eta}' = (\eta_1 \mathbf{1}_I', \eta_2 \mathbf{1}_I', \ldots, \eta_m \mathbf{1}_I'). \tag{6.6.13}$$

Note that the cluster effect η_c does not depend on the domain within cluster c.

For the polytomous case, let η_{kc} denote the cluster effect of category k of Y in cluster c. Again, in this set-up, the cluster effect does not depend on the domain $i(= 1, \ldots, I)$. Now, $\boldsymbol{\eta}(mId \times 1) = (\eta_1', \ldots, \eta_c', \ldots, \eta_m')'$ where $\eta_c(ID \times 1) = (\tilde{\eta}_c', \tilde{\eta}_c', \ldots, \tilde{\eta}_c')'$, that is in η_c, $\tilde{\eta}_c$ is repeated I times, $\tilde{\eta}_c(d \times 1) = (\eta_{1c}, \ldots, \eta_{dc})'$. Thus

$$\boldsymbol{\eta}' = (\mathbf{1}_I' \otimes (\eta_{11}, \ldots, \eta_{d1}), \ldots, \mathbf{1}_I' \otimes (\eta_{1c}, \ldots, \eta_{dc}), \ldots, \mathbf{1}_I' \otimes (\eta_{1m}, \ldots, \eta_{dm}))_{1 \times (mId)}. \tag{6.6.14}$$

The models (6.6.13) and (6.6.14) are constant cluster effect model. It has been assumed here that Var(η_{kc}) or Var(η_c) is a constant for all c. Also, it has been assumed that the logits in the same cluster are affected in the same way for all domains by a common cluster effect η_{kc} (or η_c). These assumptions may not always hold in practice.

A more flexible approach will be to allow η_{ic} in place of η_c for binary case and η_{ikc} for η_{kc} for polytomous case. Thus, for the dichotomous case, we assume $\boldsymbol{\eta}' = (\boldsymbol{\eta}_1', \ldots, \boldsymbol{\eta}_c', \ldots, \boldsymbol{\eta}_m')$ where $\boldsymbol{\eta}_c' = (\eta_{1c}, \ldots, \eta_{Ic})$ with Cov $(\eta_{ic}, \eta_{i'c}) = w_{ii'}$, Cov $(\eta_{ic}, \eta_{i'c'}) = 0$ for all $i, i', c(\neq)c'$. Here

$$\begin{aligned} \text{Var }(\mathbf{u}) &= \boldsymbol{\Sigma} = \otimes_{i=1}^{I} \otimes_{c=1}^{m} \sigma_{ic}/n_{ic} \\ \text{Var }(\boldsymbol{\eta}) &= \mathbf{W} = \otimes_{c=1}^{m} \mathbf{W}_c(I \times I) \end{aligned} \tag{6.6.15}$$

where n_{ic} is the number of observations in the ith domain of cluster c. The case of polytomous response variable can be treated similarly. In general, the model is, therefore,

$$\mathbf{L} = \mathbf{XB} + \boldsymbol{\eta} + \mathbf{U},$$

$$Var(\boldsymbol{\eta}) = \mathbf{W}, \; Var(\mathbf{U}) = \boldsymbol{\Sigma}, \; Var(\mathbf{L}) = \mathbf{W} + \boldsymbol{\Sigma} = \boldsymbol{\Omega} \tag{6.6.16}$$

where **L**, **B**, $\boldsymbol{\eta}$ and **U** are column vectors of suitable lengths and **X** is suitably defined as in (6.6.10). To test hypothesis $H_0 : \mathbf{CB} = \mathbf{r}$, the test statistic is

$$X^2 = (\mathbf{C}\hat{\mathbf{B}} - \mathbf{r})'[\mathbf{C}(\mathbf{X}'\hat{\boldsymbol{\Omega}}^{-1}\mathbf{X})^{-1}\mathbf{C}']'(\mathbf{C}\hat{\mathbf{B}} - \mathbf{r}) \tag{6.6.17}$$

which follows central chi-square with appropriate d.f. under null hypothesis.

A practical problem is that in many situations, there will be some domains in some clusters for which $n_{ic} = 0$ or a very small quantity. In such circumstances, one calculates empirical logits by collapsing across clusters,

$$l_{ik} = \log\left[\frac{\sum_c n_{ic}\hat{\pi}_{ikc}}{\sum_c n_{ic}(1-\hat{\pi}_{i0c})}\right]. \tag{6.6.18}$$

For further details in this area, the reader may refer to Holt and Ewings (1989).

Note 6.6.1 In many sample surveys, there are items that require individuals in different strata to make at least one of a number of choices. Within each strata, we assume that each individual has made all his/her choices, and the number of individuals with none of these choices is not reported. The analysis of such survey data are complex, and the categorical table with mutually exclusive categories can be sparse. Nandram (2009) used a simple Bayesian product multinomial-Dirichlet model to fit the count data both within and across the strata. Using the Bayes factor, the author shows how to test that the proportion of individuals with each choice are the same over the strata.

6.7 Exercises and Complements

6.1: Prove the relation (6.2.27).

6.2: Prove the relation (6.2.29).

6.3 Suppose we write the empirical logits in the multiresponse case of Sect. 6.6 in the following form. Let

$$\begin{aligned}\mathbf{L}_{I\times d} &= ((l_{ik})) = [\mathbf{l}_{(1)},\ldots,\mathbf{l}_{d)}]\\ \mathbf{B}_{p\times d} &= [\beta^{(1)},\ldots,\beta^{(d)}],\\ \mathbf{U}_{I\times d} &= [\mathbf{u}_{(1)},\ldots,\mathbf{u}_{(d)}]\end{aligned} \tag{i}$$

where u_{ik}'s are the random error terms. Then the model is

$$\mathbf{L} = \mathbf{XB} + \mathbf{U}. \tag{ii}$$

We assume that the rows $\mathbf{u}_i'(i = 1,\ldots,I)$'s of $\mathbf{U}$ are uncorrelated and $E(\mathbf{u}_i) = \mathbf{0}$ and Cov $(\mathbf{u}_i) = \mathbf{\Sigma}_i$. Find the least square estimate of $\mathbf{B}$.

Again, writing

$$\begin{aligned}\mathbf{L}^V &= (\mathbf{l}'_{(1)},\ldots,\mathbf{l}'_{(I)})'_{Id\times 1}\\ \mathbf{B}^V &= (\beta^{(1)'},\ldots,\beta^{(d)'})'_{pd\times 1}\\ \mathbf{U}^V &= (\mathbf{u}'_{(1)},\ldots,\mathbf{u}'_{(I)})'_{Id\times 1}\end{aligned} \tag{iii}$$

show that the model in (ii) reduces to

$$\mathbf{L}^V = \mathbf{X}^*\mathbf{B} + \mathbf{U}^V \qquad (iv)$$

where

$$\mathbf{X}^* = \mathbf{I}_{d\times d} \otimes \mathbf{X}_{I\times p},$$

$\otimes$ denoting the direct product operator. Then

$$E(\mathbf{U}^V) = \mathbf{0}, \;\; \text{Cov } (\mathbf{U}^V) = \oplus_{i=1}^{I} \boldsymbol{\Sigma}_i = \boldsymbol{\Omega} \text{ (say)} \qquad (v)$$

where $\oplus$ denotes the block diagonal matrix.

Show that the generalized least square estimator of $\mathbf{B}^V$ is

$$\begin{aligned}\hat{\mathbf{B}}^V &= (\mathbf{X}^{*'}\boldsymbol{\Omega}^{-1}\mathbf{X}^*)^{-1}(\mathbf{X}^{*'}\boldsymbol{\Omega}^{-1}\mathbf{L}^V)\\ &= [\mathbf{I}_d \otimes (\mathbf{X}'\mathbf{X})^{-1}\mathbf{X}']\mathbf{L}^V.\end{aligned} \qquad (vi)$$

(Holt and Ewings 1989)

6.4: Find out the expression for $\boldsymbol{\Sigma}_i$ and of $\boldsymbol{\Sigma}$ in (6.6.11).

Chapter 7
Analysis in the Presence of Classification Errors

Abstract So far we have assumed that there was no error in classifying the units according to their true categories. In practice, classification errors may be present and in these situations usual tests of goodness-of-fit, independence, and homogeneity become untenable. This chapter considers modifications of the usual test procedures under this context. Again, units in a cluster are likely to be related. Thus in a cluster sampling design where all the sampled clusters are completely enumerated, Pearson's usual statistic of goodness-of-fit seems unsuitable. This chapter considers modification of X_P^2 statistic under these circumstances.

Keywords Classification error · Misclassification probabilities · Double sampling · Weighted cluster sampling design · Models of association

7.1 Introduction

In the previous chapters we have tried to deal extensively with methods for analyzing categorical data generated from field surveys under complex survey designs, but did not assume the presence of any error in classifying the units in different categories. However, in practice data items themselves may be subject to classification errors and these may bias the estimated cell-probabilities and distort the properties of relevant test statistics. Rao and Thomas (1991), Heo (2002), and others considered this problem and suggested modification to usual tests of goodness-of-fit, independence, and homogeneity carried out when there is no misclassification.

Also, in cluster sampling, where each selected cluster is enumerated completely, units belonging to the same (sampled) cluster are likely to be related to each other. If the units are again grouped in different categories, their category membership is also likely to be related. Thus in cluster sampling conventional Pearson statistic of goodness-of-fit may not be suitable. For these situations Cohen (1976) and others suggested some modified procedures. In this chapter we make a brief review of these results.

P. Mukhopadhyay, *Complex Surveys*, DOI 10.1007/978-981-10-0871-9_7

7.2 Tests for Goodness-of-Fit Under Misclassification

Consider a finite population of size N whose members are divided into T classes with unknown proportions $\pi_1, \ldots, \pi_T$ $(\sum_{i=1}^{T} \pi_k = 1)$, $\pi_i = N_i/N$, N_i being the unknown number of units in the population in class i. A sample of size n is drawn from this population following a sampling design $p(s)$, for example, a stratified multistage design. Let $n_1, \ldots, n_T$ $(\sum_k n_k = n)$ denote the observed cell frequencies in the sample.

We want to test the hypothesis $H_0 : \pi_i = \pi_{i0}, i = 1, \ldots, T$ where the π_{i0} are the hypothesized proportions. This is the usual goodness-of-fit hypothesis which has been extensively dealt within Chap. 4.

Suppose now that there are errors in classification so that a unit which actually belongs to the category j is erroneously assigned to category k.

Let $\boldsymbol{\pi} = (\pi_1, \ldots, \pi_T)'$ be the unknown true cell-probabilities and $\mathbf{p} = (p_1, \ldots, p_T)'$ be the observed proportions (probabilities) in different cells. Let also $\mathbf{A} = ((a_{jk}))$ be the $T \times T$ matrix of classification probabilities whose (j, k)th element $a_{j,k}$ denotes the probability that an observation which belongs truly to the category j is erroneously assigned to the category k. It is readily seen that

$$\mathbf{p} = \mathbf{A}'\boldsymbol{\pi} \tag{7.2.1}$$

We are to test the null hypothesis $H_0 : \boldsymbol{\pi} = \boldsymbol{\pi}_0$ which by virtue of (7.2.1) is equivalent to the hypothesis

$$H_0' : \mathbf{p} = \mathbf{p}_0$$

where $\mathbf{p}_0 = \mathbf{A}'\boldsymbol{\pi}_0$.

7.2.1 *Methods for Considering Misclassification Under SRS*

We now consider tests for goodness-of-fit when the units are selected by *srswr*.

Case (i): **A** *known*: In this situation Mote and Anderson (1965) ignored the effects of misclassification and considered the usual Pearson statistics

$$X_P^2 = \sum_{i=1}^{T} (n_i - n\pi_{i0})^2/(n\pi_{i0}) \tag{7.2.2}$$

and showed that X_P^2 in general leads to the inflated type I error. They proposed an alternative test statistic

$$X_P'^2 = \sum_{i=1}^{T} (n_i - np_{i0})^2/(np_{i0}) \tag{7.2.3}$$

as a $\chi^2_{(T-1)}$ variable. They showed that this test is asymptotically correct, but its asymptotic power under misclassification is less than the asymptotic power of X_P^2 with no misclassification.

Case (ii): **A** *unknown*: In this case Mote and Anderson (1965) proposed two different models depicting relation between $\boldsymbol{\pi}$ and $\mathbf{p}$: (a) the misclassification rate is same for all the classes; (b) misclassification occurs only in adjacent categories and at a constant rate. For model (a) they considered the relation

$$p_{i0} = \pi_{i0} + \theta(1 - T\pi_{i0}) \quad (7.2.4)$$

where θ is an unknown parameter ($0 < \theta < 1/T$) and $T \geq 3$. They also assumed that at least two of the π_{i0} are not equal to $1/T$. For model (b) they considered similar relations between p_{i0} and π_{i0}'s, all involving an unknown parameter θ. These are

$$\begin{aligned} p_{10} &= \pi_{10}(1-\theta) + \pi_{20}\theta \\ p_{i0} &= \pi_{i-10}\theta + \pi_{i0}(1-2\theta) + \pi_{i+10}\theta, \quad i = 2, \ldots, T-1 \\ p_{T0} &= \pi_{T-10}\theta + \pi_{T0}(1-\theta), \end{aligned}$$

where θ is an unknown parameter.

They showed that for model (a), the *ml* equations for θ is

$$\sum^{\prime} \left[\frac{n_i}{\theta - q_{i0}} \right] = 0 \quad (7.2.5)$$

where $q_{i0} = \pi_{i0}/(T\pi_{i0} - 1)$, $\pi_{i0} \neq T^{-1}$ and the summation in (7.2.5) is over those i's for which $\pi_{i0} \neq T^{-1}$.

For model (b), the *ml* equation is also given by

$$\sum_{i=1}^{T} \left[\frac{n_i}{\theta - q_{i0}} \right] = 0 \quad (7.2.6)$$

where

$$\begin{aligned} q_{10} &= \pi_{10}/(\pi_{10} - \pi_{20}) \\ q_{i0} &= \pi_{i0}/(2\pi_{i0} - \pi_{i-1,0} - \pi_{i+1,0}), \quad i = 2, \ldots T-1 \\ q_{T0} &= \pi_{T0}/(\pi_{T0} - \pi_{T-1,0}). \end{aligned} \quad (7.2.7)$$

For both the models they obtained $\hat{\theta}$, *mle*'s for θ and thus obtained the estimates $\hat{p}_{i0} = p_{i0}(\hat{\theta})$. An asymptotically correct test of H_0 is obtained by treating

$$\hat{X}_P^{\prime 2} = \sum_{i=1}^{T} (n_i - n\hat{p}_{i0})^2/(n\hat{p}_{i0}) \quad (7.2.8)$$

as a $\chi^2_{(T-2)}$ variable.

7.2.2 Methods for General Sampling Designs

Rao and Thomas (1991) considered the effects of survey designs on goodness-of-fit tests when misclassification is present and the sampling design is arbitrary.

Case (i): **A** *known*: They considered

$$\tilde{X}_P'^2 = n \sum_{i=1}^{T} (\hat{\pi}_i - p_{i0})^2 / p_{i0} \tag{7.2.9}$$

which is obtained from (7.2.3) by substituting the survey estimates $\hat{\pi}_i$ for n_i/n. In this case the survey estimate $\hat{\boldsymbol{\pi}}$ is a consistent estimate of the category probabilities $\mathbf{p}$ under misclassification. Under some regularity conditions $\hat{\boldsymbol{\pi}}^{(1)} = (\hat{\pi}_1, \ldots, \hat{\pi}_{T-1})'$ is asymptotically distributed as $N_{T-1}(\mathbf{p}_0^{(1)}, \mathbf{V})$ under H_0 where $\mathbf{V}$ is the $(T-1) \times (T-1)$ covariance matrix of $\hat{\boldsymbol{\pi}}^{(1)}$ and $\mathbf{p}^{(1)}$ is defined similarly. Therefore, it readily follows from Rao and Scott (1981) that their first-order corrected test statistic

$$\tilde{X}_{P(c)}'^2 = \tilde{X}_P'^2 / \hat{\delta}_0' \tag{7.2.10}$$

is a $\chi^2_{(T-1)}$ variable with $\hat{\delta}_0'$, mean of the estimated eigenvalues $\hat{\delta}_i'$'s is given by

$$\hat{\delta}_0' = \frac{1}{T-1} \sum_{i=1}^{T} \frac{\hat{\pi}_i}{p_{i0}} (1 - \hat{\pi}_i) \hat{d}_i \tag{7.2.11}$$

where $\hat{d}_i$ is the deff of $\hat{\pi}_i$ (compare with (4.2.31)). $\tilde{X}_{P(c)}'^2$ has a remarkable advantage for secondary data analysis in which it depends only on the estimated cell proportions $\hat{\pi}_{i0}$ and the cell deffs $\hat{d}_i$, which are often available from published tables.

Similarly, the second-order corrected test statistic is

$$\tilde{X}_{P(cc)}'^2 = \frac{\tilde{X}_P'^2}{\hat{\delta}_0'(1 + \hat{a}'^2)} = \frac{\tilde{X}_{P(c)}'^2}{1 + \hat{a}'^2} \tag{7.2.12}$$

as $\chi^2_{(\nu)}$ where $\hat{a}'^2$ is given as in (4.2.34) and $\nu = (T-1)/(1 + \hat{a}'^2)$.

Case (ii): **A** *unknown*: For general sampling designs it is difficult to find *ml* equations for θ for both models (a) and (b). Rao and Thomas (1991), therefore, found a 'pseudo' *mle* of θ obtained from (7.2.6) and (7.2.7) by replacing n_i/n with the survey estimates $\hat{\pi}_i$. The asymptotic consistency of $\hat{\boldsymbol{\pi}}$ ensures that of $\hat{\theta}$ and hence that of $\hat{p}_{i0}(\hat{\theta})$.

The uncorrected test is given by

$$\hat{\tilde{X}}_P'^2 = n \sum_{i=1}^{T} (\hat{\pi}_i - \hat{p}_{i0})^2 / \hat{p}_{i0}. \tag{7.2.13}$$

A first-order corrected statistic is given by

$$\hat{\hat{X}}'^2_{P(c)} = \hat{\hat{X}}'^2_P / \hat{\hat{\delta}}'_0 \tag{7.2.14}$$

where the correction factor $\hat{\hat{\delta}}'_0$ is the mean of the nonzero eigenvalues of the generalized design matrix and is given by

$$\hat{\hat{\delta}}'_0 = \frac{1}{T-2}\left[\sum_{i=1}^{T} \frac{\hat{\pi}_i}{\hat{p}_{i0}}(1-\hat{\pi}_i)\hat{d}_i - \text{deff}\left(\sum \frac{\pi_{i0}}{p_{i0}}\hat{\pi}_i\right)\right] \tag{7.2.15}$$

where deff $(\sum \frac{\pi_{i0}}{p_{i0}}\hat{\pi}_i)$ depends on the estimated full covariance matrix $\hat{\mathbf{V}}$ of $(\hat{\pi}_1, \ldots, \hat{\pi}_{T-1})'$. The test that treats $\hat{\hat{X}}'^2_{P(c)}$ as $\chi^2_{(T-2)}$ therefore requires the knowledge of $\hat{\mathbf{V}}$. Omitting the deff terms in (7.2.15) and calling it as $\hat{\delta}_0(\alpha)$ we get a conservative test that depends only on the cell deffs $\hat{d}_i$'s. This is

$$\hat{\hat{X}}'^2_{P(c)}(\alpha) = \hat{\hat{X}}'^2_P / \hat{\delta}_0(\alpha).$$

Similarly, a second-order corrected test can be obtained.

7.2.3 A Model-Free Approach

Double sampling method can be used to test the hypothesis of goodness-of-fit without invoking any model of misclassification. In double- or two-phase sampling the error-prone measurements are done on a large first-phase sample of size m selected according to some sampling design and more expensive error-free measurements are then made on a smaller subsample of size n, also selected according to a specific design.

(a) SRS design at both phase: Let m_k be the number of units in category k in the first-phase sample and suppose of these units n_{jk} are found to actually belong to true category j in the second-phase sampling. Then the *mle*'s of the π_j and the misclassification probabilities a_{jk} are

$$\hat{\pi}_j = \sum_{k=1}^{T} \hat{p}_k \left(\frac{n_{jk}}{n_{0k}}\right) \tag{7.2.16}$$

and

$$\hat{a}_{jk} = \left(\frac{\hat{p}_k}{\hat{\pi}_j}\right)\left(\frac{n_{jk}}{n_{0k}}\right) \tag{7.2.17}$$

where $n_{0k} = \sum_j n_{jk}$ and $\hat{p}_k = m_k/m$. Tenenbein (1972) derived the estimated covariance matrix $\mathbf{V}^* = ((v^*_{jk}))$ of the *mle*'s $\hat{\boldsymbol{\pi}}^{(1)} = (\hat{\pi}_1, \ldots, \hat{\pi}_{T-1})'$. These are

$$v^*_{jj} = \hat{V}(\hat{\pi}_j) = \frac{1}{n}\left[\hat{\pi}_j - \hat{\pi}_j^2 \sum_{k=1}^{T} \frac{\hat{a}^2_{jk}}{\hat{p}_j}\right] + \frac{\hat{\pi}^2_{jk}}{m}\left[\sum_{k=1}^{T} \frac{\hat{a}^2_{jk}}{\hat{\pi}_k} - 1\right] \quad (7.2.18)$$

and

$$v^*_{jl} = \hat{C}ov(\hat{\pi}_j, \hat{\pi}_l) = -\frac{1}{n}\hat{\pi}_j\hat{\pi}_l \sum_{k=1}^{T} \frac{\hat{a}_{jk}\hat{a}_{lk}}{\hat{p}_k} + \frac{\hat{\pi}_j\hat{\pi}_l}{m}\left[\sum_{k=1}^{T} \frac{\hat{a}_{jk}\hat{a}_{lk}}{\hat{p}_k} - 1\right], \quad j \neq l. \quad (7.2.19)$$

Consider now the simple chi-square statistic

$$X^{*2}_P = n\sum_{i=1}^{T}(\hat{\pi}_i - \pi_{i0})^2/\pi_{i0} = (\hat{\boldsymbol{\pi}}^{(1)} - \boldsymbol{\pi}^{(1)}_0)'\boldsymbol{\Pi}^{(1)-1}_0(\hat{\boldsymbol{\pi}}^{(1)} - \boldsymbol{\pi}^{(1)}_0) \quad (7.2.20)$$

where $\boldsymbol{\Pi}^{(1)}_0 = \text{Diag.}\,(\pi_{10}, \ldots \pi_{T-10}) - \boldsymbol{\pi}^{(1)}\boldsymbol{\pi}^{(1)'}$ (see (4.2.4)), for testing $H_0 : \pi_i = \pi_{i0}, i = 1, \ldots, T$. Its asymptotic distribution under H_0 is not $\chi^2_{(T-1)}$, since the covariance matrix of $\hat{\boldsymbol{\pi}}^{(1)}$ is not $\boldsymbol{\Pi}^{(1)}_0$. Assuming that $\hat{\boldsymbol{\pi}}^{(1)}$ is approximately normal $N_{T-1}(\boldsymbol{\pi}^{(1)}, \mathbf{V}^*)$ for sufficiently large subsamples we have Rao-Scott (1981) first-order corrected test statistics

$$X^{*2}_{P(c)} = X^{*2}_P/\hat{\delta}^*_0, \quad (7.2.21)$$

as a $\chi^2_{(T-1)}$ variable where $\hat{\delta}^*_0$ is the mean of the eigenvalues of the generalized design effect matrix $\boldsymbol{\Pi}^{(1)-1}_0\hat{\mathbf{V}}^*$. We note that

$$(T-1)\hat{\delta}^*_0 = n\sum_{j=1}^{T}\frac{v^*_{jj}}{\pi_{j0}}. \quad (7.2.22)$$

A second-order corrected test is

$$X^{*2}_{P(cc)} = X^{*2}_{P(c)}/(1 + a^{*2}) \quad (7.2.23)$$

which follows $\chi^2_{\nu^*}$ under H_0 where

$$a^{*2} = \sum_i(\delta^*_i - \delta^*_0)^2/[(T-1)\delta^{*2}_0] \quad (7.2.24)$$

and $\nu^* = (T-1)/(1 + a^{*2})$.

(b) Sampling design is not SRS at at least one of the phases: Suppose that the sampling design is other than srs in at least one of the phases. Let $\hat{\boldsymbol{\pi}}^{(1)}$ represent a consistent estimator of $\boldsymbol{\pi}^{(1)}$ in the present survey design. Let $\hat{\mathbf{V}}_1$ be a consistent estimator of covariance matrix of $\hat{\boldsymbol{\pi}}^{(1)}$. Then a first-order corrected test is given by $X^2_{P(c)} = X^{*2}_P/\hat{\delta}_0$ which follows asymptotically a $\chi^2_{(T-1)}$ variable, where $\hat{\delta}_0$ is the mean of the eigenvalues of $\boldsymbol{\Pi}_0^{(1)-1}\hat{\mathbf{V}}_1$. It can be seen that

$$(T-1)\hat{\delta}_0 = n\sum_{j=1}^{T}\hat{v}_{1jj}/\pi_{j0} = \sum_{j=1}^{T}\frac{\hat{\pi}_j}{\pi_{j0}}(1-\hat{\pi}_j)\hat{d}_j \qquad (7.2.25)$$

where $v_{1jj} = \hat{V}(\hat{\pi}_j)$ and $\hat{d}_j = \hat{v}_{1jj}n/[\hat{\pi}_j(1-\hat{\pi}_j)]$ is the deff of the jth cell. Hence, the first-order corrected test depends only on the estimated proportions and their standard errors under the specified two-phase design. Similarly, a second-order correction can be made.

7.3 Tests of Independence Under Misclassification

Suppose we have two-way cross-classified data according to two variables A and B, having r and c categories, respectively. Let π_{jk} be the probability that a unit truly belongs to the (j,k)th cell $(j=1,\ldots,r;k=1,\ldots,c)$. Also $\pi_{j0}=\sum_k \pi_{jk}$, $\pi_{0k}=\sum_j \pi_{jk}$ are the row and column marginal probabilities, respectively. The hypothesis of independence of the variable A and B can be stated as

$$H_0: \pi_{jk} = \pi_{j0}\pi_{0k} \; j=1,\ldots,r;\; k=1,\ldots,c. \qquad (7.3.1)$$

We assume, however, that there are errors in classification. Thus, when a unit actually belonging to the row u is erroneously assigned row j, similarly, a unit which actually belongs to column v is by mistake allotted to column k. Again the observations may be obtained by sampling from the population according to some specific designs including *srs*.

Let a_{ujvk} be the probability that an observation which truly belongs to the (u,v) cell is erroneously assigned to the (j,k)th cell. Let p_{jk} be the probability that an observation is found in the (j,k)th cell when misclassification is present and let π_{uv} be the probability that a unit truly belongs to the (u,v)th cell. Therefore,

$$p_{jk} = \sum_{u=1}^{r}\sum_{v=1}^{c}\pi_{uv}a_{ujvk}. \qquad (7.3.2)$$

In matrix notation (7.3.2) can be written as

$$\mathbf{p} = \mathbf{A}'\boldsymbol{\pi} \qquad (7.3.3)$$

where $\mathbf{p} = (p_{11}, \ldots, p_{1c}, \ldots, p_{r1}, \ldots, p_{rc})'$, π defined in a similar manner and $\mathbf{A} = ((a_{ujvk}))$ is the $(rc) \times (rc)$ matrix of probabilities. The rows of $\mathbf{A}$ are indexed by the subscripts (u, v) listed in lexicographic order and the columns of $\mathbf{A}$ are similarly indexed by the subscripts (j, k), also listed in lexicographic order.

7.3.1 Methods for Considering Misclassification Under SRS

Case (i): **A** *known and classification errors are independent*: The usual Pearson statistics, without considering misclassification, is

$$X_P^2 = \sum_{j=1}^{r} \sum_{k=1}^{c} (n_{jk} - n_{j0}n_{0k})^2 / [n_{j0}n_{0k}/n] \tag{7.3.4}$$

where n_{jk} is the number of observed units in the (j, k)th cell and $n_{j0} = \sum_k n_{jk}$, $n_{0k} = \sum_j n_{jk}$ are, respectively, the row and column totals. If the way in which error of classification occurs in rows is independent of the way errors in classification occurs in columns, i.e., if $a_{ujvk} = b_{uj}e_{vk}$, then it has been shown that H_0 is equivalent to $H_0' : p_{jk} = p_{j0}p_{0k} \forall\ j, k$. As a result the usual chi-square test (7.3.4) which treats X_P^2 as $\chi^2_{(r-1)(c-1)}$ will be asymptotically correct, but its asymptotic power under misclassification will be less than its power with no misclassification (Mote and Anderson 1965).

Case (ii): **A** *known, but classification errors are not independent*: In this case, Pearson statistics X_P^2 in (7.3.4) though asymptotically correct leads to inflated type I error.

From Example 5.6.1 we note that the hypothesis H_0 in (7.3.1) may be expressed as $H_{2|1} : \theta_2 = \mathbf{0}$ in the log-linear model (5.2.6) (read with (5.6.1)), and therefore find a 'pseudo' *mle* of $\tilde{u}$, θ_1 and hence of π following an iterative scheme involving weighted least squares method of Scott et al. (1990). Denoting the *mle* of $\mathbf{p}$ by $\hat{\mathbf{p}}$ an asymptotically correct test of H_0 is

$$\hat{X}_P'^2 = \sum_{j=1}^{r} \sum_{k=1}^{c} (n_{jk} - n\hat{p}_{jk})^2 / (n\hat{p}_{jk}) \tag{7.3.5}$$

which follows $\chi^2_{(r-1)(c-1)}$ distribution under H_0.

Case (iii): **A** *unknown*: A double sampling method must be used to obtain the appropriate test of independence.

7.3.2 Methods for Arbitrary Survey Designs

Case (i): **A** *known and classification errors are independent*: Consider the test statistic

$$X_P^2 = n \sum_{j=1}^{r} \sum_{k=1}^{c} (\hat{\pi}_{jk} - \hat{\pi}_{j0}\hat{\pi}_{0k})^2 / (\hat{\pi}_{i0}\hat{\pi}_{0j}). \tag{7.3.6}$$

which is used for testing hypothesis of independence when classification errors are absent. It readily follows that the first-order corrected test $X^2_{P(c)}$ and the second-order corrected test $X^2_{P(cc)}$ can be used without any modification.

Case (ii): **A** *known but classification errors are not independent*: we consider the test statistic

$$\hat{\tilde{X}}_P^{\prime 2} = n \sum_{j=1}^{r} \sum_{k=1}^{c} (\hat{\pi}_{jk} - p_{jk})^2 / \hat{p}_{jk} \tag{7.3.7}$$

which is obtained from (7.3.5) by substituting the survey estimates $\hat{\pi}_{jk}$ for n_{jk}/n and using the 'pseudo' *mle* for $\hat{\mathbf{p}}$. The true probabilities $\boldsymbol{\pi}$ are estimated by $\mathbf{A}'^{-1}\hat{\mathbf{p}}$.

Rao and Thomas (1991) have shown that $\hat{\tilde{X}}_P^{\prime 2}$ is asymptotically distributed as $\sum_{i=1}^{(r-1)(c-1)} \delta_i' W_i$ where each δ_i' is independent $\chi^2_{(1)}$ variable and δ's are the eigenvalues of the deff matrix

$$\boldsymbol{\Gamma} = n(\tilde{\mathbf{X}}_2'\mathbf{PAD}_{\Pi}^{-1}\mathbf{A}'\mathbf{P}\tilde{\mathbf{X}}_2)^{-1}(\tilde{\mathbf{X}}_2'\mathbf{PAD}_{\Pi}^{-1}\mathbf{VD}_{\Pi}^{-1}\mathbf{A}'\mathbf{P}\tilde{\mathbf{X}}_2) \tag{7.3.8}$$

with

$$\tilde{\mathbf{X}}_2 = [\mathbf{I} - \mathbf{X}_1(\mathbf{X}_1'\mathbf{PAD}_{\Pi}^{-1}\mathbf{A}'\mathbf{PX}_1)^{-1}\mathbf{X}_1'\mathbf{PAD}_{\Pi}^{-1}\mathbf{A}'\mathbf{P}]\mathbf{X}_2, \tag{7.3.9}$$

$\mathbf{D}_{\Pi} =$ Diag. $(\pi_{11}, \ldots, \pi_{rc})$ and $\pi_{jk} = \pi_{j0}\pi_{0k}$.

A first-order corrected test is

$$\hat{\tilde{X}}_{P(c)}^{\prime 2} = \hat{\tilde{X}}_P^{\prime 2} / \hat{\delta}_0' \tag{7.3.10}$$

which follows $\chi^2_{(r-1)(c-1)}$ under H_0 where $\hat{\delta}_0'$ is the mean of the estimated eigenvalues. Similarly, a second-order correction can be implemented.

In this case, however, $\hat{\delta}_0'$ cannot be expressed simply in terms of estimated cell deffs and marginal row and column deffs, unlike in case of no-misclassification error.

A *unknown*: Here again, a double sampling design has to be used. As before the first-phase sample size is m and the second-phase sample size is n. Let $\hat{\boldsymbol{\pi}}$ represent a consistent estimator of $\boldsymbol{\pi}$ under the specific two-phase design and $\hat{\mathbf{V}}$ be a consistent estimator of covariance matrix of $\boldsymbol{\pi}$. A Pearsonian chi-square for testing independence is given by (7.3.4). Standard corrections can also be applied.

7.4 Test of Homogeneity

Following Rao and Thomas (1991), Heo (2002) considered test of homogeneity when errors of misclassification are present. Suppose there are two independent populations U_1 and U_2 and independent samples of sizes n_1 and n_2, respectively, are selected from these two populations. Suppose also that there is a categorical variable with T mutually exclusive and exhaustive classes with $\boldsymbol{\pi}_{i+} = (\pi_{i1}, \ldots, \pi_{iT})'$ and $\mathbf{p}_{i+} = (p_{i1}, \ldots, p_{iT})'$ as the vectors of true and observed proportions, respectively, for the ith population ($i = 1, 2$). The hypothesis of homogeneity is given by

$$H_0 : \boldsymbol{\pi}_1 = \boldsymbol{\pi}_2 = \boldsymbol{\pi}_0$$

where $\boldsymbol{\pi}_i = (\pi_{i1} \ldots, \pi_{iT-1})'(i = 1, 2)$ and $\boldsymbol{\pi}_0$ is defined accordingly.

Let X be a predictor which predicts the category Z of a unit in population U_i when the unit actually belongs to category Y. We shall denote the probability of such a misclassification by

$$P_i(Z = k|Y = j, X = x). \tag{7.4.1}$$

When misclassification errors exist, it may be important to determine the extent to which the misclassification probabilities are homogeneous within specific groups. We shall say that misclassification probabilities are homogeneous within population U_i if the expression (7.4.1) does not depend on x. In such a situation we shall also say that the misclassification errors of U_i are homogeneous.

When the misclassification probabilities are homogeneous, customary design-based estimators of the proportions of reported classification will converge to

$$\mathbf{p}_{i+} = \mathbf{A}_i' \boldsymbol{\pi}_{i+} \tag{7.4.2}$$

where $\mathbf{A}_i = ((a_{i,jk}))$ is a $T \times T$ matrix with (j, k)th element $a_{i;jk}$. The (j, k)th element of matrix $\mathbf{A}_i$ is the probability, $P_i(Z = k|Y = j)$ of a unit being classified into the kth class when its true class is j.

Suppose now that there are categorical explanatory variables and the intersection of all the explanatory variables partition the population U_i into C groups, $U_{ic}(c = 1, \ldots, C)$ and the vectors analogous to $\boldsymbol{\pi}_i$ is $\boldsymbol{\pi}_{ic} = (\pi_{ic1}, \ldots, \pi_{icT-1})'$ and $\mathbf{p}_{ic}$ is defined similarly. Here π_{icj} denotes the true proportion in category j of elements in group c of population $U_i(j = 1, \ldots, T-1; c = 1, \ldots, C; i = 1, 2)$. In this case, as in Eq. (7.4.2) we have

$$\mathbf{p}_{ic+} = \mathbf{A}_{ic}' \boldsymbol{\pi}_{ic+} \tag{7.4.3}$$

where $\mathbf{A}_{ic} = ((a_{ic;jk}))$ is the associated misclassification matrix with the (j, k)th element $a_{ic;jk} = P_{ic}(Z = k|Y = j)$ for members of the subpopulation U_{ic}.

The vector $\mathbf{p}_{ic+}$ is defined as

$$\mathbf{p}_{ic+} = M_{ic}^{-1}\left(\sum_{t\in U_{ic}} I_{t1}, \ldots, \sum_{t\in U_{ic}} I_{tT}\right)'$$

where M_{ic} is the size of U_{ic} and $I_{tj} = 1(0)$ if the person gives answer j, and 0 (otherwise). Similarly, the vector $\boldsymbol{\pi}_{ic+}$ is

$$\boldsymbol{\pi}_{ic+} = M_{ic}^{-1}\left(\sum_{t\in U_{ic}} \delta_{t1}, \ldots, \sum_{t\in U_{ic}} \delta_{tT}\right)'$$

where $\delta_{tj} = 1(0)$ if a person's true category is j (otherwise). The vector of the overall observed proportions for population U_i is, therefore,

$$\mathbf{p}_{i+} = \sum_{c=1}^{C} R_{ic}\mathbf{p}_{ic+} = \sum_{c=1}^{C} R_{ic}\mathbf{A}'_{ic}\boldsymbol{\pi}_{ic+} \tag{7.4.4}$$

where $R_{ic} = M_i^{-1}M_{ic}$ and $M_i = \sum_{c=1}^{C} M_{ic}$, the number of units in U_i. When $\mathbf{A}_{ic} = \mathbf{A}_i \ \forall\ c$, (7.4.4) reduces to

$$\begin{aligned}
\mathbf{p}_{i+} &= \mathbf{A}'_i \sum_c R_{ic}\boldsymbol{\pi}_{ic+} \\
&= \mathbf{A}'_i \sum_c M_i^{-1}\left(\sum_{t\in U_{ic}} \delta_{t1}, \ldots, \sum_{t\in U_{ic}} \delta_{cT}\right)' \\
&= \mathbf{A}'_i M_i^{-1}\left(\sum_{t\in U_i} \delta_{t1}, \ldots, \sum_{t\in U_i} \delta_{tT}\right)' \\
&= \mathbf{A}'_i \boldsymbol{\pi}_{i+}.
\end{aligned}$$

Assume now that all $\mathbf{A}_{ic}$ are non-singular matrices and are not all equal. Let $\mathbf{B}_{ic} = ((b_{ic;jk})) = (\mathbf{A}'_{ic})^{-1}$. Then from expressions (7.4.3) and (7.4.4),

$$\begin{aligned}
\boldsymbol{\pi}_{i+} &= \sum_c R_{ic}\boldsymbol{\pi}_{ic+} \\
&= \sum_{c=1}^{C} R_{ic}\mathbf{B}_{ic}\mathbf{p}_{ic+}.
\end{aligned} \tag{7.4.5}$$

When all $\mathbf{A}_{ic}$ are equal, expression (7.4.5) simplifies to

$$\begin{aligned}\boldsymbol{\pi}_{i+} &= \mathbf{A}_i^{-1}\sum_c R_{ic}\mathbf{p}_{ic+}\\ &= \mathbf{A}_i^{-1}M_i^{-1}\left(\sum_{t\in U_i} I_{t1},\ldots,\sum_{t\in U_i} I_{tT}\right)'\\ &= (\mathbf{A}_i')^{-1}\mathbf{p}_{i+}.\end{aligned}$$

Point Estimation: Consider now the following sampling design D_1. Suppose that population U_i has been stratified into H strata with N_h clusters in the hth stratum. From the hth stratum $n_h \geq 2$ clusters are selected independently with unequal probability p_{ha} (for the ath cluster) and with replacement across the strata ($a = 1,\ldots,n_h;\ h = 1,\ldots H$). Within the ath first stage unit in the hth stratum $n_{ha} \geq 1$ ultimate units or elements are selected according to some sampling method with selection probability p_{hab} (for the bth element) from N_{ha} units, $b = 1,\ldots,n_{ha};\ a = 1,\ldots,n_h;\ h = 1,\ldots,H$. The total number of ultimate units in the population is $N = \sum_{h=1}^{H}\sum_{a=1}^{N_h} N_{ha}$ and in the sample is $n = \sum_{h=1}^{H}\sum_{i=1}^{n_h} n_{ha}$.

Replacing the triple subscript (hab) by the single subscript t for convenience, we have the estimates

$$\hat{R}_{ic} = \hat{M}_i^{-1}\hat{M}_{ic} \tag{7.4.6}$$

where $\hat{M}_{ic} = \sum_{t\in s_{ic}} w_t,\ \hat{M}_i = \sum_{t\in s_i} w_t,\ w_t$ is the unit-level sample survey weight, s_{ic} is the set of sample in U_{ic}, and s_i is the set of sample units in U_i. Also,

$$\hat{\mathbf{p}}_{ic+} = \hat{M}_{ic}^{-1}\left(\sum_{t\in s_{ic}} w_t I_{t1},\ldots,\sum_{t\in s_{ic}} w_t I_{tT}\right)'. \tag{7.4.7}$$

Then from expressions (7.4.6) and (7.4.7),

$$\hat{R}_{ic}\hat{\mathbf{p}}_{ic+} = \hat{M}_i^{-1}\left(\sum_{i\in s_{ic}} w_t I_{t1},\ldots,\sum_{t\in s_{ic}} w_t I_{tT}\right)' = \hat{\mathbf{e}}_{ic},\ (\text{say}). \tag{7.4.8}$$

Note that $\hat{\mathbf{e}}_{ic}$ is a vector of sample ratios. Moreover, from expression (7.4.5)

$$\hat{\boldsymbol{\pi}}_{i+} = \sum_{c=1}^{C}\mathbf{B}_{ic}\hat{\mathbf{e}}_{ic} \tag{7.4.9}$$

where we have assumed $\mathbf{B}_{ic}$'s are known. The jth element of $\hat{\boldsymbol{\pi}}_{i+}$ is $\hat{\pi}_{ij} = \sum_{c=1}^{C} \mathbf{B}_{icj0}\hat{\mathbf{e}}_{ic}$ where $\mathbf{B}_{icj0} = (b_{ic,j1},\ldots,b_{ic,jT})$ is the jth row of the $T\times T$ matrix $\mathbf{B}_{ic0}$. The authors used a logistic regression model to estimate the misclassification probabilities $a_{ic;jk}$.

Estimation of Variance of $\hat{\pi}_{i+}$: Assume that the matrices $\mathbf{A}_{ic}$ and $\mathbf{B}_{ic}$ are known. Define the $CT \times 1$ vector $\hat{\mathbf{e}}_{i0} = (\hat{\mathbf{e}}'_{i1}, \ldots, \hat{\mathbf{e}}'_{iC})'$. We note that the vector $\hat{\mathbf{e}}_{i0}$ is a vector of sample ratios. Again, the expression (7.4.9) can be written as

$$\begin{aligned}\hat{\pi}_{i+} &= \sum_{c=1}^{C} \mathbf{B}_{ic}\hat{\mathbf{e}}_{ic}\\ &= \begin{bmatrix}\mathbf{B}_{i1}\ \mathbf{B}_{i2}\ \ldots\ \mathbf{B}_{iC}\end{bmatrix}\begin{bmatrix}\hat{\mathbf{e}}_{i1}\\ \hat{\mathbf{e}}_{i2}\\ \cdot\\ \cdot\\ .\hat{\mathbf{e}}_{iC}\end{bmatrix}\\ &= \mathbf{B}_{i000}\hat{\mathbf{e}}_{i0}\end{aligned} \tag{7.4.10}$$

where $\mathbf{B}_{i000}$ is a $T \times (CT)$ matrix with jth row equal to a $1 \times CT$ vector $\mathbf{B}_{i0j0} = (\mathbf{B}_{i1j0}, \ldots, \mathbf{B}_{iCj0})$. Thus an estimate of the variance of the approximate distribution of $\hat{\pi}_{i+}$ is

$$\hat{\mathbf{V}}(\hat{\pi}_{i+}) = \mathbf{B}_{i000}\hat{\mathbf{V}}(\hat{\mathbf{e}}_{i0})\mathbf{B}'_{i000} \tag{7.4.11}$$

with jth diagonal element $\hat{V}(\hat{\pi}_{ij}) = \mathbf{B}_{i0j0}\hat{\mathbf{V}}(\hat{\mathbf{e}}_{i0})\mathbf{B}'_{i0j0}$.

Asymptotic properties of $\hat{\pi}_i$: We have from (7.4.10) $\hat{\pi}_{i+} = \mathbf{B}_{i000}\hat{\mathbf{e}}_{i0}$. Thus $\hat{\pi}_{i+}$ is a linear function of $\hat{\mathbf{e}}_{i0}$ and $\hat{\mathbf{e}}_{i0}$ is a vector of sample ratios (vide (7.4.8)). Therefore, if under certain conditions $\hat{\mathbf{e}}_{i0}$ is a consistent estimator of the corresponding population quantities with its asymptotic distribution normal then under the same conditions $\hat{\pi}_{i+}$ is a consistent estimator of π_{i+} and $\sqrt{n_i}(\hat{\pi}_i - \pi_i)$ converges in distribution to a $(T-1)$ variate normal distribution $N_{T-1}(\mathbf{0}, \mathbf{V}_{\pi i})$.

For the design D_1 Shao (1996) gave conditions for consistency and asymptotic normality of design-based estimators of nonlinear functions, like ratios of population total. Along with these conditions Heo and Eltinge (2003) assumed the condition (C_1): $\hat{\mathbf{V}}(\hat{\pi})$ is a consistent estimator of $\mathbf{V}_{\pi i}$ where $\hat{\mathbf{V}}(\hat{\pi})$ is the upper $(T-1) \times (T-1)$ sub-matrix of $\hat{\mathbf{V}}(\hat{\pi}_{i+})$ in (7.4.11).

Thus under design D_1, and some regularity conditions including the condition C_1, the Wald test statistic for homogeneity is

$$X_{WH}(c)^2 = (\hat{\pi}_1 - \hat{\pi}_2)'\hat{\mathbf{V}}^{-1}(\hat{\pi}_1 - \hat{\pi}_2) \tag{7.4.12}$$

where $\hat{\mathbf{V}} = \hat{\mathbf{V}}_1 + \hat{\mathbf{V}}_2$ is asymptotically distributed as $\chi^2_{(T-1)}$ under H_0. The authors evaluated the power of the test (7.4.12) and applied their proposed methods to the data from the Dual Frame National Health Interview Survey (NHIS)/Random-Digit-Dialing (RDD) Methodology and Field Test Project.

7.5 Analysis Under Weighted Cluster Sample Design

A weighted cluster sample survey design is frequently used in large demographic surveys. As is well known, in such surveys each selected cluster, generally a household, is completely enumerated. Sociodemographic and health characteristic of each member of the selected household is recorded and is multiplied by a weight which is roughly proportional to the inverse of the probability of the individual being included in the sample on the basis of post-stratified geographic and demographic status of the individual. This type of weighting is necessary to estimate certain characteristics of the target population at reasonable cost in large sample survey situations.

Say a sample of n clusters is drawn by simple random sampling with replacement from a population of N clusters, the sth cluster containing M_s elementary units or units. The population size is therefore $M_0 = \sum_{s=1}^{N} M_s$ and sample size is $m = \sum_{s=1}^{n} M_s$. As we know, in cluster sampling all the M_s units in the selected sth cluster are completely enumerated.

Suppose that each member of a cluster is classified according to two characteristics in a two-way contingency table: once by a characteristic represented by r rows and once by a different characteristic represented by c columns, $r \times c = q$. Let a_i denote the sample count of members that fall in category i and A_i the population count of members in the same category, $i = 1, \ldots, q$ so that $\sum_i a_i = m$, $\sum_i A_i = M_0$. Also let π_i be the probability that a unit (member of a cluster) belongs to the category i, $\sum_{i=1}^{q} \pi_i = 1$.

We denote by $P_{ij\ldots t}$ the probability that the first member of a cluster falls in category i, the second member in category $j, \ldots$, and the last member in category t, $(i, j, \ldots, t = 1, \ldots, q)$.

When $M_s = 2$ for each s, Cohen (1976) suggested the following model of association between members of the same cluster:

$$P_{ij} = \begin{cases} \alpha\pi_i + (1-\alpha)\pi_i^2 \; (i = j) \\ (1-\alpha)\pi_i\pi_j \; (i \neq j) \end{cases} \tag{7.5.1}$$

with $P_{ij} > 0, \sum P_{ij} = 1, 0 \leq \alpha \leq 1$. If $\alpha = 0$, $P_{ij} = \pi_i\pi_j$, the two members are totally independent; if $\alpha = 1$, $P_{ij} = \pi_i (i = j)$ and $0 (i \neq j)$, the two members are completely dependent. The other degree of association between members of the same cluster is thus reflected in the value of α between these two extremes. Cohen considered clusters each containing only two siblings.

Altham (1976) extended Cohen's model to the case, $M_s = 3 \forall s$:

$$P_{ijk} = \begin{cases} \alpha\pi_i + (1-\alpha)\pi_i^3 (i = j = k) \\ (1-\alpha)\pi_i\pi_j\pi_k \; \text{(otherwise)} \end{cases} \tag{7.5.2}$$

where $P_{ijk} > 0, \sum P_{ijk} = 1$ and $0 \leq \alpha \leq 1$. Some other models have been presented by Altham (1976), Plackett and Paul (1978), and Fienberg (1979).

Chi-square test of a simple hypothesis: Usual Pearson statistic for testing goodness-of-fit hypothesis $H_0 : \pi_i = \pi_{i0}, i = 1, \ldots, q$ is

$$X_P^2 = \sum_{i=1}^{q} \frac{(a_i - m\pi_{i0})^2}{m\pi_{i0}}. \tag{7.5.3}$$

However, in a complex survey situation in which the elementary units are dependent variables (because of the relations like (7.5.1) and (7.5.2)) and weighted (as introduced in the beginning of this section and elaborated subsequently), the statistic X_P^2 is not appropriate.

When the data consist of a random sample of n clusters, each containing two members, Cohen (1976) showed that a valid test statistic for goodness-of-fit is

$$X_P^2(C) = \frac{X_P^2}{1+\alpha}, \quad 0 \le \alpha \le 1, \tag{7.5.4}$$

where $X_P^2(C)$ has a limiting $\chi^2_{(q-1)}$ distribution as $n \to \infty$. Altham (1976) extended Cohen's result to the case $M_s = M \ \forall\, s$. In this case the test statistic is

$$X_P^2(A) = \frac{X_P^2}{1+\alpha(M-1)}, \quad 0 \le \alpha \le 1. \tag{7.5.5}$$

which also converges in distribution to $\chi^2_{(q-1)}$ as $n \to \infty$.

When the members in the clusters are totally independent, i.e., $\alpha = 0$, $X_P^2(A) = 0$ and the conventional test X_P^2 in (7.5.3) can be used as if there is no clustering. When they are totally dependent, i.e., $\alpha = 1$, then $X_P^2(A) = X_P^2/M$. Here though the observed sample size is nM, the effective sample size is n.

Choi (1980) considered the following weighting scheme for the weighted cluster sample survey. Let w_{st} be the statistical weight for the tth member of the sth cluster ($t = 1, \ldots, M_s$; $s = 1, \ldots, n$), and these are assumed to be known from some previous data collection procedures. We have $1 \le w_{st} \le M_0$ and unweighted data are obtained as a special case by setting $w_{st} = 1$. We can obtain the overall weighted counts $\hat{M}_0 = \sum_{s=1}^{n} \sum_{t=1}^{M_s} w_{st}$. Defining the indicator function

$$\delta_{ist} = \begin{cases} 1 & \text{if the } (s,t) \text{ member falls in } i\text{th category} \\ 0 & \text{otherwise} \end{cases}$$

the sample count of members falling in the ith category is $a_i = \sum_{s=1}^{n} \sum_{t=1}^{M_s} \delta_{ist}$.

Denoting by $\hat{A}_i$ the weighted count of the population members that fall in the ith category, we have $\hat{A}_i = \sum_{s=1}^{n} \sum_{t=1}^{M_s} w_{st}\delta_{ist}$ and $\hat{M}_0 = \sum_{i=1}^{q} \hat{A}_i$.

In both weighted and unweighted cases, the cell count is the sum over independent clusters, each including M_s-dependent variables that may or may not fall in the ith region.

The problem of interest is to investigate the goodness-of-fit of π_{i0} to the weighted cell counts arising from the weighted cluster sampling. The conventional test statistic is

$$X_P^2(W) = \sum_{i=1}^{q} \frac{(\hat{A}_i - \hat{M}_0\pi_{i0})^2}{\hat{M}_0\pi_{i0}}. \tag{7.5.6}$$

Because of the possible dependence between the members of the same cluster as is reflected through the weighting of the individual members, the joint distribution of $\hat{A}_i, i = 1, \ldots, q$'s is not a multinomial distribution and hence $X_P^2(W)$ in (7.5.6) will not provide an appropriate test statistic. Under model (7.5.1), $\hat{Y} = (\hat{A}_1, \ldots, \hat{A}_{q-1})$ has the finite mean vector $\hat{M}_0\boldsymbol{\pi}_0$ and covariance matrix $G(\mathbf{D}_{\pi 0} - \boldsymbol{\pi}_0\boldsymbol{\pi}_0')$ where $\pi_0 = (\pi_{10}, \ldots, \pi_{q-10})'$, $\mathbf{D}_{\pi_0}$ is the diagonal matrix based on $\boldsymbol{\pi}_0$ and G is a positive number defined below. If H_0 is true, the correct test statistic under model (7.5.1) is

$$X_P^2(T) = (\hat{M}_0/G)X_P^2(W) \tag{7.5.7}$$

with

$$G = \sum_{s=1}^{n}\sum_{t=1}^{M_s} w_{st}^2 + \alpha \sum_{s=1}^{n}\sum\sum_{t \neq t'=1}^{M_s} w_{st}w_{st'}, \quad 0 \leq \alpha \leq 1.$$

Note that G measures the combination of effects of clustering and weighting in the weighted cluster sampling scheme. Also, $\hat{M}_0$ is the weighted sample count. Thus $0 \leq \hat{M}_0/G \leq 1$. We note that $G \to \infty$ if the weight become large and $\alpha \geq 0$. Thus the conventional test statistic $X_P^2(W)$ can be corrected by multiplying it by the scale factor $(\hat{M}_0/G)$ and the statistic so corrected can be used for testing of null hypothesis. The statistic $X_P^2(T)$ is asymptotically distributed as $\chi^2_{(q-1)}$ under H_0 as $n \to \infty$.

Chapter 8
Approximate MLE from Survey Data

Abstract Since under many situations from complex sample surveys, exact likelihoods are difficult to obtain and pseudo-likelihoods are used instead, we shall, in this chapter, consider some procedures and applications which are useful in obtaining approximate maximum likelihood estimates from survey data. After addressing the notion of ignorable sampling designs Sect. 8.2 considers exact MLE from survey data. The concept of weighted distributions due to Rao (Classical and contagious discrete distributions, pp 320–332, 1965b), Patil and Rao (Biometrics 34:179–189, 1978) and its application in maximum likelihood estimation of parameters from complex surveys have been dealt with in the next section. Subsequently, the notion of design-adjusted estimation due to Chambers (J R Stat Soc A 149:161–173, 1986) has been reviewed. We review in Sect. 8.5 the pseudo-likelihood approach to estimation of finite population parameters as developed by Binder (Stat Rev 51:279–292, 1983), Krieger and Pfeffermann (Proceedings of the symposium in honor of Prof. V.P. Godambe, 1991), among others. The following section addresses the mixed model framework which is a generalization of design-model framework considered by Sarndal et al. (Model assisted survey sampling, 1992) and others. Lastly, we consider the effect of sample selection on the standard principal component analysis and the use of alternative maximum likelihood and probability weighted estimators in this case.

Keywords Ignorable sampling design · Exact MLE · Weighted distributions · Design-adjusted estimators · Pseudo-likelihood · Mixed model framework · Principal component

8.1 Introduction

We have already seen that the estimation of model parameters of the distribution of categorical variables from data obtained through complex surveys are based on maximizing the pseudo-likelihood of the data, as exact likelihoods are rarely amenable to maximization. We shall, therefore, consider in this chapter some procedures and

P. Mukhopadhyay, *Complex Surveys*, DOI 10.1007/978-981-10-0871-9_8

applications which are useful in obtaining approximate maximum likelihood estimates from survey data.

Scott et al. (1990) proposed weighted least squares and quasi-likelihood estimators for categorical data. Maximum likelihood estimation (MLE) from complex surveys requires additional modeling due to information in the sample selection. This chapter reviews some of the approaches considered in the literature in this direction. After addressing the notion of ignorable sampling designs Sect. 8.2 considers exact MLE from survey data. The next section deals with the concept of weighted distributions due to Rao (1965), Patil and Rao (1978) and its application in MLE of parameters from complex surveys. The notion of design-adjusted estimation due to Chambers (1986) has been reviewed in the next section. We review in Sect. 8.5 the pseudo-likelihood approach to estimation of finite population parameters as developed by Binder (1983), Krieger and Pfeffermann (1991), among others. They utilized the sample selection probabilities to estimate the census likelihood equations. The estimated likelihood is then maximized with respect to the parameters of interest. Starting from a family of exponential distributions and the corresponding generalized linear models Binder (1983) obtained likelihood equations for a set of regression parameters. Taking cue from these equations, he obtained a set of estimating equations for the corresponding finite population parameters. By using Taylor series expansion, the author inverts these equations to obtain estimate of these finite population parameters and subsequently derives expressions for its variance and variance-estimators. We review these works in this section. The following section addresses the mixed model framework which is a generalization of design-model framework considered by Sarndal et al. (1992) and others. Lastly, we consider the effect of sample selection on the standard principal component analysis and the use of alternative maximum likelihood and probability weighted estimators in this case.

Intrested readers may refer to Chambers et al. (2012) for further discussion on MLE from survey data.

8.2 Exact MLE from Survey Data

We review in this section the MLE from survey data. To understand the complexity of the problem we first discuss the notion of *ignorable sampling designs.*

8.2.1 Ignorable Sampling Designs

Let $\mathbf{Z} = (Z_1, \dots, Z_k)'$ be a set of design variables and $\mathbf{Y} = (Y_1, \dots, Y_p)'$ a set of study variables taking value $\mathbf{z}_i = (z_{i1}, \dots, z_{ik})'$ and $\mathbf{y}_i = (y_{i1}, \dots, y_{ip})'$ respectively on unit $i = (1, \dots, N)$ in the finite population from which a sample s is selected by the sampling design $p(s)$. Let $\mathbf{z} = [\mathbf{z}_1 \dots, \mathbf{z}_N]' = ((z_{ij}))$ be the $N \times k$ matrix and $\mathbf{y} = [\mathbf{y}_1, \dots, \mathbf{y}_N]' = ((y_{ij}))$ be the $N \times p$ matrix of the respective values. Without any loss of generality we write $\mathbf{y} = [\mathbf{y}_s, \mathbf{y}_r]'$ where $\mathbf{y}_s = [\mathbf{y}_i; i \in s]$ and $\mathbf{y}_r = [\mathbf{y}_i; i \in \bar{s}]$.

Let $\mathbf{I} = (I_1, \ldots, I_N)'$, be a vector of sample inclusion indicator variables, $I_i = 1(0)$ if $i \in (\notin)s$. Thus $p(s) = P(\mathbf{I}|\mathbf{Y}, \mathbf{Z}; \rho)$ where ρ is a parameter involved in the distribution of $\mathbf{I}$. In general, the probability of observing a unit may also depend on $\mathbf{Y}$ as in retrospective sampling.

The basic problem of MLE from survey data is that unlike in simple random sampling the likelihood function cannot be derived directly from the distribution of $\mathbf{Y}$ in the population. This problem can often be resolved by modeling the joint distribution of $\mathbf{Y}$ and $\mathbf{Z}$. Our notation is that big case letters $\mathbf{Y}_i$, $\mathbf{Z}_i$ will denote variables corresponding to the ith unit and small-case letters $\mathbf{y}_i$, $\mathbf{z}_i$ their realized values.

Suppose that the values of $\mathbf{Z}$ are known for each unit in the population and $\mathbf{Y}$ is observed only for sample units. If we regard the design variable $\mathbf{Z}$ as random, the joint pdf of all the available data is

$$f(\mathbf{y}_s, \mathbf{I}, \mathbf{Z}; \theta, \phi, \rho) = \int f(\mathbf{y}_s, \mathbf{Y}_r|\mathbf{Z}; \theta_1)P(\mathbf{I}|\mathbf{y}_s, \mathbf{Y}_r, \mathbf{Z}; \rho_1)g(\mathbf{Z}; \phi)d\mathbf{Y}_r. \quad (8.2.1)$$

Ignoring the sample selection in the inference process implies that inference is based on the joint distribution of $\mathbf{Y}_r$ and $\mathbf{Z}$, that is, the probability $P(\mathbf{I}|\mathbf{y}_s, \mathbf{Y}_r, \mathbf{Z}; \rho_1)$ on the right side of (8.2.1) is ignored. Hence, the inference is based on

$$f(\mathbf{y}_s, \mathbf{Z}; \theta, \phi) = \int f(\mathbf{y}_s, \mathbf{Y}_r|\mathbf{Z}; \theta_1)g(\mathbf{Z}; \phi)d\mathbf{Y}_r. \quad (8.2.2)$$

The sample selection is said to be ignorable when inference based on (8.2.1) is equivalent to inference based on (8.2.2). This is clearly the case for sampling designs that depend only on the design variables $\mathbf{Z}$, since in this case $P(\mathbf{I}|\mathbf{Y}, \mathbf{Z}; \rho_1) = P(\mathbf{I}|\mathbf{Z}; \rho_1)$.

The complications in MLE from complex survey data based on (8.2.1) or (8.2.2) are now clear. All the design variables must be identified and known at the population level. Sample selection should be ignorable in the above sense or alternatively, the probabilities $P(\mathbf{I}|\mathbf{Y}, \mathbf{Z}; \rho_1)$ be modeled and included in the likelihood. Finally, the use of MLE requires the specification of the joint pdf $f(\mathbf{Y}, \mathbf{Z}; \theta, \phi) = f(\mathbf{Y}|\mathbf{Z}; \theta_1)g(\mathbf{Z}; \phi)$.

The above definition is due to Rubin (1976). Little (1982) extended Rubin's work to the case where data associated with the sample units is missing due to nonresponse.

Note 8.2.1 Forster and Smith (1998) considered nonresponse models for a single categorical response variable with categorical covariates whose values are always observed. They presented Bayesian methods for ignorable models and a particular nonignorable model and argued that the standard methods of model comparison are inappropriate for comparing ignorable and nonignorable models.

8.2.2 Exact MLE

Suppose that the sample selection is ignorable so that inference can be based on the joint distribution

$$f(\mathbf{y}_s, \mathbf{Z}; \theta, \phi) = f(\mathbf{y}_s|\mathbf{Z}; \theta_1)g(\mathbf{Z}; \phi).$$

The likelihood of (θ, ϕ) can then be factorized as

$$L(\theta, \phi; \mathbf{y}_s, \mathbf{Z}) = L(\theta_1; \mathbf{y}_s|\mathbf{Z})L(\phi; \mathbf{Z}). \tag{8.2.3}$$

Assuming that the parameters θ_1 and ϕ are distinct in the sense of Rubin (1976), MLE of θ_1 and ϕ can be calculated independently from the two components.

Example 8.2.1 If $(\mathbf{Y}_i, \mathbf{Z}_i)$ are independent multivariate normal, the MLE for $\mu_Y = E(\mathbf{Y})$ and $\mathbf{\Sigma}_Y = V(\mathbf{Y})$ are

$$\begin{aligned} \hat{\mu}_Y &= \bar{\mathbf{y}}_s + \hat{\beta}(\bar{\mathbf{Z}} - \bar{\mathbf{z}}_s); \\ \hat{\mathbf{\Sigma}}_{YY} &= \mathbf{s}_{YY} + \hat{\beta}(\mathbf{S}_{ZZ} - \mathbf{s}_{ZZ})\hat{\beta}' \end{aligned} \tag{8.2.4}$$

where $(\bar{\mathbf{y}}_s, \bar{\mathbf{z}}_s) = \sum_{i\in s}(\mathbf{y}_i, \mathbf{z}_i)/n$, $\bar{\mathbf{Z}} = \sum_{i=1}^N \mathbf{z}_i/N$, $\mathbf{S}_{ZZ} = \sum_{i=1}^N (\mathbf{z}_i - \bar{\mathbf{Z}})(\mathbf{z}_i - \bar{\mathbf{Z}})'/N$, $\mathbf{s}_{ZZ} = \sum_{i\in s}(\mathbf{z}_i - \bar{\mathbf{z}}_s)(\mathbf{z}_i - \bar{\mathbf{z}}_s)'/n$ and $\hat{\beta} = \sum_{i\in s}(\mathbf{y}_i - \bar{\mathbf{y}}_s)(\mathbf{z}_i - \bar{\mathbf{z}}_s)'\mathbf{s}_{ZZ}^{-1}/n$ and n is the sample size (Anderson 1957).

8.3 MLE's Derived from Weighted Distributions

Let X be a random variable with pdf $f(x)$. The weighted pdf of a random variable X^w is defined as

$$f^w(x) = w(x)f(x)/w \tag{8.3.1}$$

where $w = \int w(x)f(x)dx = E[w(X)]$ is the normalizing factor to make the total probability unity. Weighted distributions occur when realizations x of a random variable X with probability density $f(x)$ are observed and recorded with differential probability $w(x)$. Then $f^w(x)$ is the pdf of the resulting random variable X^w and w is the probability of recording an observation. The concept of weighting distributions was introduced by Rao (1965) and has been studied by Cox (1969), Patil and Rao (1978), among others.

Krieger and Pfeffermann (1991) considered the application of concept of weighted distributions in inference from complex surveys. Consider, a finite population of size N with a random measurement $X(i) = \mathbf{x}_i' = (\mathbf{y}_i', \mathbf{z}_i')$ generated independently from a common pdf $h(\mathbf{x}_i; \delta) = f(\mathbf{y}_i|\mathbf{z}_i; \theta_1)g(\mathbf{z}_i; \phi)$ on the unit i (using the notation of Sect. 8.2). Suppose the unit i has been included in the sample with probability $w(\mathbf{x}_i; \alpha)$. We denote by X_i^w the measurement recorded for unit $i \in s$. The pdf of X_i^w is then

$$h^w(\mathbf{x}_i; \alpha, \delta) = f(\mathbf{x}_i | i \in s) = \frac{P[i \in s | X(i) = \mathbf{x}_i] h(\mathbf{x}_i; \delta)}{P(i \in s)}$$

$$= \frac{w(\mathbf{x}_i; \alpha) h(\mathbf{x}_i; \delta)}{\int w(\mathbf{x}_i; \alpha) h(\mathbf{x}_i; \delta) d\mathbf{x}_i}. \tag{8.3.2}$$

This is the sampling distribution of the variable X_i^w after sampling. The concept has earlier been used by several authors including Quesenberg and Jewell (1986).

In analytic surveys interest often focuses on the vector parameter δ or functions thereof. Suppose a sample $S = \{1, \ldots, n\}$ is selected with replacement such that at any draw $k (= 1, \ldots, n)$, $P(j \in s) = w(\mathbf{x}_j; \alpha), j = 1, \ldots, N$. Then, the joint probability of the random variables $\{X_i^w, i = 1, \ldots, n; i \in S\}$ is $\Pi_{i=1}^n h^w(\mathbf{x}_i; \alpha, \delta)$ so that the likelihood function of δ is

$$L(\delta : \mathbf{X}_s, S) = \text{Const.} \frac{\Pi_{i=1}^n h(\mathbf{x}_i; \delta)}{[\int w(\mathbf{x}; \alpha) h(\mathbf{x}; \delta) d\mathbf{x}]^n} \tag{8.3.3}$$

where $\mathbf{X}_s = \{\mathbf{x}_1, \ldots, \mathbf{x}_n\}$.

Example 8.3.1 Assume $\mathbf{X}_i' = (\mathbf{Y}_i', \mathbf{Z}')$ are *iid* multivariate normal variables with mean $\mu_x' = (\mu_Y', \mu_Z')$ and covariance matrix

$$\mathbf{\Sigma}_{XX} = \begin{bmatrix} \mathbf{\Sigma}_{YY} & \mathbf{\Sigma}_{YZ} \\ \mathbf{\Sigma}_{ZY} & \mathbf{\Sigma}_{ZZ} \end{bmatrix}.$$

PPSWR selection: Let $T_i = \alpha_1' \mathbf{Y}_i + \alpha_2' \mathbf{Z}_i$ define a design-variable and suppose that a sample of n draws is selected by probability proportional to size (T) with replacement such that at each draw $k = 1, \ldots, n$, $P(j \in s) = t_j / N\bar{T}$ where $\bar{T} = \sum_{j=1}^N t_j / N$. We assume that N is large enough so that $E(T) = \mu_T \approx \bar{T}$. Suppose that the only information available is sample values $\mathbf{x}_i' = (\mathbf{y}_i', \mathbf{z}_i')$, $i \in s$ and the selection probabilities $w(\mathbf{x}_i; \boldsymbol{\alpha}) = t_i / N\bar{T}$ where $\boldsymbol{\alpha} = (\alpha_1, \alpha_2)$. Replacing $\bar{T}$ by μ_T, and using (8.3.3), the likelihood function of $\{\mu_X, \mathbf{\Sigma}_{XX}\}$ is

$$L(\mu_X, \mathbf{\Sigma}_{XX}; \mathbf{X}_S, S) = \frac{\Pi_{i=1}^n (\alpha' \mathbf{x}_i) \phi(\mathbf{x}_i; \mu_X, \mathbf{\Sigma}_{XX})}{(\alpha_1' \mu_Y + \alpha_2' \mu_Z)^n} \tag{8.3.4}$$

where $\phi(\mathbf{x} : \mu_X, \mathbf{\Sigma}_{XX})$ is the normal pdf with mean μ_X, and dispersion matrix Σ_{XX}. The likelihood in (8.3.4) also involves the unknown vector coefficients α. However, the values of α can generally be found up to a constant by regressing the sample selection probabilities $w(\mathbf{x}_i; \boldsymbol{\alpha})$ against α.

Stratified random sampling with T as stratification variable: Suppose the population is divided into L strata, the stratum U_h being of size N_h from which a *srswor* s_h of size n_h is drawn and the strata being defined by $0 < t^{(1)} \le t^{(2)} \le \cdots \le t^{(L-1)} < \infty$. Let $P_h = n_h / N_h$. By (8.3.2), the weighted pdf of X_i^w, the measurement for unit $i \in S$ is,

$$h^w(\mathbf{x}_i;\alpha,\delta)=f(\mathbf{x}_i|i\in s)=\begin{cases}P_1h(\mathbf{x}_i;\delta)/w & \text{if } t_i\le t^{(1)}\\ P_2h(\mathbf{x}_i;\delta)/w & \text{if } t^{(1)}\le t_i\le t^{(2)}\\ \ldots & \ldots\\ P_Lh(\mathbf{x}_i;\delta)/w & \text{if } t^{(L-1)}\le t_i\end{cases}. \quad (8.3.5)$$

For N_h sufficiently large the probability w can be approximated by

$$w=P(i\in s)=P_1\int_{-\infty}^{t^{(1)}}\phi(t)dt+\sum_{h=2}^{L-1}P_h\int_{t^{(h-1)}}^{t^{(h)}}\phi(t)dt+P_L\int_{t^{(L-1)}}^{\infty}\phi(t)dt \quad (8.3.6)$$

where $\phi(t)$ denotes the normal pdf of T with mean $\alpha'\mu_X$ and variance $\alpha'\mathbf{\Sigma}_{XX}\alpha$. Let $\Phi(t)$ denote the distribution function of T.

Suppose that the strata are large so that the sampling may be considered as being done with replacement within stratum. For given strata boundaries $\{t^{(h)}\}$ and the vector coefficients α, the likelihood function of δ can then be written from (8.3.3) as

$$L(\delta;\mathbf{X}_s,s)=\frac{\Pi_{i\in s}h(\mathbf{x}_i;\delta)\Pi_{h=1}^{L}P_h^{n_h}}{\{P_1\Phi(t^{(1)})+\sum_{h=2}^{L-1}P_h[\Phi(t^{(h)})-\Phi(t^{(h-1)})]+P_L[1-\Phi(t^{(L-1)})]\}^n}. \quad (8.3.7)$$

Often strata-boundaries are unknown and have to be estimated from sample data. If the values t_i, $i\in s$ are known, the vector α can be estimated from the regression of t_i on $\mathbf{x}_i$. Furthermore, if $t_{(1)}\le\cdots\le t_{(n)}$ are the ordered values of t_i's, the strata-boundaries can be estimated as $t^{(1)}=(t_{n_1}+t_{n_1+1})/2,\ldots,t^{(L-1)}=(t_{n^*}+t_{n^*+1})/2$, where $n^*=\sum_{h=1}^{L-1}n_h$. Substituting these estimators in (8.3.7) we get an approximate likelihood which can then be maximized with respect to *delta*. For further details the reader may refer to Krieger and Pfeffermann (1991).

8.4 Design-Adjusted Maximum Likelihood Estimation

Chambers (1986) considered parameter-estimation which is adjusted for sampling design. Let as usual $\mathbf{Y}_i=(Y_{i1},\ldots,Y_{ip})'$ be a $p\times 1$ vector representing the values of the response variables $y_1,\ldots,y_p$ on the unit i in the population ($i=1,\ldots,N$), $\mathbf{Y}=((Y_{ij}))$. We are interested in estimating the parameter θ in the joint distribution $F_N(.;\theta)$ of $\mathbf{Y}$, $\theta\in\Theta$. For this we proceed as follows.

Consider, a loss function $L_N(\theta)=L_N(\theta,\mathbf{Y})$ for estimating θ. For given $\mathbf{Y}$, a general approach to optimal estimation is to find an estimator θ_N such that

$$L_N(\theta_N)=\min[L_N(\theta),\mathbf{Y};\theta\in\Theta]. \quad (8.4.1)$$

The loss function may be chosen suitably. For example, if the joint distribution $F_N(.;\theta)$ is completely specified except θ, then $L_N(\theta,\mathbf{Y})$ may correspond to the inverse of the likelihood function for θ given $\mathbf{Y}$. Estimator obtained in this way will correspond to the maximum likelihood estimator of θ. If $F_N(.;\theta)$ is specified only up to its second moment, then minimization of $L_N(\theta, \mathbf{Y})$ may be made to correspond to least square method of estimation.

In practice, only $\mathbf{y}_s = \{\mathbf{Y}_i(i \in s)\}$ are observed, s being a sample of size n selected suitably. Let for given $\mathbf{y}_s$, $L_n(\theta, \mathbf{y}_s) = L_n(;\mathbf{y}_s)$ be the loss function defined on $\mathbf{y} \times \Theta$. An optimal estimator based on $\mathbf{y}_s$ is then θ_n such that

$$L_n(\theta_n) = \min[L_n(\theta); \theta \in \Theta]. \tag{8.4.2}$$

Now, a sample design can be characterized by an $N \times q$ matrix $\mathbf{Z} = ((Z_{ij}))$ of known population values of variables $z_1, \ldots, z_q$. For example, in stratified sampling z_j's may correspond to strata-indicator variables. Thus, $\mathbf{Z}_i = (Z_{i1}, \ldots, Ziq)'$ denotes the vector of q-strata indicators for the ith unit in the population. However, the actual process of sample selection or sample selection mechanism generally depends on factors other than in $\mathbf{Z}$.

Let $G_N(.;|\mathbf{Z};\lambda)$ denote the conditional distribution of $\mathbf{Y}$ given $\mathbf{Z}$. Here, λ is a parameter vector in this design model G_N. (Note that a design model is the superpopulation distribution of $\mathbf{Y}$ given the design-variable values $\mathbf{Z}$. Thus, a model of regression of $\mathbf{Y}$ on $\mathbf{Z}$ is a design model.) It is not the same as parameter vector θ involved in the joint distribution F_N of $\mathbf{Y}$. Let G_n denote the corresponding marginal conditional distribution of $\mathbf{y}_s$ given $\mathbf{Z}$. G_n is obtained from G_N by integrating out the nonobserved values $\mathbf{Y}_r = \{\mathbf{Y}_i, i \notin s\}$.

If the parameter λ is the same in both G_N and G_n, then the form of the design model G_n for the sample data does not depend on the sample selection mechanism. In this case, the sample selection mechanism is said to be *ignorable* (Rubin 1976). A sufficient condition for an ignorable sample selection mechanism is that sample selection process depends only on $\mathbf{Z}$. In this case, an optimal estimator of λ can be obtained by using the sample data, Z and G_n only. Note that in $G_n(.|\mathbf{Z})$ we examine the distribution of $\mathbf{y}_s$ given $\mathbf{Z}$.

In general, however, the sample selection mechanism may not be ignorable. In that case to draw inference about λ using the survey data we have to specify the design model for the whole population using the joint conditional distribution of $\mathbf{y}$ and $\mathbf{y}_r$ given $\mathbf{Z}$. This distribution is not the same as G_N and may be written as $H_N(;|\mathbf{Z};\lambda,\phi)$ which may involve a nuisance parameter ϕ.

In practice, it may be very difficult to specify H_N and hence H_n, while same is not the case with G_N and G_n. One therefore attempts to ensure that an ignorable selection mechanism is used whenever possible.

We shall henceforth assume in this section that the selection mechanism is ignorable. The problem is how to use the sample data, $\mathbf{Z}$ and G_n to produce a design-adjusted estimator of θ.

Let $G_N(.|\mathbf{y}, \mathbf{Z};\lambda) = G_N(.|\mathbf{y})$ denote the conditional distribution of $\mathbf{Y}$ given $\mathbf{Z}$ and the sample data $\mathbf{y}$ and $E_N[.|\mathbf{y}]$ the expectation with respect to $G_N(|\mathbf{y})$. Here, following the argument underlying the EM (expectation minimization)-algorithm of Dempster et al. (1977), instead of minimizing $L_N(\theta)$, we minimize $E_N[L_N(\theta)]$. An appropriate design-adjusted estimator of θ is the value θ_{ND} such that

$$E_N[L_N(\theta_{ND})|\mathbf{y}] = \min[E_N\{L_N(\theta)|\mathbf{y}\}; \theta \in \Theta]. \tag{8.4.3}$$

We shall try to provide some examples.

Example 8.4.1 Let the vector variable on the ith unit, $\mathbf{Y}_i$ be *iid* p-variate normal $N_p(\boldsymbol{\mu}, \boldsymbol{\Sigma}), i = 1, \ldots, N$. Given the sample data $\mathbf{y}_s$, we are required to find design-adjusted estimators of $(\mu, \boldsymbol{\Sigma})$. Let $\bar{\mathbf{Y}} = (\bar{Y}_1, \ldots, \bar{Y}_p)'$ denote the vector of population means of the variables, $\bar{Y}_j = \sum_{i=1}^N Y_{ij}/N$; $\mathbf{S} = ((S_{jk}))$ the finite population covariance matrix of $Y_1, \ldots, Y_p$; $S_{jk} = \sum_{i=1}^N (Y_{ij} - \bar{Y}_j)(Y_{ik} - \bar{Y}_k)/(N-1)$. Let $\boldsymbol{\Sigma}^{-1} = ((\Sigma^{jk}))$.

Taking the loss function as proportional to the logarithm of the inverse of the likelihood functions of $(\mu, \boldsymbol{\Sigma})$,

$$\begin{aligned} L_N(\mu, \boldsymbol{\Sigma}) &= \sum_{i=1}^N (\mathbf{Y}_i - \mu)'\boldsymbol{\Sigma}^{-1}(\mathbf{Y}_i - \mu) - N \log |\boldsymbol{\Sigma}^{-1}| \\ &= \sum_{j=1}^p \sum_{k=1}^p \Sigma^{jk}[S_{jk} + N(\bar{Y}_j - \mu_j)(\bar{Y}_k - \mu_k)] - N \log |\boldsymbol{\Sigma}^{-1}|. \end{aligned} \tag{8.4.4}$$

The corresponding design-adjusted loss function is obtained by taking expectation of (8.4.4) with respect to the conditional distribution of $\mathbf{Y}$ given $\mathbf{Z}$ and $\mathbf{y}_s$ with unknown parameters in the model being replaced by estimates from the survey data.

It is shown that the value of $\boldsymbol{\mu}$ and $\boldsymbol{\Sigma}$ minimizing $E_N[L_n(\mu, \boldsymbol{\Sigma})|\mathbf{y}_s]$ are given by

$$\begin{aligned} \mu_{ND} &= E_N(\bar{\mathbf{Y}}|\mathbf{y}_s) \\ \boldsymbol{\Sigma}_{ND} &= N^{-1}[\sum_{i \in s} \mathbf{y}_i\mathbf{y}_i' + \sum_{i \notin s} E_N(\mathbf{Y}_i\mathbf{Y}_i'|\mathbf{y}_s)] - \mu_{ND}\mu_{ND}'. \end{aligned} \tag{8.4.5}$$

Example 8.4.2 Stratified Random Sampling: Here the design variables $z_1, \ldots, z_q$ are stratum-indicator variables and $\mathbf{Z}$ is the matrix of their values; N_h, n_h, s_h will have their usual meanings. It is assumed that the study variables $\mathbf{Y}_{h1}, \ldots, \mathbf{Y}_{hN_h}$ within stratum h are *iid* $N_p(\mu_h, \boldsymbol{\Sigma}_h)$. Also such variables are assumed to be independent from stratum to stratum.

It is known that given the sample data $\mathbf{y}_{hs}$, the MLE's of μ_h and $\boldsymbol{\Sigma}_h$ are respectively, $\bar{\mathbf{y}}_h$, the vector of sample means and $\mathbf{s}_h$, the matrix of sample covariance matrix for the hth stratum. It then follows by (8.4.5) that the design-adjusted MLE of μ and $\boldsymbol{\Sigma}$ are

$$\begin{aligned} \mu_{ND} &= N^{-1}\textstyle\sum_h N_h\bar{\mathbf{y}}_h \\ \boldsymbol{\Sigma}_{ND} &= N^{-1}\textstyle\sum_h \left(\frac{N_h}{n_h}\right) \sum_{s_h} (\mathbf{Y}_i - \mu_{ND})(\mathbf{Y}_i - \mu_{ND})'. \end{aligned} \tag{8.4.6}$$

It is seen that both μ_{ND} and $\boldsymbol{\Sigma}_{ND}$ are design-based estimates of μ and $\boldsymbol{\Sigma}$ when the sample is selected randomly within the stratum. Under proportional allocation of sample size, both the estimators reduce to the conventional (unadjusted) estimators.

Example 8.4.3 Sampling with auxiliary information: Let $q = 1$ and assume that the design variable z which is a measure of size of a unit is linearly related to each of the p study variables $y_1, \ldots, y_p$. Assume that conditional on $\mathbf{Z}$, $\mathbf{Y}_i$ are independently distributed as $N_p(\mu_i, \mathbf{\Sigma})$ where $\mu_i = \boldsymbol{\alpha} + \beta Z_i$; $\boldsymbol{\alpha} = (\alpha_1, \ldots, \alpha_p)'$; $\beta = (\beta_1, \ldots, \beta_p)'$.

Let $\bar{\mathbf{y}}$ denote the vector of sample means of p study variables, $\bar{z}$ and s_{zz} the sample mean and sample variance of z. The MLE of $\boldsymbol{\alpha}$, β and $\mathbf{\Sigma}$ are

$$\begin{aligned} \boldsymbol{\alpha}_n &= \bar{\mathbf{y}} - \beta_n \bar{z}, \\ \beta_n &= n^{-1} \sum_{i \in s} (\mathbf{Y}_i - \bar{\mathbf{y}})(Z_i - \bar{z})/s_{zz}, \\ \mathbf{\Sigma}_n &= n^{-1} \sum_{i \in s} (\mathbf{Y}_i - \boldsymbol{\alpha}_n - \boldsymbol{\beta}_n Z_i)(\mathbf{Y}_i - \boldsymbol{\alpha}_n - \boldsymbol{\beta}_n Z_i)'. \end{aligned} \tag{8.4.7}$$

From (8.4.5), the design-adjusted MLE of $\mu = N^{-1} \sum_{i=1}^{N} \mu_i$ is then

$$\mu_{ND} = \bar{\mathbf{y}} + (\bar{Z} - \bar{z})\beta_n \tag{8.4.8}$$

where $\bar{Z}$ is the population mean. It is seen that (8.4.8) is the regression estimator of the population mean vector $\bar{\mathbf{Y}}$.

Let S_{zz}, s_{zz} be respectively the population variance and sample variance of z and $\mathbf{s}_{yy}$ the covariance matrix of $y_1, \ldots, y_p$ based on the sample data. Then from (8.4.5) it follows that

$$\mathbf{\Sigma}_{ND} = \mathbf{s}_{yy} + \beta_n \beta_n' (S_{zz} - s_{zz}). \tag{8.4.9}$$

Clearly, if the sampling design is balanced on z up to the second order, that is, if $\bar{z} = \bar{Z}$ and $s_{zz} = S_{zz}$, then both the design-adjusted estimators (8.4.8) and (8.4.9) reduce to the respective conventional estimators.

Example 8.4.4 Design-adjusted estimation of regression coefficient: Let $\mathbf{Y}$ denote a N-vector of population values of the study variable y and $\mathbf{X}$ the $N \times p$ matrix of the population values of the independent variables $x_1, \ldots, x_p$. Consider, the regression model

$$E(\mathbf{Y}|\mathbf{X}) = \mathbf{X}\beta,\ Var(\mathbf{Y}|\mathbf{X}) = \sigma^2 \mathbf{A} \tag{8.4.10}$$

where β is the $p \times 1$ vector of regression coefficients, $\sigma > 0$ is a unknown scale parameter and $\mathbf{A}$ is a diagonal matrix with strictly positive diagonal terms.

Let $\mathbf{y}, \mathbf{x}, \mathbf{a}$ denote the sample components of $\mathbf{Y}, \mathbf{X}, \mathbf{A}$, respectively. The corresponding nonsampled components are $\mathbf{Y}_r$, $\mathbf{X}_r$ and $\mathbf{A}_r$. If the model (8.4.10) applies even if the population values are replaced by their sample components, then β can be estimated in a straightforward manner using the sample data only. In practice, the model does not often remain valid for the sample data, due to the effect of sampling design (vide, for example, Holt et al. 1980).

Let $\mathbf{Z}$ be the $N \times q$ matrix of values of the variables used in the sampling design. Let $\mathbf{X}_j$ denote the $N \times 1$ column vector of values x_{ij} taken by the independent variable x_j on the population units ($j = 1, \ldots, p$; $i = 1, \ldots, N$). We consider the following design model, linking the superpopulation model (8.4.10) with the design-variable values $\mathbf{Z}$. We assume that

$$\begin{aligned} E(\mathbf{Y}|\mathbf{Z}) &= \mathbf{Z}\boldsymbol{\lambda}_0,\ E(\mathbf{X}_j|\mathbf{Z}) = \mathbf{Z}\lambda_j \\ Var(\mathbf{Y}|\mathbf{Z}) &= \chi_{00}\mathbf{I}_N,\ Var(\mathbf{X}_j|\mathbf{Z}) = \chi_{jj}\mathbf{I}_N, \\ \text{Cov}(\mathbf{Y}, \mathbf{X}_j|\mathbf{Z}) &= \chi_{0j}\mathbf{I}_N, \text{Cov}(\mathbf{X}_j, \mathbf{X}_k|\mathbf{Z}) = \mathbf{X}_{jk}\mathbf{I}_N \\ &(j, k = 0, 1, \ldots, p). \end{aligned} \tag{8.4.11}$$

Here $\lambda_j (j = 0, 1, \ldots, p)$ are unknown $q \times 1$ vectors and $\chi_{jk} (j, k = 0, 1, \ldots, p)$ are unknown scalar parameters.

Given (8.4.10), the standard loss function for the estimation of β is the quadratic loss function,

$$L_N(\beta) = (\mathbf{Y} - \mathbf{X}\beta)'\mathbf{A}^{-1}(\mathbf{Y} - \mathbf{X}\beta). \tag{8.4.12}$$

The corresponding design-adjusted loss function is obtained by taking expectation of (8.4.12) with respect to the conditional distribution of $\mathbf{Y}$ given $\mathbf{Z}$ and $\mathbf{y}$ with unknown parameters in the model being replaced by estimates from the survey data. Thus,

$$\begin{aligned} E_N[L_N(\beta)|\mathbf{y}, \mathbf{x}] = \mathbf{y}'\mathbf{a}^{-1}\mathbf{y} + E_N(\mathbf{Y}_r'\mathbf{A}_r^{-1}\mathbf{Y}_r|\mathbf{y}, \mathbf{x}) - 2\beta'\mathbf{x}'\mathbf{a}^{-1}\mathbf{y} \\ - 2\beta' E_N(\mathbf{X}_r'\mathbf{A}_r^{-1}\mathbf{X}_r|\mathbf{y}, \mathbf{x}) + \beta'\mathbf{x}'\mathbf{a}^{-1}\mathbf{x}\beta \\ + \beta' E_N(\mathbf{X}_r'\mathbf{A}_r^{-1}\mathbf{X}_r|\mathbf{y}, \mathbf{x})\beta \end{aligned} \tag{8.4.13}$$

where $\mathbf{a}$ is the diagonal submatrix of $\mathbf{A}$ corresponding to s. Therefore, design-adjusted estimator of β is

$$\beta_{ND} = [\mathbf{x}'\mathbf{a}^{-1}\mathbf{x} + E_N(\mathbf{X}_r'\mathbf{A}_r^{-1}\mathbf{X}_r|\mathbf{y}, \mathbf{x}]^{-1}[\mathbf{x}'\mathbf{a}^{-1}\mathbf{y} + E_N(\mathbf{X}_r'\mathbf{A}_r^{-1}\mathbf{Y}_r|\mathbf{y}, \mathbf{x})] \tag{8.4.14}$$

provided the inverse above exists.

Special cases:
(a) Stratified sampling design: Suppose $p = 2$ and $\mathbf{X}_1 = \mathbf{1}_N$. We will denote by $\mathbf{X}_2$, the second column of the design matrix $(\mathbf{X}_1, \mathbf{X}_2)_{N\times 2}$. It is assumed that $\mathbf{A} = \mathbf{I}_N$.

Let $\bar{\mathbf{y}} = (\bar{\mathbf{y}}_1, \ldots, \bar{\mathbf{y}}_q)'$ and $\bar{\mathbf{x}} = (\bar{\mathbf{x}}_1, \ldots, \bar{\mathbf{x}}_q)'$ denote the vectors of stratum sample means of y, x respectively. These are the best linear unbiased estimators of λ_0, λ_2 in (8.4.11). Let $s_{jk}, j, k = 0, 2$ denote the usual unbiased estimates of the scale-parameters $\chi_{j,k}$ in (8.4.11). Thus,

$$s_{02} = \frac{1}{n - q}\sum_{h=1}^{q}\sum_{i\in s_h}(Y_{ih} - \bar{y}_h)(X_{ih} - \bar{x}_h)$$

where Y_{ih}, X_{ih} are the values of Y, X respectively for the ith sample element in stratum h. Let $\beta_{ND} = (\beta_{ND1}, \beta_{ND2})'$. Substituting in (8.4.14),

$$\beta_{ND1} = N^{-1}\sum_{h=1}^{q} N_h(\bar{y}_h - \beta_{ND2}\bar{x}_h) = \mu_{NDY} - \beta_{ND2}\mu_{NDX}$$
$$\beta_{ND2} = \frac{(N-q)s_{02} + \sum_{h=1}^{q} N_h(\bar{y}_h - \mu_{NDY})(\bar{x}_h - \mu_{NDX})}{(N-q)s_{22} + \sum_{h=1}^{q} N_h(\bar{x}_h - \mu_{NDX})^2} \quad (8.4.15)$$

where $\mu_{NDY} = N^{-1}\sum_{h=1}^{q} N_h\bar{y}_h$ and $\mu_{NDX} = N^{-1}\sum_{h=1}^{q} N_h\bar{x}_h$.

8.4.1 Design-Adjusted Regression Estimation with Selectivity Bias

Chambers (1988) developed a method of design-adjusted regression when there is nonresponse in the observed sample data.

Let Y_i represent the value of the study variable $'y'$ and $\mathbf{X}_i = (X_{i0}, X_{i1}, \ldots, X_{ip})'$, a vector of values of the $(p+1)$ auxiliary variables $\mathbf{x} = (x_0, x_1, \ldots, x_p)'$, $x_{i0} = 1$ for unit $i(= 1, \ldots, N)$ in the population. The problem of interest is the estimation of $(p+1)$-regressor-vector β defined by the model

$$\mathbf{Y} = \mathbf{X}\beta + \mathbf{U} \quad (8.4.16)$$

where $\mathbf{Y} = (Y_1, \ldots, Y_N)'$, $\mathbf{X} = ((X_{ij}))$, $\mathbf{U} = (U_1, \ldots, U_N)'$. It is assumed that $E(\mathbf{U}) = \mathbf{0}$, Cov $(\mathbf{U}) = \sigma^2\mathbf{I}_N$.

It is also assumed that with each unit i, there is a censoring variable V_i with a threshold value C_i such that if i is selected in the sample s of size n, response Y_i is available if only $V_i \le C_i$. If $V_i > C_i$, the value Y_i is censored (that is, some imputed value is substituted as its value) if $i \in s$. Neither V_i nor C_i is observable.

We denote by $\mathbf{y}_1, \mathbf{x}_1, \mathbf{v}_1, \mathbf{c}_1$ the vector of Y_i values, $n_1 \times (p+1)$ matrix of $\mathbf{x}$-values, the vector of V_i's and C_i's for the uncensored sample units; the corresponding quantities for the n_2 censored sample units are $\mathbf{y}_2, \mathbf{x}_2, \mathbf{v}_2, \mathbf{c}_2$; $n_1 + n_2 = n$. After s is selected, $\mathbf{y}_1, \mathbf{x}_1, \mathbf{x}_2$ are observed. Let $\mathbf{Y}_r, \mathbf{X}_r$ denote the nonsample components of $\mathbf{Y}$ and $\mathbf{X}$, respectively. For estimating β consider a unweighted square error loss function

$$L_N(\beta : \mathbf{Y}, \mathbf{X}) = (\mathbf{Y} - \mathbf{X}\beta)'(\mathbf{Y} - \mathbf{X}\beta). \quad (8.4.17)$$

Let $E_s(.)$ denote expectation conditional on the observed sample and with respect to a model for the joint distribution of y and $\mathbf{x}$ which also accommodates the effect of censoring mechanism described above. The design-based estimator of β is then obtained by minimizing $E_s[L_N(\beta : \mathbf{Y}, \mathbf{X})]$ with unknown parameters replaced by their estimators. From (8.4.16) and (8.4.17), the design-adjusted estimator of β is

$$\hat{\beta} = \mathbf{A}^{-1}\mathbf{B} \tag{8.4.18}$$

where

$$\begin{aligned} \mathbf{A} &= \mathbf{x}_1'\mathbf{x}_1 + \mathbf{x}_2'\mathbf{x}_2 + \hat{E}_s(\mathbf{X}_r'\mathbf{X}_r), \\ \mathbf{B} &= \mathbf{x}_1'\mathbf{y}_1 + \mathbf{x}_2'\hat{E}_s(\mathbf{y}_2) + \hat{E}_s(\mathbf{X}_r'\mathbf{Y}_r). \end{aligned} \tag{8.4.19}$$

Censoring model without design adjustment
Suppose the population corresponds to the sample so that $\mathbf{Y}_r, \mathbf{X}_r$ are nonexistent. Hence, from (8.4.18),

$$\hat{\beta} = (\mathbf{x}\mathbf{x}')^{-1}[\mathbf{x}_1'\mathbf{y}_1 + \mathbf{x}_2'\hat{E}_s(\mathbf{y}_2)]. \tag{8.4.20}$$

We assume that the censoring model is such that (U_i, V_i) are *iid* bivariate normal $N_2(0, 0; \sigma^2, 1, \text{cov} = \omega)$, $i = 1, \ldots, N$. It then follows that

$$U_i = \omega V_i + e_i \tag{8.4.21}$$

where the random variable e_i is distributed independently of V_i with $E(e_i) = 0$, $V(e_i) = \sigma^2 - \omega^2$. This implies

$$Y_i = \mathbf{X}_i'\beta + U_i = \mathbf{X}_i'\beta + \omega V_i + e_i \tag{8.4.22}$$

and therefore,

$$E(Y_i|\mathbf{X}_i, V_i > C_i) = \mathbf{X}_i'\beta + \omega E(V_i > C_i). \tag{8.4.23}$$

Define

$$\mathbf{k}_2 = E(\mathbf{v}_2|\mathbf{v}_2 > \mathbf{c}_2). \tag{8.4.24}$$

Then $E(\mathbf{y}_2|\mathbf{x}_2, \mathbf{v}_2) = \mathbf{x}_2\beta + \omega\mathbf{k}_2$. Hence, from (8.4.20),

$$\hat{\beta} = (\mathbf{x}'\mathbf{x})^{-1}[\mathbf{x}_1'\mathbf{y}_1 + \mathbf{x}_2'(\mathbf{x}_2\hat{\beta} + \hat{\omega}\hat{k}_2)]. \tag{8.4.25}$$

Given $\mathbf{c}_2$, $\mathbf{k}_2$ can be calculated if V_i's are normally distributed. In this case, the components of $\mathbf{k}_2$ are K_i where

$$K_i = f(C_i)[1 - F(C_i)]^{-1}$$

and f, F are respectively density function and distribution function of a $N(0, 1)$ variable. In practice, $\mathbf{c}_2$ is not known.

If it can be assumed that $C_i = \mathbf{W}_i'\lambda$ where $\mathbf{W}_i$ is a known vector, then probability $V_i \leq C_i$ is $F(\mathbf{W}_i'\lambda)$ and standard probit analysis can be used to estimate λ. An estimate of $\mathbf{c}_2$, and hence of $\mathbf{k}_2$ then follows.
To estimate ω, we note that if the censoring variables V_i were observed, ω would be estimated by $n^{-1}\sum_{i\in s}(Y_i - \mathbf{X}_i'\beta)V_i$. An estimate of ω would then be

$$\hat{\omega} = \hat{E}_s\{n^{-1}\sum_{i\in s}(Y_i - \mathbf{X}_i'\beta)V_i|\mathbf{y}_1, \mathbf{x}\}. \tag{8.4.26}$$

Now, for a censored sample unit,

$$E(Y_i - \mathbf{X}_i'\beta)Y_i = \omega E(V_i^2) = \omega[1 + C_if(C_i)\{1 - F(C_i)\}^{-1}]. \tag{8.4.27}$$

This leads to the equation

$$\hat{\omega} = n^{-1}[\hat{\omega}\{n_2 + \sum_2 C_iK_i\} - \sum_1(Y_i - \mathbf{X}_i'\hat{\beta})H_i] \tag{8.4.28}$$

where $\sum_1$, $\sum_2$ denote summation over the noncensored and censored sample units respectively, $K_i = f(C_i)[1 - f(C_i)]^{-1}$ and $H_i = f(C_i)/F(C_i)$. Given estimated values of C_i from an initial probit analysis, equations (8.4.25) and (8.4.28) can be used to define an iterative scheme to estimate β and ω.

Censoring model with design adjustment
Consider a sample drawn by stratified sampling design with sample data subject to censoring as above. Let $\mathbf{X} = (1, \mathbf{Z})'$ with the ith row $\mathbf{X}_i' = (1 \;\; \mathbf{Z}_i')$ where $\mathbf{Z}_i = (x_{1i}, \ldots, x_{pi})'$. Assume that within each stratum h, $(Y_i, \mathbf{Z}_i)$ are *iid* and $(U_i, V_i)'$ are also *iid* $N_2(0, 0; \sigma^2, 1, \omega)$ where, of course, σ^2, ω are allowed to vary across strata. Now,

$$\hat{\beta} = [\mathbf{x}'\mathbf{x} + \hat{E}_s(\mathbf{X}_r'\mathbf{X}_r)]^{-1}[\mathbf{x}_1'\mathbf{y}_1 + \mathbf{x}_2'\hat{E}_s(\mathbf{y}_2) + \hat{E}_s(\mathbf{X}_r'\mathbf{Y}_r)]$$

where $\hat{E}_s$ denotes conditional expectation under stratified model and the censoring process. Let $\hat{\omega}_h$ denote estimate of ω in stratum h, obtained as in equation (8.4.26). For a censored sample unit in stratum h, $\hat{E}_s(Y_i) = \mathbf{X}_i'\hat{\beta} + \hat{\omega}_hK_i$. Also

$$\hat{E}_s(\mathbf{X}_r'\mathbf{X}_r) = \sum_h(N_h - n_h)\begin{bmatrix} 1 & \bar{\mathbf{z}}_h' \\ \bar{\mathbf{z}}_h & \mathbf{s}_{hZZ} + \bar{\mathbf{z}}_h\bar{\mathbf{z}}_h' \end{bmatrix} \tag{8.4.29}$$

where $\bar{\mathbf{z}}_h, \mathbf{s}_{hzz}, N_h, n_h$ denote respectively the vector of sample means, sample covariance matrix of $\mathbf{z}$-values, population size, and sample size of stratum h.

In calculation of $\hat{E}_s(\mathbf{X}_r'\mathbf{Y}_r)$, the sample-based estimator of the stratified superpopulation model parameters need to be adjusted for the bias induced by censoring. Let $\bar{\mathbf{y}}_{h1}$ denote the mean of the Y_i values for the noncensored sample units and $\bar{k}_{h2}$ the mean of the n_{h2} values of K_i for the censored sample units in stratum h. It is then shown that

$$\hat{E}_s(\mathbf{X}_r'\mathbf{Y}_r) = \sum_h (N_h - n_h)[\hat{E}_s(\bar{\mathbf{y}}_h)\{\hat{E}_s(\mathbf{s}_{hYZ}) + \bar{\mathbf{z}}_h\hat{E}_s(\bar{y}_h)\}']' \tag{8.4.30}$$

where

$$\begin{aligned}\hat{E}_s(\bar{y}_h) &= n_h^{-1}[n_{h1}\bar{y}_{h1} + n_{h2}\{\bar{\mathbf{x}}_{h2}'\hat{\beta} + \hat{\omega}_h\bar{k}_{h2}\}]\\ \hat{E}_s(\mathbf{s}_{hYZ}) &= n_h^{-1}\left[\sum\nolimits_{h1} Y_i(\mathbf{Z}_i - \bar{\mathbf{z}}_h) + \sum\nolimits_{h2}(\mathbf{X}_i'\hat{\beta} + \hat{\omega}_h K_i)(\mathbf{Z}_i - \bar{\mathbf{z}}_h)\right].\end{aligned} \tag{8.4.31}$$

Again, in estimating C_i using the model $C_i = \mathbf{W}_i'\lambda$, the marginal distribution of W_i will vary across strata, as also the censoring probability. Consequently, estimation of censoring model parameter λ must also be made design-adjusted.

8.5 The Pseudo-Likelihood Approach to MLE from Complex Surveys

Binder (1983), Chambless and Boyle (1985), Krieger and Pfeffermann (1991), among others, utilized the sample-selection probabilities to estimate the census likelihood equations. The estimated likelihoods are then maximized with respect to the vector parameter of interest.

Suppose that the population values $\mathbf{Y}_i$ are independent draws from a common distribution $f(\mathbf{Y};\theta)$ and let $l_N(\theta;\mathbf{Y}) = \sum_{i=1}^N \log f(\mathbf{Y}_i;\theta)$ define the census log-likelihood function. Under some regularity conditions, the MLE $\hat{\theta}$ solves the likelihood equations

$$\mathbf{U}(\hat{\theta}) = \frac{\partial l_N(\theta;\mathbf{Y})}{\partial\theta} = \sum_{i=1}^N \mathbf{u}(\theta;\mathbf{y}_i) = \mathbf{0} \tag{8.5.1}$$

where $\mathbf{u}(\theta;\mathbf{y}_i) = (\partial \log f(\mathbf{Y}_i;\theta))/(\partial\theta)$. The pseud-MLE of θ is defined as the solution of the equations $\hat{\mathbf{U}}(\theta) = \mathbf{0}$, where $\hat{\mathbf{U}}(\theta)$ is a design-consistent estimate of $\mathbf{U}(\theta)$ in the sense that $p\lim_{n\to\infty,N\to\infty}[\hat{\mathbf{U}}(\theta) - \mathbf{U}(\theta)] = \mathbf{0}\ \forall\ \theta \in \Theta$. The commonly used estimator of $\mathbf{U}(\theta)$ is the Horvitz–Thompson (1952) estimator so that the pseudo-MLE of θ is the solution of

$$\hat{\mathbf{U}}(\theta) = \sum_{i\in s} w_i\mathbf{u}_i(\theta;\mathbf{y}_i) = \mathbf{0} \tag{8.5.2}$$

where $w_i^{-1} = \pi_i$ is the inclusion probability of unit i in the sample s.

For the multivariate normal model with parameters $\boldsymbol{\mu}_Y$ and $\boldsymbol{\Sigma}_Y$, pseudo-MLE's obtained by using (8.5.2) are

$$\tilde{\mu}_Y = \sum_{i \in s} w_i \mathbf{y}_i / \sum_{i \in s} w_i$$
$$\tilde{\mathbf{\Sigma}}_Y = \sum_{i \in s} w_i (\mathbf{y}_i - \tilde{\mu}_Y)(\mathbf{y}_i - \tilde{\mu}_Y)' / \sum_{i \in s} w_i. \quad (8.5.3)$$

Properties of the pseudo-MLE
Suppose the score function $\mathbf{u}$ is such that $E_M[\mathbf{u}(\theta; \mathbf{y}_i)] = \mathbf{0}$ where E_M denotes expectation with respect to model M. Then under some regularity conditions, the pseudo-likelihood equations (8.5.2) are closest to the census likelihood equations (8.5.1) in the mean-square-error (MSE) sense, among all design-unbiased estimating equations. The MSE is taken, here, both with respect to randomization distribution of sample-selection and the model. This result follows by the property of Horvitz–Thompson estimator.

Another property of the use of probability-weighted statistics (8.5.2) instead of the simple unweighted statistics is that the resulting estimators are in general design-consistent for the estimators obtained by solving the census likelihood equations, irrespective of whether the model is correct and/or the sampling design depends on the $\mathbf{y}$-values or not (Pfeffermann 1992).

Various models under the pseudo-likelihood approach have been studied by Binder (1983), Chambles and Boyle (1985), Roberts et al. (1987), Skinner et al. (1989), among others. We shall consider here the results due to Binder (1983).

8.5.1 Analysis Based on Generalized Linear Model

Defining Finite Population Parameters

The generalized linear models described by Nelder and Wedderburn (1972) is defined as follows. Suppose the observation y_k is a random sample from an exponential family of distributions with parameters θ_k and ϕ:

$$p(y_k; \theta_k, \phi) = \exp[\alpha(\phi)\{y_k\theta_k - g(\theta_k) + h(y_k)\} + \gamma(\phi, y_k)], k = 1, 2, \ldots, \quad (8.5.4)$$

where $\alpha(\phi) > 0$. For this distribution $E(y_k) = g'(\theta_k) = \mu(\theta_k)$ and $V(y_k) = \mu'(\theta_k)/\alpha(\phi)$. θ_k is called the *natural parameter* of the distribution.

We assume that for each k we have access to the value $\mathbf{x}_k = (x_{k1}, \ldots, x_{kp})'$ of a vector of auxiliary variables $\mathbf{x} = (x_1, \ldots, x_p)'$. Let $\theta_k = f(\mathbf{x}_k'\beta)$ where $f(.)$ is a known differentiable function and $\beta = (\beta_1, \ldots, \beta_p)'$ a set of unknown parameters. For observations $(y_k, \mathbf{x}_k')$, $k = 1, \ldots, N$ the log-likelihood function of β is

$$L(\beta|\mathbf{y}, \mathbf{X}) = \log p(\mathbf{y}; \theta, \phi)$$
$$= \alpha(\phi) \sum_{k=1}^{N} \{y_k\theta_k - g(\theta_k) + h(y_k)\} + \sum_{k=1}^{N} \gamma(\phi, y_k),$$

where $\mathbf{y} = (y_1, \ldots, y_N)'$, $\theta = (\theta_1, \ldots, \theta_N)'$ and $\mathbf{X}_{N\times p} = ((x_{ki}))$. Now $\partial\theta_k/\partial\beta_i = f'(\mathbf{x}_k'\beta)x_{ki}$. Hence, the likelihood equations are

$$\begin{aligned}\frac{\partial L}{\partial\beta_i} &= \sum_{k=1}^{N}\left(\frac{\partial L}{\partial\theta_k}\right)\cdot\left(\frac{\partial\theta_k}{\partial\beta_i}\right)\\ &= \alpha(\phi)\sum_{k=1}^{N}\{y_k - g'(\theta_k)\}f'(\mathbf{x}_k'\beta)x_{ki}\\ &= \alpha(\phi)\sum_{k=1}^{N}[y_k - \mu\{f(\mathbf{x}_k'\beta)\}]f'(\mathbf{x}_k'\beta)x_{ki} = 0,\ i = 1, \ldots, p.\end{aligned} \tag{8.5.5}$$

If $\mu(.)$ and $f'(.)$ are strictly monotonic functions, then the solution to Eq. (8.5.5) is unique, provided the matrix $\mathbf{X}$ is of full rank.

In the finite population context we assume that for the kth unit, we observe the data $(Y_k, X_{k1}, \ldots, X_{kp}) = (Y_k, \mathbf{X}_k')$. We define the population vector $\mathbf{B} = (B_1, \ldots, B_p)'$ as the solution of the equations

$$\sum_{k=1}^{N}[Y_k - \mu\{f(\mathbf{X}_k'\mathbf{B})\}]f'(\mathbf{X}_k'\mathbf{B})X_{ki} = 0\ (i = 1, \ldots, p). \tag{8.5.6}$$

Expression (8.5.6) can be more generally written as

$$\sum_{k=1}^{N}\mathbf{u}(Y_k, X_k; \mathbf{B}) = \mathbf{0} \tag{8.5.7}$$

where $\mathbf{u}$ is a p-dimensional vector-valued function of the data $(Y_k, \mathbf{X}_k)$ and the parameter $\mathbf{B}$.

We now introduce a slight change of notation. Let us write $\mathbf{Z}_k = (Z_{k1}, \ldots, Z_{kq})'$ as the data vector for the kth unit, $\mathbf{Z} = ((Z_{ki}))$ as the $N \times q$ matrix of data values for all the units in the population. Instead of $\mathbf{B}$ we shall write the finite population parameters as $\psi = (\psi_1, \ldots, \psi_p)'$. Following (8.5.7) the finite population parameters are now defined by the expression of the form

$$\mathbf{W}_N(\psi) = \sum_{k=1}^{N}\mathbf{u}(\mathbf{Z}_k; \psi) - \mathbf{v}(\psi) = \mathbf{0} \tag{8.5.8}$$

where $\mathbf{v}(\psi)$ is completely known for a given ψ, but $\mathbf{u}(\mathbf{Z}_k; \psi)$ is known only for those units which occur in the sample.

We now estimate ψ, derive asymptotic variance of $\hat{\psi}$ by Taylor series expansion and find an estimator of this asymptotic variance.

Estimation of ψ

The term $\sum_{k=1}^{N} \mathbf{u}(\mathbf{Z}_k; \psi) = \mathbf{U}(\psi)$ (say) is the population total of functions $\mathbf{u}(\mathbf{Z}_k; \psi)$. We represent the sample-based estimator of this total as $\hat{\mathbf{U}}(\psi)$. Assume that $\hat{\mathbf{U}}(\psi)$ is asymptotically normally distributed with mean $\mathbf{U}(\psi)$ and variance $\mathbf{\Sigma}_{\mathbf{U}}(\psi)$. Let

$$\dot{\mathbf{W}}(\psi) = \hat{\mathbf{U}}(\psi) - \mathbf{v}(\psi). \tag{8.5.9}$$

(Chambless and Boyle (1985) derived asymptotic normality of $\hat{\mathbf{U}}(\boldsymbol{\psi})$ using an approach due to Fuller (1975).) Following (8.5.8), the estimator $\hat{\psi}$ is defined by the solution of the estimating equation

$$\dot{\mathbf{W}}(\hat{\psi}) = \hat{\mathbf{U}}(\hat{\psi}) - \mathbf{v}(\hat{\psi}) = \mathbf{0}. \tag{8.5.10}$$

Example 8.5.1 In the linear regression case, we have the estimator of parameter regression coefficient $\boldsymbol{\beta}$ given by the solution of the equations

$$\mathbf{S}_{xx}\hat{\beta} = \mathbf{S}_{xy} \tag{8.5.11}$$

where $\mathbf{S}_{xx}$ and $\mathbf{S}_{xy}$ are respectively estimates of $\mathbf{X}'\mathbf{X}$ and $\mathbf{X}'\mathbf{Y}$. Considering that $\mathbf{S}_{xy} - \mathbf{S}_{xx}\hat{\beta} = \sum_{k=1}^{N}(y_k - \mathbf{x}_k'\hat{\beta})\mathbf{x}_k$ we set $\mathbf{u}(y_k, \mathbf{x}_k; \beta) = (y_k - \mathbf{x}_k'\beta)\mathbf{x}_k$ and $\mathbf{v}_k = \mathbf{0}$. □

Variance of $\hat{\boldsymbol{\psi}}$

To obtain the variance of $\hat{\psi}$, we take a Taylor-series expansion of $\dot{\mathbf{W}}(\hat{\psi})$ around $\hat{\psi} = \psi_0$, the true value of ψ. We obtain

$$\mathbf{0} = \dot{\mathbf{W}}(\hat{\psi}) \approx \dot{\mathbf{W}}(\psi_0) + \left[\frac{\partial \dot{\mathbf{W}}(\psi)}{\partial \psi}\right]_{\psi_o} (\hat{\psi} - \psi_0)$$

so that

$$\dot{\mathbf{W}}(\psi_0) \approx -\left[\frac{\partial \dot{\mathbf{W}}(\psi)}{\partial \psi}\right]_{\psi_0} (\hat{\psi} - \psi_0).$$

Calculating the variance of both sides, we obtain in the limit,

$$\mathbf{\Sigma}_{\mathbf{U}}(\psi_0) = \left[\frac{\partial \mathbf{W}_N(\psi_0)}{\partial \psi_0}\right] V(\hat{\psi}) \left[\frac{\partial \mathbf{W}_N(\psi_0)}{\partial \psi_0}\right]' \tag{8.5.12}$$

since,

$$\frac{\partial \dot{\mathbf{W}}(\psi_0)}{\partial \psi_0} = \frac{\partial \{\hat{\mathbf{U}}(\psi_0) - \mathbf{v}(\psi_0)\}}{\partial \psi_0} \approx \frac{\partial \mathbf{W}_N(\psi_0)}{\partial \psi_0}.$$

From (8.5.12), it follows that

$$V(\hat{\psi}) = \left[\frac{\partial \mathbf{W}_N(\psi_0)}{\partial \psi_0}\right]^{-1} \mathbf{\Sigma}_{\mathbf{U}}(\psi_0) \left\{\left[\frac{\partial \mathbf{W}_N(\psi_0)}{\partial \psi_0}\right]'\right\}^{-1} \tag{8.5.13}$$

provided the matrix $\frac{\partial \mathbf{W}_N(\psi_0)}{\partial \psi_0}$ is of full rank. An estimate of $V(\hat{\psi})$ is

$$\hat{V}(\hat{\psi}) = \left[\frac{\partial \dot{\mathbf{W}}(\hat{\psi})}{\partial \hat{\psi}}\right]^{-1} \hat{\boldsymbol{\Sigma}}_{\mathbf{U}}(\hat{\psi}) \left\{\left[\frac{\partial \dot{\mathbf{W}}(\hat{\psi})}{\partial \hat{\psi}}\right]'\right\}^{-1}. \tag{8.5.14}$$

8.5.2 *Estimation for Linear Models*

We shall now apply the above-mentioned results to different models.

Generalized Linear Model

We shall now consider, the estimation of parameters of the general linear models described in (8.5.4) and their variances. We concentrate on the following important case. Suppose $f(t) = t$. Then $\theta_k = \mathbf{x}_k'\beta = \sum_{j=1}^p x_{kj}\beta_j$. The parameter of interest is $\mathbf{B} = (B_1, \ldots, B_p)'$ where B_j is the finite population parameter corresponding to the superpopulation parameter β_j and is thus defined by (vide equations (8.5.6))

$$\sum_{k=1}^N \left[Y_k - \mu\left(\sum_{j=1}^p B_j X_{kj}\right)\right] X_{ki} = 0 (i = 1, \ldots, p). \tag{8.5.15}$$

Therefore, from (8.5.8),

$$\mathbf{W}_N(\mathbf{B}) = \sum_{k=1}^N \left[Y_k - \mu(\mathbf{X}_k'\mathbf{B})\right] \mathbf{X}_k = \mathbf{0}. \tag{8.5.16}$$

If $\hat{\mathbf{G}}$ is an estimate of $\sum_{k=1}^n Y_k \mathbf{X}_k$ and $\hat{\mathbf{R}}(\mathbf{B})$ is an estimate of $\sum_{k=1}^N \mu(\mathbf{X}_k'\mathbf{B})\mathbf{X}_k$, then the estimate of $\mathbf{B}$ is given by the solution of the equation

$$\hat{\mathbf{R}}(\hat{\mathbf{B}}) = \hat{\mathbf{G}}. \tag{8.5.17}$$

Here,

$$\frac{\partial \mathbf{W}_N(\mathbf{B})}{\partial \mathbf{B}} = -\sum_{k=1}^N \mu'(\mathbf{X}_k'\mathbf{B})\mathbf{X}_k'\mathbf{X}_k = -\mathbf{X}'\boldsymbol{\Lambda}(\mathbf{B})\mathbf{X}$$

where $\boldsymbol{\Lambda}(\mathbf{B}) =$ Diag.$(\mu'(\mathbf{X}_1'\mathbf{B}), \ldots, \mu'(\mathbf{X}_N'\mathbf{B}))$. Therefore, we obtain from (8.5.13),

$$V(\hat{\mathbf{B}}) = [\mathbf{X}'\boldsymbol{\Lambda}(\mathbf{B})\mathbf{X}]^{-1}\boldsymbol{\Sigma}(\mathbf{B})[\mathbf{X}'\boldsymbol{\Lambda}(\mathbf{B})\mathbf{X}]^{-1} \tag{8.5.18}$$

where $\boldsymbol{\Sigma}(\mathbf{B})$ is the variance of the estimator of the total $\sum_{k=1}^N [Y_k - \mu(\mathbf{X}_k'\mathbf{B})]\mathbf{X}_k$ based on the values $\{e_i, \mathbf{x}_i\}$, $i \in$ sample and $e_i = Y_i - \mu(\mathbf{X}_i'\mathbf{B})$.

Let $\hat{\mathbf{\Sigma}}(\hat{\mathbf{B}})$ be a consistent estimator of $\mathbf{\Sigma}(\mathbf{B})$, based on observations $\{\hat{e}_i, \mathbf{x}_i\}$, $\hat{e}_i = Y_i - \mu(\mathbf{X}_i'\hat{\mathbf{B}})$, $i \in$ sample. Also, let $\hat{\mathbf{J}}(\hat{\mathbf{B}})$ be a consistent estimator of $\mathbf{X}'\mathbf{\Lambda}(\mathbf{B})\mathbf{X}$. From (8.5.14), we have

$$\hat{V}(\hat{\mathbf{B}}) = [\hat{\mathbf{J}}^{-1}(\hat{\mathbf{B}})]\hat{\mathbf{\Sigma}}(\hat{\mathbf{B}})[\hat{\mathbf{J}}^{-1}(\hat{\mathbf{B}})]. \tag{8.5.19}$$

Again, from (8.5.17), $\hat{\mathbf{R}}(\hat{\mathbf{B}}) - \hat{\mathbf{G}} = \mathbf{0}$. If $\mathbf{B}_0$ is the true value of $\mathbf{B}$, expanding $\hat{\mathbf{R}}(\hat{\mathbf{B}})$ around $\hat{\mathbf{R}}(\mathbf{B}_0)$, we have

$$\mathbf{0} = \hat{\mathbf{R}} - \hat{\mathbf{G}} = \hat{\mathbf{R}}(\mathbf{B}_0) + \left[\frac{\partial \hat{\mathbf{R}}(\mathbf{B})}{\partial \mathbf{B}}\right]_{\mathbf{B}_0} (\hat{\mathbf{B}} - \mathbf{B}_0) - \hat{\mathbf{G}}.$$

Hence,

$$\hat{\mathbf{B}} = \mathbf{B}_0 - \left(\frac{\partial \hat{\mathbf{R}}(\mathbf{B}_0)}{\partial \mathbf{B}_0}\right)^{-1} [\hat{\mathbf{R}}(\mathbf{B}_0) - \hat{\mathbf{G}}].$$

Thus, if $\hat{\mathbf{B}}$ is solved from (8.5.17), by Newton–Raphson method,

$$\hat{\mathbf{B}}^{(j+1)} = \hat{\mathbf{B}}^{(j)} - \left(\frac{\partial \hat{\mathbf{R}}(\hat{\mathbf{B}}^{(j)})}{\partial \hat{\mathbf{B}}^{(j)}}\right)^{-1} [\hat{\mathbf{R}}(\hat{\mathbf{B}}^{(j)}) - \hat{\mathbf{G}}]$$

where $\hat{\mathbf{B}}^{(t)}$ is the value of $\mathbf{B}$ at the tth iteration. Again, $(\partial \hat{\mathbf{R}}(\hat{\mathbf{B}}))/(\partial \hat{\mathbf{B}}) = \hat{\mathbf{J}}(\hat{\mathbf{B}})$. Hence,

$$\hat{\mathbf{B}}^{(j+1)} = \hat{\mathbf{B}}^{(j)} - \hat{\mathbf{J}}^{-1}(\hat{\mathbf{B}}^{(j)})[\hat{\mathbf{R}}(\hat{\mathbf{B}}^{(j)}) - \hat{\mathbf{G}}].$$

Classical Linear Model

Here $f(t) = t$ and $\mu(\theta) = \theta$. Thus, $\theta_k = \mathbf{x}_k'\beta$, $E(Y_k) = \theta_k$. Therefore, $\mathbf{W}_N(\mathbf{B}) = \sum_{k=1}^{N}[Y_k - \mathbf{X}_k'\mathbf{B}]\mathbf{X}_k = \mathbf{X}'\mathbf{Y} - \mathbf{X}'\mathbf{X}\mathbf{B}$. Again, $(\partial \mathbf{W}_N(\mathbf{B}))/(\partial \mathbf{B}) = -\mathbf{X}'\mathbf{X}$. Hence, $\hat{\mathbf{V}}(\hat{\mathbf{B}}) = (\mathbf{S}_{xx})^{-1}\hat{\mathbf{\Sigma}}(\hat{\mathbf{B}})(\mathbf{S}_{xx})^{-1}$, where $\mathbf{S}_{xx}$ is an estimate of $\mathbf{X}'\mathbf{X}$ (assuming that the observations on $\mathbf{x}$ in the population are centered against the population mean vector), $\hat{\mathbf{\Sigma}}(\hat{\mathbf{B}})$ is an estimate of the variance of an estimator of total $\sum_{k=1}^{N}(Y_k - \mathbf{X}_k'\mathbf{B})$ based on observations $(\hat{e}_k, \mathbf{X}_k)$, $\hat{e}_k = Y_k - \mathbf{X}_k'\hat{\mathbf{B}}$, $k \in$ sample.

Coefficient of multiple determination
Consider the multiple regression model

$$\mathbf{Y}(N \times 1) = \mathbf{X}\beta + \mathbf{U}(N \times 1) \tag{i}$$

where $\mathbf{X} = (\mathbf{1}_N, \mathbf{X}_1)$ is $N \times (r+1)$ and $\beta = (\beta_0, \beta_1, \ldots, \beta_r)' = (\beta_0, \mathbf{B}')'$. The model (i) can be written as

$$\tilde{\mathbf{Y}} = \tilde{\mathbf{X}}\mathbf{B} + \tilde{\mathbf{U}} \tag{ii}$$

where $\tilde{\mathbf{Y}} = \mathbf{Y} - \mathbf{1}_N\bar{Y}$, $\bar{Y} = \mathbf{1}'_N\mathbf{Y}/N$, $\tilde{\mathbf{X}} = \mathbf{X}_1 - \mathbf{1}_N\bar{\mathbf{x}}'$, $\bar{\mathbf{x}}' = \mathbf{1}'_N\mathbf{X}_1/N$, $\tilde{\mathbf{U}} = \mathbf{U} - \mathbf{1}_N\bar{u}$, $\bar{u} = \mathbf{1}'_N\mathbf{U}/N$. Let

$$\mathbf{S} = (N-1)^{-1}\begin{bmatrix}\tilde{\mathbf{Y}}'\\ \tilde{\mathbf{X}}'\end{bmatrix}[\tilde{\mathbf{Y}}\ \tilde{\mathbf{X}}] = \begin{bmatrix}S_{yy} & \mathbf{S}_{yx}\\ \mathbf{S}_{xy} & \mathbf{S}_{yx}\end{bmatrix}.$$

The ordinary least square estimator of $\mathbf{B}$ is

$$\hat{\mathbf{B}} = (\tilde{\mathbf{X}}'\tilde{\mathbf{X}})^{-1}\tilde{\mathbf{X}}'\tilde{\mathbf{Y}} = \mathbf{S}_{xx}^{-1}\mathbf{S}_{xy}.$$

The simple correlation between $\tilde{\mathbf{Y}}$ and $\hat{\tilde{\mathbf{Y}}} = \tilde{\mathbf{X}}\hat{\mathbf{B}}$ is called the coefficient of multiple determination of y on $\mathbf{x}$ and is given by

$$\begin{aligned} R^2_{y.\mathbf{x}} = R^2 &= \frac{\mathbf{S}_{yx}\mathbf{S}_{xx}^{-1}\mathbf{S}_{xy}}{S_{yy}} \\ &= \frac{\hat{\tilde{\mathbf{Y}}}'\hat{\tilde{\mathbf{Y}}}}{\tilde{\mathbf{Y}}'\tilde{\mathbf{Y}}}. \end{aligned} \qquad (iii)$$

The estimating equations for the finite population parameters, $\bar{Y}$, $\mathbf{B}$, and R^2 is

$$\mathbf{W}_N = \begin{bmatrix} \mathbf{Y}'\mathbf{1} - N\bar{Y} \\ \mathbf{X}'\mathbf{Y} - \mathbf{X}'\mathbf{X}\mathbf{B} \\ (\mathbf{Y}'\mathbf{Y} - N\bar{Y}^2)(R^2 - 1) + \mathbf{Y}'\mathbf{Y} - \mathbf{Y}'\mathbf{X}\mathbf{B} \end{bmatrix} = \begin{bmatrix} 0 \\ \mathbf{0} \\ 0 \end{bmatrix}. \qquad (8.5.20)$$

Here

$$\left[\frac{\partial \mathbf{W}}{\partial(\bar{Y}, \mathbf{B}, R^2)}\right]^{-1} = \begin{bmatrix} N^{-1} & \mathbf{0} & 0 \\ 0 & -(\mathbf{X}'\mathbf{X})^{-1} & 0 \\ \frac{2\bar{Y}(1-R^2)}{SSY} & -\frac{\mathbf{B}'}{SSY} & (SSY)^{-1} \end{bmatrix}$$

where $SSY = \tilde{\mathbf{Y}}'\tilde{\mathbf{Y}}$. Hence, expression of $\hat{V}(\hat{\psi})$ can be written by (8.5.14). Here $\hat{\mathbf{\Sigma}}_U$ is the estimate of covariance matrix of $W_N(\psi)$ in (8.5.20). To write $W_N(\psi)$ in the form $\sum_{k=1}^N \mathbf{u}(\mathbf{Z}_k; \bar{Y}, \mathbf{B}, R^2) = \sum_{k=1}^N \mathbf{c}_k$ (say), where $\mathbf{c}_k = (c_{k1}, \ldots, c_{kr+2})'$ we note that $c_{k1} = Y_k$, $(c_{k2}, \ldots, c_{kr+1})' = (Y_k - \mathbf{x}'_k\hat{\mathbf{B}}_k)\mathbf{x}_k$ and $c_{kr+2} = (\hat{R}^2 Y_k - \mathbf{x}'_k\hat{\mathbf{B}})Y_k$.

Logistic Regression for Binary Response Variable

Here the dependent variable y is dichotomous, taking values 0 and 1. The functions $g(\theta) = \log(1 + e^{\theta})$ and $f(t) = t$. Thus,

$$\mu(\mathbf{X}'_k\mathbf{B}) = \frac{\exp(\mathbf{X}'_k\mathbf{B})}{1+\exp(\mathbf{X}'_k\mathbf{B})}$$

$$\mu'(\mathbf{X}'_k\mathbf{B}) = \mu(\mathbf{X}'_k\mathbf{B})(1 - \mu(\mathbf{X}'_k\mathbf{B})).$$

Logistic Regression for Polytomous Response Variable

Suppose that each member of the population belongs to exactly one of the q distinct categories with proportion $\pi_i(\beta)$ for the ith category, We assume that associated with category i, there is a vector $\mathbf{x}_i = (x_{i1}, \ldots, x_{ir})'$ of values of auxiliary variables $\mathbf{x}' = (x_1, \ldots, x_r)'$ so that approximately

$$\pi_i(\beta) = \frac{\exp(\mathbf{x}_i'\beta)}{\sum\limits_{s=1}^{q} \exp(\mathbf{x}_s'\beta)}$$

proportion of individuals falls in the ith category. Let $\pi(\beta) = (\pi_1(\beta), \ldots, \pi_q(\beta))'$ and $\mathbf{N} = (N_1, \ldots, N_q)'$ where N_i denotes the number of units in category i in the population. If $\mathbf{N}$ is obtained by a multinomial model, the log-likelihood is

$$\log L = \text{ Constant } + \sum_{i=1}^{q} N_i \log\{\pi_i(\beta)\}.$$

Now

$$\frac{\partial \pi_i}{\partial \beta_j} = \pi_i(\beta)\left[x_{ij} - \sum_{s=1}^{q} x_{sj}\pi_s(\beta)\right].$$

Hence

$$\begin{aligned}\frac{\partial \log L}{\partial \beta_j} &= \sum_{i=1}^{q} \frac{\partial \log L}{\partial \pi_i} \cdot \frac{\partial \pi_i}{\partial \beta_j} \\ &= \sum_{i=1}^{q} N_i x_{ij} - \sum_{i=1}^{N} N_i \sum_{s=1}^{q} x_{sj}\pi_s(\beta),\ j = 1, \ldots, r.\end{aligned} \quad (8.5.21)$$

Therefore, the maximum likelihood equations for the finite population parameters $\mathbf{B} = (B_1, \ldots, B_r)'$ is

$$\mathbf{W}_N(\mathbf{B}) = \mathbf{X}'\mathbf{N} - \mathbf{X}'\pi(\mathbf{B})[\mathbf{1}'\mathbf{N}] = \mathbf{0} \quad (8.5.22)$$

where $\mathbf{X}$ is the $q \times r$ matrix of x_{ij} values. An estimator $\hat{\mathbf{B}}$ of $\mathbf{B}$ therefore satisfies

$$\dot{\mathbf{W}}(\hat{\mathbf{B}}) = \mathbf{X}'\hat{\mathbf{N}} - \mathbf{X}'\pi(\hat{\mathbf{B}})[\mathbf{1}'\hat{\mathbf{N}}] = \mathbf{0} \quad (8.5.23)$$

where $\hat{\mathbf{N}}$ is a consistent asymptotically normal estimator of $\mathbf{N}$ with covariance matrix $V(\hat{\mathbf{N}})$. Let $\mathbf{D}_\pi(\mathbf{B}) =$ Diag. $(\pi_1(\mathbf{B}), \ldots, \pi_q(\mathbf{B}))$, $\mathbf{H}_\pi(\mathbf{B}) = \mathbf{D}_\pi(\mathbf{B}) - \pi(\mathbf{B})\pi(\mathbf{B})'$. Then

$$\frac{\partial \mathbf{W}(\mathbf{B})}{\partial \mathbf{B}} = -(\mathbf{1}'\mathbf{N})\mathbf{X}'\mathbf{H}(\mathbf{B})\mathbf{X}.$$

Again,

$$\Sigma_U(\mathbf{B}) = V\{\hat{\mathbf{W}}(\mathbf{B})\} = V\{\mathbf{X}'[\mathbf{I}_q - \pi\mathbf{1}']\hat{\mathbf{N}}\}$$
$$= \mathbf{X}'[\mathbf{I} - \pi\mathbf{1}']V(\hat{\mathbf{N}})[\mathbf{I} - \pi\mathbf{1}']\mathbf{X}. \quad (8.5.24)$$

Therefore, by (8.5.13),

$$V(\hat{\mathbf{B}}) = (\mathbf{1}'\mathbf{N})^{-2}(\mathbf{X}'\mathbf{H}\mathbf{X})^{-1}\mathbf{X}'[\mathbf{I} - \pi\mathbf{1}']V(\hat{\mathbf{N}})[\mathbf{I} - \mathbf{1}\pi']\mathbf{X}(\mathbf{X}'\mathbf{H}\mathbf{X})^{-1}. \quad (8.5.25)$$

The author illustrates the application of the techniques with respect to the data from Canada Health Survey, 1978–1979.

8.6 A Mixed (Design-Model) Framework

In the traditional $p\xi$ (design-model) framework inference is based on $P(s)\xi(\mathbf{Y})$ distribution where $P(s)$ is the probability of selecting a sample s and $\xi(\mathbf{Y})$ is the superpopulation model distribution of the response variable vector $\mathbf{Y} = (Y_1, \ldots, Y_N)'$, Y_i being the response from unit i (Sarndal et al. 1992; Mukhopadhyay 1996, 2007). In the mixed framework proposed by Rizzo (1992) inference is based on the joint distribution $P(\mathbf{I}, \mathbf{Y})$ where $\mathbf{I}$ is an indicator random vector with its value $I_i = 1(0)$ according as unit $i \in s$ (otherwise). The distribution of $\mathbf{I}$ may be based on the covariates that are related to $\mathbf{Y}$ so the joint distribution $P(\mathbf{I}, \mathbf{Y})$ does not necessarily factor into the product of marginal distributions. This framework which uses predictive regression estimators as functions of the probability of selection as covariates provide an alternative to classical p-weighted estimators for finite population parameters in complex surveys.

Let $\mathbf{X}$ be a set of covariates correlated with the response variable $\mathbf{Y}$ and let $\mathbf{Z}$ be a set of variables contained in $\mathbf{X}$. The survey design is based on $\mathbf{Z}$. Let $\Pi_\infty = \{\{\pi_i\} \cup \{\pi_{ij}\} \cup \{\pi_{ijk}\} \cup \ldots\}$ where π_i, π_{ij}, etc. are the first order, second order, etc. inclusion probabilities. Clearly, Π_∞ is random since $\mathbf{Z}$ is so. We focuss on inference about population mean $\bar{Y} = N^{-1}\sum_{i=1}^N Y_i$ or some other aspect of the distribution of $\mathbf{Y}$. The relationship between $\mathbf{X}$ and $\mathbf{Y}$, for example, is not of our direct interest.

After the unit i is selected, observation on variable Y_i is made. Thus, the sample of Y_i is drawn from the distribution $P(Y_i|I_i = 1)$. From this, we have to go back to the original superpopulation model distribution of Y_i which is $P(Y_i)$.

Earlier authors considered $P(\mathbf{Y}|\mathbf{X}, \mathbf{I})(= P(\mathbf{Y}|\mathbf{X}))$ or $P(\mathbf{Y}|\mathbf{Z}, \mathbf{I})$. Instead of the whole distribution, Chambers (1986), Pfeffermann (1988), among others, considered $E(\mathbf{Y}|\mathbf{Z})$.

Rizzo considers the specification of $P(Y_i|\Pi_\infty, \mathbf{I})$ or $P(Y_i|\Pi_\infty)$, since given Π_∞, $\mathbf{I}$ is ignorable (because distribution of $\mathbf{I}$ depends on Π_∞ alone). This approach is stated to be more precise and simple than $P(Y_i|\mathbf{X})$-approach.

If Π_∞ is sufficiently complex, then even this specification may be difficult. Denoting by y_i the value of Y_i on the unit $i \in s$, we make the following simplifying assumptions.

(i) $\mathcal{E}(Y_i|\Pi_\infty) = \mathcal{E}(Y_i|\pi_i)$;, i.e., the expectation of y_i is related only to the π_i associated with it, rather than the other π_i's and with joint probabilities;
(ii) $Var(Y_i|\Pi_\infty) = Var(Y_i|\pi_i)$;
(iii) $\text{Cov}(Y_i, Y_j|\Pi_\infty) = \text{Cov}(Y_i, Y_j|\pi_i, \pi_j, \Delta_{ij})$ where $\Delta_{ij} = \pi_{ij} - \pi_i\pi_j$.

If the assumption (i) is reasonable, then it may be assumed in most cases that

$$\mathcal{E}(Y_i|\pi_i) = \mathbf{\Pi}_i\gamma \tag{8.6.1}$$

where $\mathbf{\Pi}_i = (f_1(\pi_i), \ldots, f_k(\pi_i))$, a row vector of known functions of π_i and γ is a $k \times 1$ vector of regression coefficients, i.e., conditional expectation of Y_i is a linear combination of π_i.

Similarly, we assume

$$Var(Y_i|\pi_i) = \mathbf{P}_i\gamma_m \tag{8.6.2}$$

where $\mathbf{P}_i = (p_1(\pi_i), \ldots, p_m(\pi_i))$, a row vector and γ_m a $m \times 1$ vector of parameters.

Assuming that Δ_{ij} is noninformative (i.e., it does not depend on y's), assumption (iii) may be modified as $(iii)'$: $\text{Cov}(Y_i, Y_j|\Pi_\infty) = \text{Cov}(Y_i, Y_j|\pi_i, \pi_j)$. Further, assuming that in the original superpopulation model Y_i's are independently distributed, it may be reasonable to assume that this covariance, given Π_∞, is zero.

Consider the following estimators:

(a) The Horvitz–Thompson estimator (HTE), $\hat{\bar{Y}}_{HT} = \frac{1}{N}\sum_{i\in s}\frac{y_i}{\pi_i}$;
(b) An adjusted version of HTE, $\hat{\bar{Y}}_{Ha} = \frac{1}{Nd_\pi}\sum_{i\in s}\frac{y_i}{\pi_i}$ where $d_\pi = \frac{1}{N}\sum_{i\in s}\frac{1}{\pi_i}$;
(c) The simple predictive regression estimator:

$$\begin{aligned}\hat{\bar{Y}}_P &= \frac{1}{N}\left[\sum_{i\notin s}\hat{\mathcal{E}}(Y_i|\pi) + \sum_{i\in s} y_i\right] \\ &= \frac{1}{N}\left[\sum_{i\notin s}\mathbf{\Pi}_i\hat{\gamma} + \sum_{i\in s} y_i\right]\end{aligned} \tag{8.6.3}$$

where $\hat{\gamma}$ is the least square estimator (LSE) of γ. Thus from (8.6.1),

$$\hat{\gamma} = (\mathbf{\Pi}_s'\mathbf{\Pi}_s)^{-1}\mathbf{\Pi}_s'\mathbf{y}_s$$

where $\mathbf{\Pi}_s = (\mathbf{\Pi}_i; i \in s)_{n\times k}$.

$\hat{\bar{Y}}_P$ is Royall's (1970) predictor using the conditional expectation $\mathcal{E}(Y_i|\pi_i)$. Royall used $\mathcal{E}(Y_i|\mathbf{x}_i)$ where $\mathbf{X}' = (\mathbf{x}_1, \ldots, \mathbf{x}_N)$.

(d) An estimator with lower variance than $\hat{\hat{Y}}_P$ is

$$\hat{\hat{Y}}_G = \frac{1}{N}[\sum_{i \notin s} \mathbf{\Pi}_i \tilde{\gamma} + \sum_{i \in s} y_i] \tag{8.6.4}$$

where

$$\tilde{\gamma} = (\mathbf{\Pi}_s' \tilde{\mathbf{V}}_s^{-1} \mathbf{\Pi}_s)^{-1} (\mathbf{\Pi}_s' \tilde{\mathbf{V}}_s^{-1} \mathbf{y}_s)$$

where

$$\tilde{\mathbf{V}}_s = \text{ Diag. } (\mathbf{P}_i \tilde{\gamma}_m : i \in s),$$

$$\tilde{\gamma}_m = (\mathbf{P}_s' \mathbf{P}_s)^{-1} (\mathbf{P}_s' \mathbf{e}_s^{(2)}),$$

$\mathbf{P}_s = (\mathbf{P}_i;\ i \in s)$ and $\mathbf{e}_s^{(2)} = (\hat{e}_i^2;\ i \in s)$, $\hat{e}_i^2 = (y_i - \mathbf{\Pi}_i \hat{\gamma})^2$. We have assumed here that $\mathbf{P}_i \tilde{\gamma}_m > 0$.

We now consider the moment properties of these estimators. Suppose, the overall superpopulation model for $\mathbf{Y}$ is: $Y_1, \ldots, Y_N$ are independently distributed with

$$Y_i = f(\mathbf{x}_i, \beta) + \epsilon_i \tag{8.6.5}$$

where ϵ's are independently distributed and ϵ_i is independent of $\mathbf{x}_i$ and f is some function of $\mathbf{x}_i$, possibly nonlinear and/or nonadditive. A general model of this form can be accommodated, because we have to only specify $P(\mathbf{Y}|\Pi_\infty)$ in this approach. For finding the estimator we have to only specify $\mathcal{E}(Y_i|\Pi_\infty)$ and under assumption (i), $\mathcal{E}(Y_i|\pi_i)$.

From (8.6.1) we can write

$$Y_i = \mathbf{\Pi}_i \gamma + e_i \tag{8.6.6}$$

where $\mathcal{E}(e_i|\pi_i) = 0$. From assumptions (i), (ii), (iii)', (8.6.1), (8.6.2) and the fact that $\mathcal{E}(\mathbf{Y}|\Pi_\infty, \mathbf{I}) = \mathcal{E}(\mathbf{Y}|\Pi_\infty)$ the following results follow.

$$\begin{aligned} \mathcal{E}(e_i|\pi_i, I_i = 1) &= \mathcal{E}(e_i|\pi_i, I_i = 0) = 0 \\ Var(e_i|\pi_i, I_i = 1) &= Var(e_i|\pi_i) = \mathbf{P}_i \gamma_m \\ \text{Cov}(e_i, e_j|\pi_i, \pi_j, I_i = 1, I_j = 1) &= \text{Cov}(e_i, e_j|\pi_i, \pi_j) = 0. \end{aligned} \tag{8.6.7}$$

The prediction error of $\hat{\hat{Y}}_P$ can be written as

$$\hat{\hat{Y}}_P - \bar{Y} = \frac{1}{N}[\sum_{i \notin s} \mathbf{\Pi}_i (\frac{1}{n} \mathbf{\Pi}_s' \mathbf{\Pi}_s)^{-1} (\frac{1}{n} \mathbf{\Pi}_s' \mathbf{e}_s) - \sum_{i \notin s} e_i] \tag{8.6.8}$$

where $\mathbf{e}_s = \{e_i : i \in s\}$. It immediately follows that

$$\mathcal{E}(\hat{\hat{Y}}_P - \bar{Y}|\Pi_\infty, \mathbf{I}) = 0.$$

Thus $\hat{\hat{Y}}_P$ is conditionally and hence unconditionally unbiased predictor of $\bar{Y}$.

Under some regularity conditions $\hat{\hat{Y}}_G$ is asymptotically unbiased. When the variance ratios $Var(Y_i|\pi_i)/Var(Y_j|\pi_j) = \mathbf{P}_i\gamma_m/\mathbf{P}_j\gamma_m$ are known, $\hat{\hat{Y}}_G$ will be exactly unbiased both conditionally and hence unconditionally.

It can be easily checked that

$$N(\hat{\hat{Y}}_{Ha} - \bar{Y}) = \sum_{i\in s} \mathbf{\Pi}_i\gamma[\frac{1}{d_\pi \pi_i} - 1] - \sum_{i\in \bar{s}} \mathbf{\Pi}_i\gamma + \sum_{i\in s} e_i[\frac{1}{d_\pi \pi_i} - 1] - \sum_{i\notin s} e_i. \tag{8.6.9}$$

Its conditional expectation given $\Pi_\infty, \mathbf{I}$ gives zero for the last two terms, but the first two terms have nonzero expectation. The bias tends to be small and is unconditionally zero when expectation is taken over all samples with d_π constant.

Let w_i be the ith element of the $1 \times n$ row vector $\sum_{i\in s} \mathbf{\Pi}_i[(\frac{1}{n})\mathbf{\Pi}_s'\mathbf{\Pi}_s]^{-1}[(\frac{1}{n})\mathbf{\Pi}_s]$. Then

$$Var(\hat{\hat{Y}}_P - \bar{Y}) = \frac{1}{N^2}\sum_{i\in s} w_i^2[\mathbf{P}_i\gamma_m] + \frac{1}{N}\sum_{i\notin s} \mathbf{P}_i\gamma_m. \tag{8.6.10}$$

A consistent variance estimator can be obtained by replacing γ_m by $\tilde{\gamma}_m$. It is shown that under some conditions,

$$Var(\hat{\hat{Y}}_G - \bar{Y}) = \left(\frac{1}{N}\sum_{i\notin s} \mathbf{\Pi}_i\right)\left(\frac{1}{n}\mathbf{\Pi}_s'\mathbf{V}_s^{-1}\mathbf{\Pi}_s\right)^{-1}\left(\frac{1}{N}\sum_{i\notin s} \mathbf{\Pi}_s\right)' + \frac{1}{N^2}\sum_{i\notin s} \mathbf{P}_i\gamma_m + 0(n^{-1}). \tag{8.6.11}$$

Superpopulation variables for the sample are e_i's. In the case of cluster sampling we have therefore to look into covariances among e_i's. Now,

$$\begin{aligned}\mathrm{Cov}(e_i, e_j|\Pi_\infty) &= \sqrt{Var(e_i|\Pi_\infty)}\sqrt{Var(e_j|\Pi_\infty)}corr(e_i, e_j|\Pi_\infty)\\ &= \sqrt{\mathbf{P}_i\gamma_m}\sqrt{\mathbf{P}_j\gamma_m}corr(e_i, e_j|\Pi_\infty).\end{aligned}$$

We now make the simplifying assumption

$$corr(e_i, e_j|\Pi_\infty) = corr(e_i, e_j|\pi_i, \pi_j, \Delta_{ij}). \tag{8.6.12}$$

Let α_{ij} be an indicator function whose value is 1(0) according as $(i, j) \in s$ (otherwise). Therefore, α_{ij} is a function of $\mathbf{Z}$ rather that of Π_∞. Assume that

$$corr(e_i, e_j|\pi_i, \pi_j, \Delta_{ij}, \alpha_{ij}) = corr(e_i, e_j|\alpha_{ij}) = \rho\alpha_{ij} \tag{8.6.13}$$

where $\rho > 0$. With these assumptions $\mathbf{V} = Var(\mathbf{e}|\Pi_\infty)$ can be given as (a) $V_{ij} = 0$ if i, j are in different clusters; (b) $V_{ij} = \sqrt{\mathbf{P}_i\gamma_m}\sqrt{\mathbf{P}_j\gamma_m}\rho$ if i, j are in same cluster; (c) $V_{ij} = \mathbf{P}_i\gamma_m$ if $i = j$. With this specification one can recalculate the estimators and their variances.

8.7 Effect of Survey Design on Multivariate Analysis of Principal Components

Sknner et al. (1986) examined the effects of sample selection on standard principal component analysis and the use of alternative maximum likelihood and probability-weighted estimators.

Associated with each unit i in the population there is a $p \times 1$ vector $\mathbf{y}_i$ of unknown values, some of which are to be measured in the survey. Also, there is a $q \times 1$ vector $\mathbf{z}_i$ of values of auxiliary variables, assumed to be known for each unit in the population.

Let

$$\mathbf{y} = (\mathbf{y}_1, \ldots, \mathbf{y}_N)_{p\times N}, \quad \mathbf{z} = (\mathbf{z}_1, \ldots, \mathbf{z}_N)_{q\times N}, \tag{8.7.1}$$

$s = (i_1, \ldots, i_n)$, a sample of n distinct units. The data consist of $(\mathbf{y}_s, s, \mathbf{z})$ where $\mathbf{y}_s = (\mathbf{y}_{i_1}, \ldots, \mathbf{y}_{i_n})$.

We assume that $(\mathbf{y}_i', \mathbf{z}_i')'$ is a realization of a random matrix $(\mathbf{Y}_i', \mathbf{Z}_i')'$ where $\mathbf{Y}_i = (Y_{i1}, \ldots, Y_{ip})'$ and $\mathbf{Z}_i = (Z_{i1}, \ldots, Z_{iq})'$. Also assume that the vector $(\mathbf{Y}_i', \mathbf{Z}_i')'$ is a random sample from a multivariate normal distribution

$$\begin{bmatrix}\mathbf{Y}_i\\ \mathbf{Z}_i\end{bmatrix} \sim N_{p+q}\left[\begin{bmatrix}\mu_y\\ \mu_z\end{bmatrix}, \begin{bmatrix}\mathbf{\Sigma}_{yy} & \mathbf{\Sigma}_{yz}\\ \mathbf{\Sigma}_{zy} & \mathbf{\Sigma}_{zz}\end{bmatrix}\right], \quad i = 1, \ldots, N \tag{8.7.2}$$

where $\mathbf{\Sigma}_{zz}$ is assumed to be positive definite. First, we shall try to estimate the covariance matrix $\mathbf{\Sigma}_{yy}$.

Let θ be the vector of parameters involved in the distribution in (8.7.2). Then, the likelihood function may be written as

$$L(\theta|\mathbf{y}_s, s, \mathbf{z}_s) \propto f(s|z)f(\mathbf{y}_s|z; \theta)f(\mathbf{z}; \theta). \tag{8.7.3}$$

The sample selection mechanism $f(s|\mathbf{z})$ does not depend on θ, and hence, it can be ignored in likelihood inference about θ.

Let $\pi_i(\mathbf{z})$ be the inclusion probability of the unit i in the sample, given $\mathbf{z}$ and

$$
\begin{aligned}
\beta_{yz} &= \boldsymbol{\Sigma}_{yz}\boldsymbol{\Sigma}_{zz}^{-1},\ w_i = (N\pi_i(\mathbf{z}))^{-1},\\
w(s) &= \textstyle\sum_{i\in s} w_i,\ \hat{\mu}_{y:\pi w} = \sum_{i\in s} w_i y_i,\\
\mathbf{x}_i &= (\mathbf{y}_i', \mathbf{z}_i')',\ \bar{\mathbf{x}}_s = n^{-1}\textstyle\sum_{i\in s}\mathbf{x}_i = (\bar{\mathbf{y}}_s', \bar{\mathbf{z}}_s')'\\
\bar{\mathbf{x}} &= N^{-1}\textstyle\sum_{i=1}^{N}\mathbf{x}_i = (\bar{\mathbf{y}}', \bar{\mathbf{z}}')'\\
\begin{bmatrix} \mathbf{s}_{yys} & \mathbf{s}_{yzs} \\ \mathbf{s}_{zys} & \mathbf{s}_{zzs} \end{bmatrix} &= \mathbf{s}_{xxs} = (n-1)^{-1}\textstyle\sum_{i\in s}(\mathbf{x}_i - \bar{\mathbf{x}}_s)(\mathbf{x}_i - \bar{\mathbf{x}}_s)'\\
\begin{bmatrix} \tilde{\mathbf{s}}_{yys} & \tilde{\mathbf{s}}_{yzs} \\ \tilde{\mathbf{s}}_{zys} & \tilde{\mathbf{s}}_{zzs} \end{bmatrix} &= n^{-1}(n-1)\mathbf{s}_{xxs}\\
\mathbf{b}_{yzs} &= \mathbf{s}_{yzs}\mathbf{s}_{zzs}^{-1}\\
\begin{bmatrix} \mathbf{S}_{yy} & \mathbf{S}_{yz} \\ \mathbf{S}_{zy} & \mathbf{S}_{zz} \end{bmatrix} &= \mathbf{S}_{xx} = N^{-1}\sum_{i=1}^{N}(\mathbf{x}_i - \bar{\mathbf{x}})(\mathbf{x}_i - \bar{\mathbf{x}})'.
\end{aligned}
\tag{8.7.4}
$$

We shall consider, the following three point estimators of $\boldsymbol{\Sigma}_{yy}$:

$$
\begin{aligned}
&\text{(i)}\,\hat{\boldsymbol{\Sigma}}_{yy:srs} = \mathbf{s}_{yys};\\
&\text{(ii)}\,\hat{\boldsymbol{\Sigma}}_{yy:ML} = \tilde{\mathbf{s}}_{yys} + \mathbf{b}_{yzs}(\mathbf{S}_{zz} - \tilde{\mathbf{s}}_{zzs})\mathbf{b}_{yzs}';\\
&\text{(iii)}\,\hat{\boldsymbol{\Sigma}}_{yy:\pi w} = \textstyle\sum_{i\in s} w_i\mathbf{y}_i\mathbf{y}_i' - w(s)^{-1}\hat{\mu}_{y;\pi w}\hat{\mu}_{y:\pi w}'.
\end{aligned}
\tag{8.7.5}
$$

The first estimator is the conventional estimator applicable to simple random sampling. The second estimator is obtained by maximizing the likelihood in (8.7.3) (Anderson 1957). The third estimator is the π-weighted estimator of $\boldsymbol{\Sigma}_{yy}$ and is approximately design unbiased.

Hence, the first estimator ignores the sampling design and prior information. The second estimator takes account of the multivariate normal model in (8.7.2). The third estimator takes account of the sampling design through w_i's.

To examine the properties of the estimators we consider the conditional model expectation of the estimators given the sample s and prior information $\mathbf{z}$. Now,

$$
\begin{aligned}
E(\hat{\boldsymbol{\Sigma}}_{yy:srs}|s,\mathbf{z}) &= \boldsymbol{\Sigma}_{yy} + \beta_{yz}(\mathbf{s}_{zzs} - \boldsymbol{\Sigma}_{zz})\beta_{yz}'\\
E(\hat{\boldsymbol{\Sigma}}_{yy:ML}|s,\mathbf{z}) &= \alpha\boldsymbol{\Sigma}_{yy} + \beta_{yz}(\mathbf{S}_{zz} - \alpha\boldsymbol{\Sigma}_{zz})\beta_{yz}'\\
E(\hat{\boldsymbol{\Sigma}}_{yy:\pi w}|s,\mathbf{z}) &= \alpha_w\boldsymbol{\Sigma}_{yy} + \beta_{yz}(\hat{\boldsymbol{\Sigma}}_{zz:\pi w} - \alpha_w\boldsymbol{\Sigma}_{zz})\beta_{yz}'
\end{aligned}
\tag{8.7.6}
$$

where

$$
\begin{aligned}
\alpha &= [n - q - 1 + \text{ trace }(\mathbf{S}_{zz}\mathbf{s}_{zzs}^{-1})]/n,\\
\alpha_w &= w(s) - \textstyle\sum_{i\in s} w_i^2/w(s),\\
\hat{\boldsymbol{\Sigma}}_{zz:\pi w} &= \textstyle\sum_{i\in s} w_i\mathbf{z}_i\mathbf{z}_i' - w(s)\sum_{i\in s} w_i\mathbf{z}_i\sum_{i\in s} w_i\mathbf{z}_i'.
\end{aligned}
\tag{8.7.7}
$$

Since the samples are generally selected using the variable $\mathbf{z}$, the estimators are not in general unbiased, even asymptotically. Hence, $\hat{\Sigma}_{yy,srs}$ may in general have a bias of $0(1)$ and be unconditionally inconsistent for $\mathbf{\Sigma}_{yy}$. In contrast $\hat{\Sigma}_{yy,ML}$ has a conditional bias of $0(1)$, assuming that $\alpha = 1 + 0_p(n^{-1})$, and is generally so unconditionally for Σ_{yy}. The π-weighted estimator $\hat{\Sigma}_{yy,\pi w}$ has a conditional bias of $0_p(n^{-1/2})$, but is unconditionally consistent for $\mathbf{\Sigma}_{yy}$.

8.7.1 Estimation of Principal Components

Assuming that the eigenvalues $\lambda_1 > \lambda_2 > \cdots > \lambda_p$ of $\mathbf{\Sigma}_{yy}$ are distinct, the corresponding normalized eigenvectors $\gamma_1, \ldots, \gamma_p$ are uniquely defined by

$$\begin{aligned} \mathbf{\Sigma}_{yy}\gamma_i &= \lambda_i\gamma_i,\ i = 1, \ldots, p \\ \gamma_i'\gamma_k &= 1(i = k); = 0\ (\text{otherwise})\,. \end{aligned} \tag{8.7.8}$$

The principal components of $\mathbf{Y}$ then consist of the linear combinations $\chi_j = \gamma_j'\mathbf{Y}(j = 1, \ldots, p)$. Here, we are interested in estimating the λ_i's and γ_j's. For these we use the eigenvalues and the eigenvectors of the three estimators $\hat{\mathbf{\Sigma}}_{yy}$ given above. To evaluate the properties of these estimators, we use the linear Taylor series expansion about $\mathbf{\Sigma}_{yy}$. If $\hat{\lambda}_1 > \cdots > \hat{\lambda}_p$ are the eigenvalues of $\hat{\mathbf{\Sigma}}_{yy}$ and $\hat{\gamma}_1, \ldots, \hat{\gamma}_p$ are the corresponding eigenvectors, then it follows from Girshick (1939) that

$$\begin{aligned} \hat{\lambda}_i &\approx \lambda_i + \gamma_i'(\hat{\mathbf{\Sigma}}_{yy} - \mathbf{\Sigma}_{yy})\gamma_i \\ \hat{\gamma}_i &\approx \gamma_i + \textstyle\sum_{k(\neq i)=1}^{p} w_{ik}\gamma_k, \end{aligned} \tag{8.7.9}$$

where $w_{ik} = \gamma_i'(\hat{\mathbf{\Sigma}}_{yy} - \mathbf{\Sigma}_{yy})\gamma_k/(\hat{\lambda}_i - \hat{\lambda}_k)$.

The authors consider the conditional model expectation of $\hat{\lambda}_{j:srs}$ and $\hat{\gamma}_{j;srs}$ given $s, \mathbf{z}$. They also justify the validity of approximations in (8.7.9) through simulation studies.

Appendix A
Asymptotic Properties of Multinomial Distribution

Abstract Since multinomial distribution is one of the main pillars on which the models for analysis of categorical data collected from complex surveys thrive, the Appendix makes a review of the asymptotic properties of the multinomial distribution and asymptotic distribution of Parson chi-square statistic X_P^2 for goodness-of-fit based on this distribution. General theory of multinomial estimation and testing in case the population proportions $\pi_1, \ldots, \pi_{t-1}$ depend on several parameters $\theta_1, \ldots, \theta_s (s < t - 1)$, also unknown, is then introduced. Different minimum-distance methods of estimation, like, X_P^2, likelihood ratio statistic G^2, Freeman–Tukey (Ann Math Stat 21:607–611, 1950) statistic $(FT)^2$ and Neyman's (Contribution to the theory of χ^2 tests, pp 239–273, 1949) statistic X_N^2 have been defined and their asymptotic distribution studied under the full model as well as nested models in the light of, among others, Birch's (Ann Math Stat 35:817–824, 1964) illuminating results. Finally, Neyman's (Contribution to the theory of χ^2 tests, pp 239–273, 1949) and Wald's (Trans Am Math Soc 54:429–482, 1943) procedures for testing general hypotheses relating to population proportions have been revisited.

Keywords Multinomial distribution · Pearson's statistic X_P^2 · Likelihood ratio · Freeman-Tukey statistic · Neyman's statistic · Wald statistic · Nested models

A.1 Introduction

This chapter reviews asymptotic properties of the multinomial distribution and associated tests of goodness-of-fit. Sections A.2 and A.3 deal respectively with asymptotic normality of the multinomial distribution and the asymptotic distribution of Pearsonian goodness-of-fit statistic X_P^2 based on observations in a multinomial sampling. The following section addresses the general theory of multinomial estimation and testing and considers all the four goodness-of-fit statistics, X_P^2, Wilk's likelihood ratio statistic G^2, Freeman-Tukey statistic F^2, and Neyman's statistic X_N^2. This section also considers Birch's (1964) result on expansion of maximum likelihood estimators

P. Mukhopadhyay, *Complex Surveys*, DOI 10.1007/978-981-10-0871-9

(MLE's) of the parameters around their true values. Asymptotic distribution of all these four statistics is considered in the next section. The subsequent sections consider nested models and procedures for testing general hypotheses.

A.2 Multinomial Distribution

Let $\mathbf{X}_n = (X_{n1}, \ldots, X_{nt})'$ have the multinomial distribution $\mathcal{M}(n, \pi)$ with parameter (n, π) where $\pi = (\pi_1, \ldots, \pi_t)'$, $\sum_{i=1}^t \pi_i = 1$, i.e.,

$$P[X_{n1} = x_{n1}, \ldots, X_{nt} = x_{nt}] = n!\Pi_{i=1}^t \frac{\pi_i^{x_{ni}}}{x_{ni}!}.$$

Then it is known,

$$\begin{aligned} E(\mathbf{X}_n) &= n\pi \\ \text{Cov}\ (\mathbf{X}_n) &= n(\mathbf{D}_\pi - \pi\pi') \end{aligned} \tag{A.2.1}$$

where $\mathbf{D}_\pi = \text{Diag.}\ (\pi_1, \ldots, \pi_t)$.

Let $\hat{\mathbf{p}} = n^{-1}\mathbf{X}_n$ be the vector of sample proportions and $\mathbf{U}_n = \sqrt{n}(\hat{\mathbf{p}} - \pi)$. Then

$$\begin{aligned} E(\mathbf{U}_n) &= \mathbf{0} \\ \text{Cov}\ (\mathbf{U}_n) &= \mathbf{D}_\pi - \pi\pi'. \end{aligned} \tag{A.2.2}$$

It is known that $\hat{\mathbf{p}}$ is the maximum likelihood estimator (MLE) of π and $\hat{\mathbf{p}}$ converges in probability to π.

A.2.1 *Asymptotic Normality*

We have the following theorem.

Theorem A.2.1 *For large n, $\mathbf{U}_n$ converges in law to $\mathbf{U}$(i.e. $\mathcal{L}(\mathbf{U}_n) \rightarrow \mathcal{L}(\mathbf{U})$) where $\mathbf{U}$ has the multivariate normal distribution with mean $\mathbf{0}$ and covariance matrix $\mathbf{D}_\pi - \pi\pi'$.*

Proof Proof follows by the moment generating function and the continuity theorem.

Note A.2.1 The covariance matrix in (A.2.1) (and in (A.2.2)) is singular, because $\hat{\mathbf{p}}$ satisfies the linear constraint $\sum_{i=1}^t \hat{p}_i = 1$.

A.2.2 Asymptotic Normality When $\pi = \pi^0 + \mu n^{-1/2}$

The model is useful in the testing of goodness-of-fit hypothesis $H_0 : \pi = \pi^0$, when the model being tested is wrong, but not very wrong. Here, π^0 and μ are $t \times 1$ vector of constants.

In this case,

$$\begin{aligned} E(\mathbf{X}_n) &= n\pi^0 + \sqrt{n}\mu \\ \text{Cov}\,(\mathbf{X}_n) &= n(\mathbf{D}_{\pi^0} - \pi^0\pi^{0'}) + \sqrt{n}(\mathbf{D}_\mu - 2\pi^0\mu') + \mu\mu'. \end{aligned} \tag{A.2.3}$$

Since $\sum_{i=1}^t \pi_i^0 = \sum_{i=1}^t \pi_i = 1$, we must have $\sum_{i=1}^t \mu_i = 0$. Setting $\mathbf{U}_n = \sqrt{n}(\hat{\mathbf{p}} - \pi^0)$ we have

$$\begin{aligned} E(\mathbf{U}_n) &= \mu \\ \text{Cov}\,(\mathbf{U}_n) &= n^{-1}\,\text{Cov}\,(\mathbf{X}_n) \\ &= \mathbf{D}_{\pi^0} - \pi^0\pi^{0'} + n^{-1/2}(\mathbf{D}_\mu - 2\pi^0\mu') + n^{-1}\mu\mu' \end{aligned} \tag{A.2.4}$$

Theorem A.2.2 *The variable* $\mathbf{U}_n$ *converges in distribution to* $\mathbf{U}$ *where* $\mathbf{U}$ *has the multivariate normal* $(\mu, \mathbf{D}_{\pi^0} - \pi^0\pi^{0'})$ *distribution.*

Proof Proof follows by the moment generating function and the continuity theorem.

A.3 Distribution of Pearson Chi-Square Under a Simple Hypothesis

For testing the hypothesis of goodness-of-fit $H_0 : \pi = \pi^0$ (a known vector), the Pearson chi-square statistic is

$$X_P^2 = \sum_{i=1}^t \frac{(X_{ni} - n\pi_i^0)^2}{n\pi_i^0} \tag{A.3.1}$$

where $(X_{n1}, \ldots, X_{nt})'$ have a multinomial distribution (n, π^0) under H_0. Now X_P^2 can be written as

$$X_P^2 = \mathbf{U}_n'(\mathbf{D}_{\pi^0})^{-1}\mathbf{U}_n \tag{A.3.2}$$

where

$$\mathbf{U}_n = \sqrt{n}(\hat{\mathbf{p}} - \pi^0), \quad \hat{\mathbf{p}} = n^{-1}\mathbf{X}_n.$$

We now recall the following lemma.

Lemma A.3.1 *Let $\mathcal{L}(\mathbf{X}_n) \rightarrow \mathcal{L}(\mathbf{X})$ and let $g(.)$ be a continuous function. Then $\mathcal{L}\{g(\mathbf{X}_n)\} \rightarrow \mathcal{L}\{g(\mathbf{X})\}$.*

For a proof, see Rao (1965), p. 104.

Since $\mathcal{L}(\mathbf{U}_n) \rightarrow \mathcal{L}(\mathbf{U})$, where $\mathbf{U}$ has a multivariate normal distribution with parameters $(\mathbf{0}, \mathbf{D}_{\pi^0} - \pi^0\pi^{0'})$ and since $\mathbf{x}'(\mathbf{D}_{\pi^0})^{-1}\mathbf{x}$ is a continuous function of $\mathbf{x}$, by using Lemma A.3.1, $\mathcal{L}(X_P^2) = \mathcal{L}(\mathbf{U}_n'(\mathbf{D}_{\pi^0})^{-1}\mathbf{U}_n) \rightarrow \mathcal{L}(\mathbf{U}'(\mathbf{D}_{\pi^0})^{-1}\mathbf{U})$. The problem then reduces to the problem of finding the distribution of the quadratic form of a multivariate normal random vector. To this effect, we have the following result.

Lemma A.3.2 *If $\mathbf{X} = (X_1, \ldots, X_T)'$ has the multivariate normal distribution $N(\mathbf{0}, \mathbf{\Sigma})$ and $Y = \mathbf{X}'\mathbf{A}\mathbf{X}$, where $\mathbf{A}$ is a symmetric matrix, then Y is distributed in large sample as $\sum_{i=1}^{T} \lambda_i Z_i^2$ where $Z_1, \ldots, Z_T$ are independent $N(0, 1)$ variables and $\lambda_1, \ldots, \lambda_T$ are the eigenvalues of $(\mathbf{A}^{1/2})'\mathbf{\Sigma}(\mathbf{A})^{1/2}$.*

For a proof, see Rao (1965), p. 149.

By Lemma A.3.2 it follows that $\mathbf{U}'\mathbf{D}_{\pi^0}^{-1}\mathbf{U}$ is distributed as $\sum_{i=1}^{t} \lambda_i Z_i^2$ where the λ_i's are eigenvalues of

$$\mathbf{B} = \mathbf{D}_{\pi^0}^{-1/2}(\mathbf{D}_{\pi^0} - \pi^0\pi^{0'})\mathbf{D}_{\pi^0}^{-1/2} = \mathbf{I} - \sqrt{\pi^0}\sqrt{\pi^{0'}} \tag{A.3.3}$$

where $\sqrt{\pi^0} = (\sqrt{\pi_1^0}, \ldots, \sqrt{\pi_t^0})'$. It is easy to verify that $\mathbf{B}$ is a symmetric idempotent matrix. Hence, its eigenvalues are either 0 or 1. Also, the number of eigenvalues of $\mathbf{B}$ is equal to the trace of $\mathbf{B}$. Now

$$\begin{aligned} \text{tr. } (\mathbf{B}) &= \text{tr. } (\mathbf{I}) - \text{tr. } (\sqrt{\pi^0}\sqrt{\pi^{0'}}) \\ &= t - 1. \end{aligned}$$

Hence, $t-1$ of the eigenvalues each equal 1 and one eigenvalue is 0. Therefore, $\mathcal{L}[\mathbf{U}'\mathbf{D}_{\pi^0}^{-1}\mathbf{U}] = \mathcal{L}[\sum_{i=1}^{t-1} Z_i^2] = \chi^2_{(t-1)}$. Therefore, under the null hypothesis, asymptotic distribution of X_P^2 is $\chi^2_{(t-1)}$.

Distribution of X_P^2 under an alternative hypothesis

We want to find the distribution of Pearson chi-square X_P^2 in (A.3.1) when $\pi = \pi^0 + n^{-1/2}\mu$.

It can be shown that

$$\mathcal{L}(X_P^2) \rightarrow \mathcal{L}[\mathbf{U}'\mathbf{D}_{\pi^0}^{-1}\mathbf{U}] \tag{A.3.4}$$

where $\mathbf{U}$ has the multivariate normal distribution with parameters $(\mu, \mathbf{D}_{\pi^0} - \pi^0\pi^{0'})$. It can be shown that $Y = \mathbf{U}'\mathbf{D}_{\pi^0}^{-1}\mathbf{U}$ has a noncentral chi-square distribution with $t-1$ degrees of freedom (d.f.) and non-centrality parameter (*ncp*)

$$\psi^2 = \mu' \mathbf{D}_{\pi^0}^{-1} \mu.$$

If we write $\boldsymbol{\mu}$ as

$$\boldsymbol{\mu} = \sqrt{n}(\boldsymbol{\pi} - \boldsymbol{\pi}^0)$$

then the result is sometimes stated with the *ncp* given as

$$\psi^2 = n(\boldsymbol{\pi} - \boldsymbol{\pi}^0)' \mathbf{D}_{\boldsymbol{\pi}^0}^{-1} (\boldsymbol{\pi} - \boldsymbol{\pi}^0).$$

A.4 General Theory for Multinomial Estimation and Testing

In general, the cell probabilities $\pi_1(\theta), \ldots, \pi_t(\theta)$ will involve unknown parameters $\theta = (\theta_1, \ldots, \theta_s)'$, $s < t - 1$. We shall consider the problem of estimation of π_i's in this situation. (In Sect. A.2 we assumed that π_i's are known quantities, namely π_i^0's.)

Let $\mathcal{S}_t$ be the set of all t-dimensional probability vectors

$$\mathcal{S}_t = \left\{ \mathbf{p} : p_i \geq 0 \text{ and } \sum_{i=1}^{t} p_i = 1 \right\}. \tag{A.4.1}$$

We shall denote by $\mathbf{p}$ a generic point in $\mathcal{S}_t$ and by π the special point in $\mathcal{S}_t$ denoting the true cell probabilities.

In $\mathcal{S}_t$, the multinomial random vector is the vector of sample cell proportions $\hat{\mathbf{p}}$, rather than the vector of cell-counts $\mathbf{X}_n = (X_{n1}, \ldots, X_{nt})'$. Clearly, the point $\hat{\mathbf{p}}$ also lies in $\mathcal{S}_t$.

Let $\pi = \mathbf{f}(\theta)$, that is, $\pi_1 = f_1(\theta), \ldots, \pi_t = f_t(\theta)$, where $\theta = (\theta_1, \ldots, \theta_s)'$, $s < t - 1$ is a set of unknown parameters. We are required to estimate θ's and hence π_i's. The relation $\pi = \mathbf{f}(\theta)$ is the assumed model under this multinomial sampling.

The vector θ is a vector of parameters and we assume that $\theta \in \Theta$, a subset of the s-dimensional Euclidean space $\mathcal{R}^s$. As θ ranges over Θ, $\mathbf{f}(\theta)$ ranges over a subset $\mathcal{M}$ of $\mathcal{S}_t$. Any model of categorical data structure can therefore be defined by the assumption $\pi \in \mathcal{M}$ or by the pair $(\mathbf{f}(\theta), \Theta)$. If the model is correct, there exists a parameter-value $\phi \in \Theta$ such that $\pi = \mathbf{f}(\phi)$ and $\pi \in \mathcal{M}$. If the model is not correct there does not exist any such ϕ and $\pi \notin \mathcal{M}$.

A.4.1 Estimation

We shall write $\hat{\pi}, \hat{\theta}$ as generic symbols for estimators of π, θ respectively. Usually, we require $\hat{\pi}$ to be close to $\tilde{\mathbf{p}}$, a design-based consistent estimator of π. Under multinomial sampling $\tilde{\mathbf{p}} = \hat{\mathbf{p}} = \mathbf{X}_n/n$. Alternatively, we choose a suitable estimator $\hat{\theta}(\in \Theta)$ of θ and find $\hat{\pi} = \mathbf{f}(\hat{\theta})$.

Minimum distance method of estimation
The observed point $\hat{\mathbf{p}}$ is a natural estimator of π, when we do not restrict that $\pi \in \mathcal{M}$. Usually, we shall require $\hat{\pi}$ to be close to $\hat{\mathbf{p}}$ and that $\hat{\pi} \in \mathcal{M}$, to reflect that the model is true.

We consider a suitable distance function $K(\mathbf{x}, \mathbf{y})$ where $\mathbf{x}$ and $\mathbf{y}$ are two points in $\mathcal{S}_t$. The function should have the following properties:

(i) $K(\mathbf{x}, \mathbf{y}) \geq 0$;
(ii) $K(\mathbf{x}, \mathbf{y}) = 0$ *iff* $\mathbf{x} = \mathbf{y}$;
(iii) If $||\mathbf{x} - \mathbf{y}||$ is increased sufficiently, then $K(\mathbf{x}, \mathbf{y})$ is also increased.

We find that value of θ, $\hat{\theta}$ (say) in Θ for which the distance $K(\hat{\mathbf{p}}, \mathbf{f}(\hat{\theta}))$ is minimum and take $\hat{\pi} = \pi(\hat{\theta})$ as the minimum K-distance estimate of π. Thus, the minimum K-distance estimate of ϕ is $\hat{\theta}$ where

$$K(\hat{\mathbf{p}}, \mathbf{f}(\hat{\theta})) = \min{}_{\theta \in \Theta} K(\hat{\mathbf{p}}, \mathbf{f}(\theta)), \tag{A.4.2}$$

$\hat{\pi}$ is obtained by putting $\theta = \hat{\theta}$ in the function $\pi(\theta)$. Some regularity conditions are required on $\mathbf{f}$ and Θ to ensure that such a vector of functions $\mathbf{f}$ exists (vide Sect. A.4.3).

Some distance measures between two points $\mathbf{x}, \mathbf{y}$ in $\mathcal{S}_t$ are

(i)

$$X_D^2(\mathbf{x}, \mathbf{y}) = \sum_{i=1}^{t} \frac{(x_i - y_i)^2}{y_i}; \tag{A.4.3a}$$

(ii)

$$G_D^2(\mathbf{x}, \mathbf{y}) = 2\sum_{i=1}^{t} x_i \log\left(\frac{x_i}{y_i}\right); \tag{A.4.3b}$$

(iii)

$$F_D^2 = 4\sum_{i=1}^{t} (\sqrt{x_i} - \sqrt{y_i})^2; \tag{A.4.3c}$$

(iv)

$$X_{ND}^2 = \sum_{i=1}^{t} \frac{(x_i - y_i)^2}{x_i}. \tag{A.4.3d}$$

These distance functions give, respectively, the following four test statistics, based on the distance between $\hat{\mathbf{p}}$, vector of sample proportions (or any design-based consistent estimator of π) and $\hat{\pi} = \pi(\hat{\theta})$:

(a)

$$\tilde{X}_P^2 = nX_D^2(\hat{\mathbf{p}}, \pi(\hat{\theta})) = n\sum_{i=1}^{t}\frac{(\hat{p}_i - \pi_i(\hat{\theta}))^2}{\pi_i(\hat{\theta})}; \tag{A.4.4}$$

(b)

$$G^2 = nG_D^2(\hat{\mathbf{p}}, \pi(\hat{\theta})) = 2n\sum_{i=1}^{t}\hat{p}_i\log\left(\frac{\hat{p}_i}{\pi_i(\hat{\theta})}\right); \tag{A.4.5}$$

(c)

$$(FT)^2 = nF_D^2(\hat{\mathbf{p}}, \pi(\hat{\theta})) = 4n\sum_{i=1}^{t}\left(\sqrt{p_i} - \sqrt{\pi_i(\hat{\theta})}\right)^2; \tag{A.4.6}$$

(d)

$$X_N^2 = nX_{ND}^2(\hat{\mathbf{p}}, \pi(\hat{\theta})) = n\sum_{i=1}^{t}\frac{(\hat{p}_i - \pi_i(\hat{\theta}))^2}{\hat{p}_i}. \tag{A.4.7}$$

In all these functions, $\hat{\theta}$ is that value of θ in Θ for which the corresponding distance function has the minimum value. However, other methods of estimation of θ, and hence of π, such as maximum likelihood method and method of moments may be used.

When $\hat{\pi} = \pi(\hat{\theta})$ is the fitted value found by maximum likelihood method or by minimum chi-square method, $\tilde{X}_P^2$ is the Pearson chi-square statistic X_P^2.

When $\hat{\pi}$ is the fitted value by maximum likelihood method, G^2 is the Wilk's likelihood ratio statistic. In this case, minimum G^2 statistic is the same as the MLE of π (proof given in Eq. (A.4.8)).

The statistic $(FT)^2$ is called the Freeman-Tukey (1950) goodness-of-fit statistic, when $\hat{\pi}$ is the fitted value found by the maximum likelihood method. Fienberg (1979) reviewed the literature and properties of these statistics.

The statistic X_N^2 is the modified chi-square statistic and was suggested by Neyman (1949). In this section, we shall not deal with X_N^2 any more.

A.4.2 Asymptotic Distribution of MLE of θ

If the model $\pi = \mathbf{f}(\theta)$ is correct, the likelihood function is

$$L(\theta|\mathbf{x}) = \binom{n}{x_1, \ldots, x_t} \Pi_{i=1}^{t}\{f_i(\theta)\}^{x_i}.$$

Therefore,

$$\begin{aligned} -2\log L(\theta|\mathbf{x}) &= -2\log \binom{n}{x_1,\ldots,x_t} - 2\sum_{i=1}^{t} x_i \log \pi_i \\ &= nG^2(\hat{\mathbf{p}}, \pi) - 2\log \binom{n}{x_1,\ldots,x_t} - 2n\sum_{i=1}^{t} \hat{p}_i \log(\hat{p}_i). \end{aligned} \tag{A.4.8}$$

Hence, maximizing the likelihood function $L(\theta|\mathbf{x})$ is equivalent to minimizing the distance function $G^2(\hat{\mathbf{p}}, \mathbf{f}(\theta))$. We shall denote by $\tilde{\theta}$ and $\tilde{\pi} = \pi(\tilde{\theta})$ the MLE of θ and π respectively.

Expansion of the MLE $\tilde{\theta}$ around the true value ϕ
Let ϕ be the true value of θ when the model is correct. We now consider the asymptotic expansion of the MLE, $\tilde{\theta}$, around the true value ϕ when the model is correct. This important result is due to Birth (1964). Assume that $s < t - 1$. All these results hold under some regularity conditions listed in Sect. A.4.3.

Theorem A.4.1 *Assume that $\pi = \mathbf{f}(\phi)$ and that $\mathbf{f}(\phi)$ lies in $\mathcal{M}$. Let $\mathbf{A}$ be the $t \times s$ matrix whose (i, j)th element is*

$$a_{ij} = \frac{1}{\sqrt{\pi_i}}\left(\frac{\partial f_i(\phi)}{\partial \theta_j}\right). \tag{A.4.9}$$

Then, under regularity conditions of Sect. A.4.3, as $\hat{\mathbf{p}} \to \pi$,

$$\tilde{\theta} = \phi + (\mathbf{A}'\mathbf{A})^{-1}\mathbf{A}'\mathbf{D}_{\pi}^{-1/2}(\hat{\mathbf{p}} - \pi) + 0_p(n^{-1/2}). \tag{A.4.10}$$

An important consequence of Theorem A.4.1 is the asymptotic distribution of $\tilde{\theta}$ under the hypothesis that the model is correct.

Theorem A.4.2 *Under the conditions of Theorem A.4.1, the asymptotic distribution of $\sqrt{n}(\tilde{\theta} - \phi)$ is*

$$N(\mathbf{0}, (\mathbf{A}'\mathbf{A})^{-1}). \tag{A.4.11}$$

An estimate of the covariance matrix of $\tilde{\theta}$ is

$$\hat{\mathbf{D}}(\tilde{\theta}) = n^{-1}(\mathbf{A}'(\tilde{\theta})\mathbf{A}(\tilde{\theta}))^{-1}. \tag{A.4.12}$$

Using Theorem A.4.2, we can obtain the asymptotic distribution of the fitted values $\mathbf{f}(\tilde{\theta}) = \tilde{\pi}$ *under the assumption that the model is correct. Now, by Taylor expansion up to the first order term,*

$$\mathbf{f}(\tilde{\theta}) = \mathbf{f}(\phi) + \left(\frac{\partial \mathbf{f}}{\partial \theta}\right)(\tilde{\theta} - \phi) + 0_p(n^{-1/2}) \tag{A.4.13}$$

where

$$\left(\frac{\partial \mathbf{f}}{\partial \theta}\right) = \left(\left(\frac{\partial f_i(\phi)}{\partial \theta_j}\right)\right).$$

Hence, we have

$$\sqrt{n}(\mathbf{f}(\tilde{\theta}) - \mathbf{f}(\phi)) = \sqrt{n}\left(\frac{\partial \mathbf{f}}{\partial \theta}\right)(\tilde{\theta} - \phi) + 0_p(1). \tag{A.4.14}$$

It therefore, follows by (A.4.10) and (A.4.11) that

$$\mathcal{L}[\sqrt{n}(\mathbf{f}(\tilde{\theta}) - \mathbf{f}(\phi))] = \mathcal{L}[\sqrt{n}(\tilde{\pi} - \pi)] \rightarrow N\left(\mathbf{0}, \left(\frac{\partial \mathbf{f}}{\partial \theta}\right)(\mathbf{A}'\mathbf{A})^{-1}\left(\frac{\partial \mathbf{f}}{\partial \theta}\right)'\right). \tag{A.4.15}$$

A.4.3 Regularity Conditions

Clearly, θ is a function of $\mathbf{p}$, $\theta = \theta(\mathbf{p}) \in \Theta$. The problem is that there may be more than one value of $\theta(\mathbf{p})$ for some value of $\mathbf{p}$. This ambiguity vanishes if $\mathbf{p}$ is sufficiently close to $\mathcal{M}$ and the following regularity conditions hold.

These conditions are due to Birch (1964) and are generally satisfied. Assume that the model is correct so that $\pi = \mathbf{f}(\phi)$. Also, assume $s < t - 1$.

(1) The point ϕ is an interior point of Θ so that ϕ is not on the boundary of Θ and there is a s-dimensional neighborhood of ϕ that is entirely contained in Θ.
(2) $\pi_i = f_i(\phi) > 0 \; \forall \, i = 1, \ldots, t$. Thus, π_i is an interior point of $\mathcal{S}_t$ and does not lie on the boundary of $\mathcal{S}_t$.
(3) The mapping $\mathbf{f} : \Theta \rightarrow \mathcal{S}_t$ is totally differentiable at ϕ, so that the partial derivative of f_i with respect to θ_j exists at ϕ and $\mathbf{f}(\theta)$ has a linear approximation at ϕ given by

$$f_i(\theta) = f_i(\phi) + \sum_{j=1}^{s}(\theta_j - \phi_j)\frac{\partial f_i(\phi)}{\partial \theta_j} + o(||\theta - \phi||)$$

as $\theta \rightarrow \phi$.

(4) The Jacobian matrix ($\frac{\partial \mathbf{f}}{\partial \theta}$) is of full rank, i.e., of rank s. Thus, $\mathbf{f}(\theta)$ maps a small neighborhood of ϕ (in Θ) into a small t-dimensional neighborhood of $\mathbf{f}(\phi)$ in $\mathcal{M}$.
(5) The inverse mapping $\mathbf{f}^{-1} : \mathcal{M} \to \Theta$ is continuous at $\mathbf{f}(\phi) = \pi$. In particular, for every $\epsilon > 0$, there exists a $\delta > 0$, such that if $||\theta - \phi|| \geq \epsilon$, then $||\mathbf{f}(\theta) - \mathbf{f}(\phi)|| \geq \delta$.
(6) The mapping $\mathbf{f} : \Theta \to \mathcal{S}_t$ is continuous at every point θ in Θ.

A.5 Asymptotic Distribution of the Goodness-of-Fit Statistics

When **x** and **y** are close together, the values of the three distance functions (A.4.3a)–(A.4.3c) are nearly identical. As a result, if the model is correct and π is estimated in a reasonable way (not necessarily by the maximum likelihood method), the three goodness-of-fit statistics (A.4.4)–(A.4.6) will have the same limiting distribution. When the model is not correct, the three goodness-of-fit statistics do not have the same limiting distribution and may yield very different results.

We state below two relevant theorems without proof, for which the interested readers may refer to Bishop et al. (1975).

Theorem A.5.1 *Let $\hat{\pi}$ be any estimate of π (not necessarily MLE), $\pi_i > 0 \, \forall \, i$ such that $\hat{\mathbf{p}}$ and $\hat{\pi}$ have asymptotically a joint normal distribution*

$$\mathcal{L}\left[\sqrt{n}\begin{bmatrix}\hat{\mathbf{p}} - \pi \\ \hat{\pi} - \pi\end{bmatrix}\right] \to N(\mathbf{0}, \mathbf{\Sigma})$$

for some dispersion matrix

$$\mathbf{\Sigma} = \begin{bmatrix}\mathbf{\Sigma}_{11} & \mathbf{\Sigma}_{12} \\ \Sigma_{21} & \mathbf{\Sigma}_{22}\end{bmatrix}.$$

Then $nX_D^2(\hat{\mathbf{p}}, \hat{\pi}), nG_D^2(\hat{\mathbf{p}}, \hat{\pi}), nF_D^2(\hat{\mathbf{p}}, \hat{\pi})$ all have the same limiting distribution.

Theorem A.5.2 *Under the assumption of Theorem A.5.1, $\mathcal{L}[nX_D^2(\hat{\mathbf{p}}, \hat{\pi})] \to \sum_{i=1}^{t-1} \lambda_i Z_i^2$ where the Z_i^2's are independent chi-square variables with one d.f. and the λ_i's are the eigenvalues of*

$$\mathbf{D}_\pi^{-1/2}[\mathbf{\Sigma}_{11} - \Sigma_{12} - \mathbf{\Sigma}_{21} + \mathbf{\Sigma}_{22}]\mathbf{D}_\pi^{-1/2}. \quad \text{(A.5.1)}$$

This result follows from Theorem A.5.1 and Lemma A.3.2.

A.5.1 Limiting Distribution of X_P^2 When π is Estimated by the Maximum Likelihood Method

We shall now find the asymptotic distribution of the Pearson chi-square statistic X_P^2 under the assumption that the model is correct, i.e., $\pi \in \mathcal{M}$ and π is estimated by the maximum likelihood method.

Theorem A.5.3 *Assume that the regularity conditions (1)–(6) of Sect. A.4.3 hold and that $\pi \in \mathcal{M}$. If $\tilde{\pi} = \mathbf{f}(\tilde{\theta})$ where $\tilde{\theta}$ is the MLE of θ and if $X_P^2 = nX_D^2(\hat{\mathbf{p}}, \tilde{\pi})$, then*

$$\mathcal{L}[X_P^2] \to \chi^2_{(t-s-1)}. \tag{A.5.2}$$

Proof We first find the joint distribution of $\tilde{\pi}$ and $\hat{\mathbf{p}}$. From Theorem A.4.1,

$$\tilde{\theta} - \phi = (\mathbf{A}'\mathbf{A})^{-1}\mathbf{A}'\mathbf{D}_\pi^{-1}(\hat{\mathbf{p}} - \pi) + 0_p(n^{-1/2}).$$

From regularity condition (3), $\mathbf{f}$ has the following expansion as $\theta \to \phi$,

$$\mathbf{f}(\theta) - \mathbf{f}(\phi) = \left(\frac{\partial \mathbf{f}}{\partial \theta}\right)(\theta - \phi) + 0(||\theta - \phi||).$$

It then follows that

$$\begin{aligned} \tilde{\pi} - \pi &= \left(\tfrac{\partial \mathbf{f}}{\partial \theta}\right)(\mathbf{A}'\mathbf{A})^{-1}\mathbf{A}'\mathbf{D}_\pi^{-1/2}(\hat{\mathbf{p}} - \pi) + 0_p(n^{-1/2}) \\ &= \mathbf{L}(\hat{\mathbf{p}} - \pi) + 0_p(n^{-1/2}) \end{aligned} \tag{A.5.3}$$

where

$$\mathbf{L} = \mathbf{D}_\pi^{-1/2}\mathbf{A}(\mathbf{A}'\mathbf{A})^{-1}\mathbf{A}'\mathbf{D}_\pi^{-1/2}$$

since

$$\left(\frac{\partial \mathbf{f}}{\partial \theta}\right) = \mathbf{D}_\pi^{1/2}\mathbf{A}.$$

Therefore,

$$\begin{bmatrix} \hat{\mathbf{p}} - \pi \\ \tilde{\pi} - \pi \end{bmatrix} = \begin{bmatrix} \mathbf{I} \\ \mathbf{L} \end{bmatrix}(\hat{\mathbf{p}} - \pi) + 0_p(n^{-1/2}).$$

Hence

$$\sqrt{n}\begin{bmatrix} \hat{\mathbf{p}} - \pi \\ \tilde{\pi} - \pi \end{bmatrix} \to N(\mathbf{0}, \mathbf{\Sigma}) \tag{A.5.4}$$

where

$$\mathbf{\Sigma} = \begin{bmatrix} \mathbf{D}_\pi - \pi\pi' & (\mathbf{D}_\pi - \pi\pi')\mathbf{L}' \\ \mathbf{L}(\mathbf{D}_\pi - \pi\pi') & \mathbf{L}(\mathbf{D}_\pi - \pi\pi')\mathbf{L}' \end{bmatrix}. \tag{A.5.5}$$

The relevant matrix whose eigenvalues we are required to find by Theorem A.5.2 is

$$\mathbf{B} = \mathbf{D}_\pi^{-1/2}[\mathbf{\Sigma}_{11} - \mathbf{\Sigma}_{12} - \mathbf{\Sigma}_{21} + \mathbf{\Sigma}_{22}]\mathbf{D}_\pi^{-1/2}$$

which simplifies to

$$\mathbf{B} = \mathbf{I} - \sqrt{\pi}\sqrt{\pi'} - \mathbf{A}(\mathbf{A}'\mathbf{A})^{-1}\mathbf{A}' \tag{A.5.6}$$

where $\sqrt{\pi} = (\sqrt{\pi_1}, \ldots, \sqrt{\pi_t})'$. It is easy to find that tr $\mathbf{B} = t - s - 1$, when the result follows.

Note A.5.1 The only property of $\tilde{\theta}$ used in the theorem is the expansion (A.4.10). This property is satisfied by a host of estimators including minimum chi-square estimator. Estimators satisfying this property are called best asymptotically normal (BAN) estimators.

A.6 Nested Models

When we consider several models, such as, log-linear model, logit model for π, we may need to consider G^2 or some other goodness-of-fit statistics for several subclasses of these models. We may enquire if any relationship exists among these statistics for several models. In general, there is no simple relationship between the values of G^2 for two different models in $\mathcal{S}_t$ except in the case of nested models.

Two models $\mathcal{M}_1, \mathcal{M}_2$ are said to be nested or more precisely, $\mathcal{M}_2$ is said to be nested within $\mathcal{M}_1$, if $\mathcal{M}_2$ is completely contained in $\mathcal{M}_1$, when they are viewed as subsets of $\mathcal{S}_t$. Usually, the situation arises when the parameter vector of $\mathcal{M}_1$ is partitioned into two sets of components, say, (θ, ψ) and $\mathcal{M}_2$ is obtained by putting the value of ψ equal to a fixed value, say, $\mathbf{0}$. Thus, the parameter vector of $\mathcal{M}_2$ is $(\theta, \mathbf{0})$. Let MLE of (θ, ψ) be denoted as $(\tilde{\theta}, \tilde{\psi})$ and the corresponding value $\tilde{\pi} = \pi(\tilde{\theta}, \tilde{\psi})$. For $\mathcal{M}_2$, we denote the MLE of θ as $\tilde{\tilde{\theta}}$ and the corresponding value of $\mathbf{f}(\tilde{\tilde{\theta}}, \mathbf{0})$ by $\tilde{\tilde{\pi}}$. Thus we get

$$G^2(\mathcal{M}_1) = nG_D^2(\hat{\mathbf{p}}, \tilde{\pi}),\ G^2(\mathcal{M}_2) = nG_D^2(\hat{\mathbf{p}}, \tilde{\tilde{\pi}}) \tag{A.6.1}$$

where $G^2(\mathcal{M}_i)$ is the G^2-goodness-of-fit statistics for the model $\mathcal{M}_i$, $i = 1, 2$. Since $\mathcal{M}_2$ is contained in $\mathcal{M}_1$,

$$G^2(\mathcal{M}_2) \geq G^2(\mathcal{M}_1),$$

because $G^2(\mathcal{M}_1)$ is minimized over a bigger parameter set. We have the following theorem.

Theorem A.6.1 *If the regularity conditions (1)–(6) hold and if the true value of* π *is* $\pi = \mathbf{f}(\phi, \mathbf{0})$, *then*

$$\mathcal{L}[G^2(\mathcal{M}_2) - G^2(\mathcal{M}_1)] \rightarrow \chi^2_{(s_1-s_2)},$$

where s_1 *is the dimension of* (θ, ψ) *and* s_2 *is the dimension of* (θ).

A.7 Testing General Hypotheses

In this section, we shall consider the problem of testing general hypotheses relating to the cell probabilities. There are two approaches to the problem, one given by Neyman (1949) and the other is due to Wald (1943).

(a) *Neyman's Approach*: Suppose that each of the S populations is divided into R categories. A simple random sample of size n_{0j} is drawn from the jth population, $j = 1, \ldots, S$. Let n_{ij} be the observed frequency in the ith category of the jth population with π_{ij} as the cell probability, $\sum_i \pi_{ij} = 1 \; \forall \; j$. Let

$$q_{ij} = \frac{n_{ij}}{n_{0j}}, \; r_j = \frac{n_{0j}}{N}, \; N = \sum_j n_{0j},$$

$$\pi = (\pi_{11}, \ldots, \pi_{(R-1),1}, \ldots, \pi_{1S}, \ldots, \pi_{(R-1),S)})'.$$

Suppose we want to test the hypotheses

$$H_0 : F_k(\pi) = 0, \; k = 1, \ldots, T(T \leq (R-1)S) \qquad \text{(A.7.1)}$$

where the F_k's are T independent functions of π_{ij}'s.

It is assumed that F_k's possess continuous partial derivatives up to the second order with respect to π's and that the $T \times \{(R-1)S\}$ matrix $((\frac{\partial F_k}{\partial \pi_{ij}}))$ is of full rank T. Let $\hat{\pi}_{ij}$ be any best asymptotically normal (BAN) estimator of π_{ij} satisfying the conditions (A.7.1). The minimum chi-square estimator, the modified minimum chi-square estimator (that is, the one minimizing X_N^2 given below) and the MLE's are all BAN estimators (Neyman 1949). It is then well-known that H_0 may be tested by using X_P^2, X_N^2 or the likelihood ratio statistic

$$X_P^2 = \sum_{j=1}^{S}\sum_{i=1}^{R} \frac{(n_{ij}-n_{0j}\hat{\pi}_{ij})^2}{n_{0j}\hat{\pi}_{ij}},$$

$$X_N^2 = \sum_{j=1}^{S}\sum_{i=1}^{R} \frac{(n_{ij}-n_{0j}\hat{\pi}_{ij})^2}{n_{ij}}, \quad \text{(A.7.2)}$$

$$G^2 = 2\sum_{j=1}^{S}\sum_{i=1}^{R} n_{ij}[\ln n_{ij} - \ln n_{0j}\hat{\pi}_{ij}].$$

Neyman has shown that if there is at least one solution of (A.7.1) such that $\pi_{ij} > 0 \; \forall \; i, j$, then each of the statistics (A.7.2), using any system of BAN estimators, has asymptotically a $\chi^2_{(T)}$ distribution under H_0 as $N \to \infty$ with r_j's fixed. Also, these tests are asymptotically equivalent.

In general, equations giving these estimates are difficult to solve and iterative methods have to be used. However, if the constraints (A.7.1) are linear in π's, the minimum X_N^2 estimate can be calculated fairly easily by solving only the linear equations. If the functions F_k's are not linear, the minimum X_N^2 can still be obtained by solving linear equations using linearized constraints

$$F_k^*(\pi) = F_k(\mathbf{q}) + \sum_{j=1}^{S}\sum_{i=1}^{R-1}\left(\frac{\partial F_k}{\partial \pi_{ij}}\right)(\pi_{ij} - q_{ij}) = 0, \; k = 1, \ldots, T. \quad \text{(A.7.3)}$$

These estimates are also the BAN estimates.

(b) *Wald's Approach*: Wald (1943) considered the following general problem. Let $\psi(\mathbf{x}_1, \ldots, \mathbf{x}_N; \theta_1, \ldots, \theta_u)$ be the joint probability distribution of N independently and identically distributed (*iid*) random variables $\mathbf{X}_m, m = 1, \ldots, N$, involving unknown parameters $\theta_1, \ldots, \theta_u$, where $\theta = (\theta_1, \ldots, \theta_u)' \in \Theta \subseteq \mathcal{R}^u$. It is assumed that ψ possesses continuous partial derivatives up to the second order with respect to θ's and the square matrix

$$\mathbf{B}(\theta) = \left(\left(-\frac{1}{N}E_\theta \frac{\partial^2 \log \psi}{\partial\theta_\alpha \partial\theta_\beta}\right)\right), \; \alpha, \beta = 1, \ldots, u \quad \text{(A.7.4)}$$

is positive definite $\forall \, \theta$ in Θ. The hypothesis to be tested is $H_\omega : \theta$ belongs to a subset ω of Θ where ω is defined by T independent constraints

$$F_k(\theta) = 0, k = 1, \ldots, T(\leq u). \quad \text{(A.7.5)}$$

It is assumed that F_k's possess continuous partial derivatives up to second order with respect to θ's. Let

$$\mathbf{h}(\theta) = (F_1(\theta), \ldots, F_T(\theta))'$$
$$\mathbf{H}(\theta) = \left(\left(\frac{\partial F_k(\theta)}{\partial \theta_\alpha}\right)\right) \quad (k = 1, \ldots, T; \alpha = 1, \ldots, u). \tag{A.7.6}$$

For testing H_ω, assuming some regularity conditions, Wald proposed the statistic

$$W = N\mathbf{h}(\hat{\theta})'[\mathbf{H}(\hat{\theta})\mathbf{B}^{-1}(\hat{\theta})\mathbf{H}'(\hat{\theta})]^{-1}\mathbf{h}(\hat{\theta}) \tag{A.7.7}$$

where $\hat{\theta}$ is the MLE of θ. The statistic W has a limiting $\chi^2_{(T)}$ distribution under H_ω. The test has been shown to be asymptotically power-equivalent to the likelihood ratio test in the sense that if W_N and L_N are the respective critical regions,

$$\lim_{N\to\infty} \{P(W_N|\theta) - P(L_N|\theta)\} = 0$$

uniformly in $\theta \in \Theta$.

Bhapkar (1966) has pointed out that X_N^2 statistic in the linear and nonlinear case (using linearized constraints) is equivalent to Wald's statistic W, as adopted to the categorical data arising from a single multinomial population, as well as for the general case of independent random samples from several populations.

We shall now apply Wald's statistic W to the categorical data problem stated at the beginning of this section. Consider independent samples drawn from S populations. Let

$$X_{mj}^{(i)} = \begin{cases} 1 & \text{if the } m\text{th observation in the } j\text{th sample belongs to category } i \\ 0 & \text{otherwise, } i = 1, \ldots, R;\ m = 1, \ldots, n_{0j};\ j = 1, \ldots, S. \end{cases}$$

Let

$$x_{mj} = \left(x_{mj}^{(1)}, \ldots, x_{mj}^{(R)}\right)'.$$

The probability distribution of X's is given by

$$\psi(\mathbf{x}_{11}, \ldots, \mathbf{x}_{n_{0S}}; \pi) = \Pi_{j=1}^{S} \Pi_{i=1}^{R} \pi_{ij}^{n_{ij}} \tag{A.7.8}$$

since $\sum_m x_{mj}^{(i)} = n_{ij}$. Taking $\theta = \pi$ with $u = (R-1)S$, it is easy to verify that

$$\mathbf{B}(\pi) = \frac{1}{N}\begin{bmatrix} n_{01}(\mathbf{D}_1^{-1} + \pi_{R1}^{-1}\mathbf{L}) & \mathbf{0} & \ldots & \mathbf{0} \\ \mathbf{0} & n_{02}(\mathbf{D}_2^{-1} + \pi_{R2}^{-1}\mathbf{L}) & \ldots & \mathbf{0} \\ . & . & \ldots & . \\ \mathbf{0} & \mathbf{0} & \ldots & n_{0S}(\mathbf{D}_S^{-1} + \pi_{RS}^{-1}\mathbf{L}) \end{bmatrix} \tag{A.7.9}$$

where $\mathbf{D}_j = \text{Diag}\,(\pi_{1j}, \ldots, \pi_{(R-1)j})$ and $\mathbf{L} = \mathbf{I}_{(R-1)\times(R-1)}$. Then $\mathbf{B}^{-1}(\pi) = N\mathbf{G}(\pi)$ where

$$\mathbf{G}(\pi) = \begin{bmatrix} n_{01}^{-1}(\mathbf{D}_1 - \pi_1\pi_1') & \mathbf{0} & \dots & \mathbf{0} \\ \mathbf{0} & n_{02}^{-1}(\mathbf{D}_2 - \pi_2\pi_2') & \dots & \mathbf{0} \\ . & . & \dots & . \\ \mathbf{0} & \mathbf{0} & \dots & n_{0S}^{-1}(\mathbf{D}_S - \pi_S\pi_S') \end{bmatrix} \quad \text{(A.7.10)}$$

where $\pi_j = (\pi_{1j}, \dots, \pi_{(R-1)j})'$.

Wald's statistic (A.7.7) for testing the hypotheses (A.7.5) takes the form

$$[\mathbf{h}(\mathbf{q})]'[\mathbf{H}(\mathbf{q})\mathbf{G}(\mathbf{q})\mathbf{H}'(\mathbf{q})]^{-1}\mathbf{h}(\mathbf{q}) \quad \text{(A.7.11)}$$

since the MLE of π is $\mathbf{q} = (q_{11}, \dots, q_{(R-1)1}, \dots, q_{1S}, \dots, q_{(R-1)S})'$. Let

$$\begin{aligned} \left(\frac{\partial F_k(\pi)}{\partial \pi_{ij}}\right)_{\pi=\mathbf{q}} &= a_{kij}, i = 1, \dots, R-1; j = 1, \dots, S \\ \mathbf{a}_{kj} &= (a_{k1j}, a_{k2j}, \dots, a_{k(R-1)j})' \\ \mathbf{a}_k &= (\mathbf{a}_{k1}', \mathbf{a}_{k2}', \dots, \mathbf{a}_{kS}')'. \end{aligned} \quad \text{(A.7.12)}$$

Then

$$\mathbf{H}'(\mathbf{q}) = (\mathbf{a}_1, \mathbf{a}_2, \dots, \mathbf{a}_T). \quad \text{(A.7.13)}$$

By (A.7.10), the (k, k')th term of $\mathbf{H}(\mathbf{q})\mathbf{G}(\mathbf{q})\mathbf{H}'(\mathbf{q})$ is

$$\mathbf{a}_k'\mathbf{G}(\mathbf{q})\mathbf{a}_{k'} = \left[\sum_j n_{0j}^{-1}\mathbf{a}_{kj}'(\mathbf{Q}_j - \mathbf{q}_j\mathbf{q}_j')\mathbf{a}_{k'j}\right]$$

where

$$\mathbf{Q}_j = \text{Diag. } (q_{1j}, \dots, q_{(R-1)j}), \mathbf{q}_j = (q_{1j}, \dots, q_{(R-1)j})'.$$

Example A.7.1 A population is divided into L strata and within each stratum is classified by double classification rule into $R \times C$ cells. Let $\pi_{ijk}(i = 1, \dots, R; j = 1, \dots, C; k = 1, \dots, L)$ be the cell-probabilities with marginal probabilities as $\pi_{i00} = \sum_{j,k} \pi_{ijk}, \pi_{ij0} = \sum_k \pi_{ijk}$, etc. Assume that $\pi_k = \pi_{00k} = \sum_{i,j} \pi_{ijk}$ are known.

The hypothesis of general independence is

$$H_0 : \pi_{ij0} = \pi_{i00}\pi_{0j0} \; (i = 1, \dots R; j = 1, \dots, C). \quad \text{(A.7.14)}$$

The hypothesis of independence within strata is

$$H_0' : \pi_{ijk} = \pi_{i0k}\pi_{0jk} \ (i = 1, \ldots, R; \ j = 1, \ldots, C; \ k = 1, \ldots, L). \quad \text{(A.7.15)}$$

The two hypotheses are not generally equivalent. In many cases, the stratification variable is not of primary interest, but only a technical device used in the designing of the survey. The hypothesis of interest is then overall independence of the two characteristics without regard to stratification. It can be shown that the necessary and sufficient condition for the equivalence of these two hypotheses is

$$\pi_{ijk} = \frac{\pi_{ij0}\pi_{0jk}\pi_{i0k}}{\pi_{i00}\pi_{0j0}\pi_{00k}} \ \forall \, i, \, j, \, k. \quad \text{(A.7.16)}$$

Let a sample of size n_k be selected from the kth population by *srswor*, $\sum_k n_k = n$. Let n_{ijk} be the cell frequencies. Assuming that the sampling fraction in each stratum is small, the distribution of $\{n_{ijk}\}$ will be the product of L multinomial distributions

$$f[\{n_{ijk}\}|\{\pi_{ijk}\}] = \Pi_k \left[\frac{n_k!}{\Pi_{i,j} n_{ijk}!} \Pi_{i,j} (\frac{\pi_{ijk}}{\pi_k})^{n_{ijk}} \right]. \quad \text{(A.7.17)}$$

For testing H_0, Bhapkar (1961, 1966) has proposed minimization of Neyman statistic

$$X_N^2 = \sum_k \frac{n_k}{\pi_k} \sum_j \sum_k \frac{(\hat{\pi}_{ijk} - \pi_{ijk})^2}{\hat{\pi}_{ijk}} \quad \text{(A.7.18)}$$

with respect to π_{ijk}, where $\hat{\pi}_{ijk} = (\frac{n_{ijk}}{n_k})\pi_k$, subject to linearization of (A.7.14). Let

$$\begin{aligned}
\pi' &= (\pi_{111}, \ldots, \pi_{RCL})' \\
\hat{\pi}' &= (\hat{\pi}_{111}, \ldots, \hat{\pi}_{RCL})' \\
\mathbf{h}(\pi) &= (h_{11}(\pi), \ldots, h_{(R-1)(C-1)}(\pi))' \\
h_{ij}(\pi) &= \pi_{ij0} - \pi_{i00}\pi_{0j0} \qquad \text{(A.7.19)} \\
\mathbf{B}(\pi) &= \tfrac{1}{n} \left(\left(E_\pi \left(\frac{\partial^2 f}{\partial \pi_{ijk} \partial \pi_{i'j'k'}} \right) \right) \right)_{\{(RC-1)L\} \times \{(RC-1)L\}} \\
\mathbf{H}(\pi) &= \left(\left(\frac{\partial h_{ij}(\pi)}{\partial \pi_{xyz}} \right) \right)_{\{(R-1)(C-1)\} \times \{(RC-1)L\}}.
\end{aligned}$$

Bhapkar has shown that the minimization of (A.7.18) subject to a linearization of (A.7.15) is given by

$$W = n[\mathbf{h}(\hat{\pi})]'[\mathbf{H}(\hat{\pi})\mathbf{B}^{-1}(\hat{\pi})\mathbf{H}'(\hat{\pi})]^{-1}[\mathbf{h}(\hat{\pi})]. \quad \text{(A.7.20)}$$

Nathan (1969) obtained approximate MLE $\tilde{\pi}_{ijk}$ by iteration procedure under H_0 and suggested the log-likelihood ratio statistic

$$G = \sum_{i,j,k} \ln \left[\frac{\hat{\pi}_{ijk}}{\tilde{\pi}_{ijk}} \right] \qquad \text{(A.7.21)}$$

and similar Pearson chi-square and Neyman chi-square statistic. All these statistics and (A.7.20) are asymptotically distributed as a $\chi^2_{(R-1)(C-1)}$ random variable under H_0. Nathan (1972) has shown that under alternative hypothesis they have noncentral distribution with $(R-1)(C-1)$ d.f. and non-centrality parameter

$$\lambda = n[\mathbf{h}(\pi)]'\mathbf{H}(\pi)[\mathbf{B}^{-1}(\pi)\mathbf{H}'(\pi)]^{-1}[\mathbf{h}(\pi)]. \qquad \text{(A.7.22)}$$

If proportional allocation is used so that $n_k = n\pi_k$, $\hat{\pi}_{ij0} = \frac{n_{ij0}}{n}$ is an unbiased estimator of π_{ij0}. In this case, Wilk's likelihood ratio test, based on overall frequency reduces to

$$G^2 = 2\left[\sum_{i,j} n_{ij0} \ln(n_{ij0}) - \sum_{i} n_{i00} \ln(n_{i00}) - \sum_{j} n_{0j0} \ln(n_{0j0}) + n\ln(n)\right]. \qquad \text{(A.7.23)}$$

Pearson's chi-square test reduces to

$$X_P^2 = \sum_{i,j} \frac{(n_{ij0} - n_{i00}n_{0j0}/n)^2}{(n_{i00}n_{0j0}/n)}. \qquad \text{(A.7.24)}$$

In the case of proportional allocation, Nathan (1975) has shown that the asymptotic power of the overall tests defined in (A.7.23) and (A.7.24) is never greater than that of the detailed tests based on all the frequencies defined in (A.7.18) and (A.7.21).

References

Ahmad T (1997) A resampling technique for complex survey data. J Indian Soc Agric Stat L3:364–379

Altham PAE (1976) Discrete variable analysis for individuals grouped into families. Biometrika 63:263–269

Anderson TW (1957) Maximum likelihood estimates for a multinomial normal distribution when some observations are missing. J Am Stat Assoc 52:200–203

Anderson TW (1973) Asymptotically efficient estimates of covariance matrices with linear structure. Ann Math Stat 1:135–141

Ballin M, Scantt M, Vicard P (2010) Estimation of contingency tables in complex survey sampling using probabilistic expert systems. J Stat Plan Inference 149(6):1501–1512

Basu DK (1958) On sampling with and without replacement. Sankhya 20:287–294

Basu DK (1971) An essay on logical foundation of survey sampling, part I. In: Godambe VP, Sprott DR (eds) Foundations of statistical inferences. Holt, Rinehart and Winston, Toronto, pp 203–242

Basu DK, Ghosh JK (1967) Sufficient statistics in sampling from a finite population. Bull Int Stat Inst 42(2):85–89

Bedrick EJ (1983) Chi-squared tests for cross-classified tables. Biometrika 70:591–595

Benhim E, Rao JNK (2004) Analysis of categorical data for complex sample surveys using inverse sampling. In: Proceeding of survey methodology section. American Statistical Association

Bhapkar VP (1961) Some tests for categorical data. Ann Math Stat 32:72–83

Bhapkar VP (1966) A note on the equivalence of two test criteria for hypotheses in categorical data. J Am Stat Assoc 61:228–235

Bickel PJ, Freedman DA (1984) Asymptotic normality and the bootstrap in stratified sampling. Ann Stat 12:470–482

Binder DA (1983) On the variance of asymptotically normal estimators from complex surveys. Int Stat Rev 51:279–292

Birch MW (1964) A new proof of the Pearson-Fisher theorem. Ann Math Stat 35:817–824

Bishop YMM, Fienberg SG, Holland PW (1975) Discrete multivariate analysis: theory and practice. The MIT Press, Massachusetts

Booth JG, Butler RW, Hall PG (1991) Bootstrap methods for finite population. Australian national university technical report SMS-96-91, Canberra

Brewer KRW, Hanif M (1983) Sampling with unequal probabilities. Lecture notes in statistic. Springer, New York

Brier SS (1980) Analysis of contingency tables under cluster sampling. Biometrika 67:591–586

P. Mukhopadhyay, *Complex Surveys*, DOI 10.1007/978-981-10-0871-9

Brillinger DR (1966) The application of the jackknife to the analysis of sample surveys. Commentary 8:74–80
Canty AJ, Davison AC (1999) Resampling-based variance estimation for labor force surveys. Stat 48:379–391
Cassel CM, Sarndal CE, Wretman JH (1976) Some results on generalized difference estimator and generalized regression estimator for finite populations. Biometrics 63:614–620
Chakraborty MC (1963) On the use of incidence matrix for designs in sampling for finite universe. J Ind Stat Assoc 1:78–85
Chakraborty RP, Rao JNK (1968) The bias and stability of the jackknife variance estimators in ratio estimation. J Am Stat Assoc 63:748
Chambers RL (1986) Design-adjusted parameter estimation. J R Stat Soc A 149:161–173
Chambers RL (1988) Design-adjusted regression estimation with selectivity bias. Appl Stat 37:323–334
Chambers RL, Steel DG, Wang S, Welsh A (2012) Maximum likelihood estimation for sample surveys. Chapman and Hall
Chambless EL, Boyle KE (1985) Maximum likelihood methods for complex survey data: logistic regression and discrete proportional hazard models. Commun Stat Theory Methods 14:1377–1392
Chao MT (1982) A general purpose unequal probability sampling plan. Biometrika 69:653–656
Chao MT, Lo SH (1985) A bootstrap method for finite population. Sankhya A 47:399–405
Chaudhuri A, Vos JWE (1988) Unified theory and strategies of survey sampling. North-Holland, Amsterdam
Choi JW (1980) Analysis of categorical data in weighted cluster sample survey. Proc Am Stat Assoc 573–578
Cochran WG (1977) Sampling techniques, 3rd edn. Wiley, New York
Cohen J (1976) The distribution of chi-square statistic under cluster samping from contingency tables. J Am Stat Assoc 71:665–670
Cox DR (1969) Some sampling problems in technology. In: Johnson NL, Smith H Jr (eds) New developments in survey sampling. Wiley, New York, pp 506–527
Cressie N, Read TRC (1984) Multinomial goodness of fit tests. J R Stat Soc B 46:440–464
Davison AC, Hinkley DV, Schechtman E (1986) Efficient bootstrap simulation. Biometrika 73:555–566
Deming WE (1956) On simplification of sampling designs through replications with equal probabilities and without replacement and without stages. J Am Stat Assoc 51:24–53
Dempster AP, Laird NM, Rubin DB (1977) Maximum likelihood from incomplete data via the EM algorithm (with discussion). J R Stat Soc B 39:1–38
Donner A (1989) Statistical methods in opthalamology: an adjusted chi-square approach. Biometrics 45:605–611
Durbin J (1953) Some results in sampling theory when the units are selected with unequal probabilities. J R Stat Soc B 15:262–269
Durbin J (1967) Estimation of sampling error in multistage surveys. Appl Stat 16:152–164
Efron B (1979) Bootstrap methods: another look at Jackknife. Ann Stat 7:1–26
Efron B (1982) The jackknife, the bootstrap and other nonsampling plans. SIAM, monograph no. 38, Philadelphia
Fattorini L (2006) Applying the Horvitz-Thompson criterion in complex designs: a computer-intensive perspective for estimating inclusion-probabilities. Biometrika 93(2):269–278
Fay RE (1979) On adjusting the Pearson chi-square statistic for cluster sampling. In Proceedings of the social statistics section, American statistical association, pp 665–670
Fay RE (1984) Replication approaches to the log-linear analysis of data from complex surveys. In Recent developments in the analysis of large scale data sets. Eurostat News, Luxemberg, pp 95–118
Fay RE (1985) A jack-knifed chi-squared test for complex samples. J Am Stat Assoc 80:148–157

Fellegi IP (1980) Approximate tests of independence and goodness of fit based on stratified multi-stage samples. J Am Stat Assoc 75:261–268
Fienberg SE (1979) The use of chi-squared statistics for categorical data problems. J R Stat Soc B B41:51–64
Fienberg SE (1980) The analysis of cross-classified data. MA Press, Cambridge
Forster JJ, Smith PWF (1998) Model-based inference for categorical survey data subject to non-ignorable non-response. J R Stat Soc B 60(1):57–70
Francisco CA, Fuller WA (1991) Quantile estimation with a complex survey design. Ann Stat 19(1):454–469
Frankel MR (1971) Inference from survey samples. Institute for social research. University of Michigan, Ann Arbour
Freeman MF, Tukay JW (1950) Transformations related to the angular and the square root. Ann Math Stat 21:607–611
Fuller WA (1975) Regression analysis for sample survey. Sankhya C 37:117–132
Goodman LA (1978) Analyzing qualitative categorical data. Cambridge University Press, Cambridge
Graubard BI, Korn EL (1993) Hypothesis testing with complex survey data: the use of classical quadratic test statistics with particular reference to regression problems. J Am Stat Assoc 88:629–641
Gray HL, Schucany WR (1972) The generalized jackknife statistics. Marcell Drekker, New York
Gross S (1980) Median estimation in sample surveys. Proceedings of the section on survey research, methods, American statistical association
Gross WF (1984) χ^2-tests with survey data. J R Stat Soc B 46:270–272
Ha'jek J (1959) Optimum strategies and other problems in probability sampling. Casopis Pest. Mat. 84:387–423
Ha'jek J (1960) Limiting distribution in simple random sampling from a finite population. Publ Math Inst Hung Acad Sci 5:361–374
Hall P (1989) On efficient bootstrap simulation. Biometrika 76(3):613–617
Hansen MH, Madow WG (1953) Sample survey methods and theory, vols I and II. Wiley, New York
Hanurav TV (1962) An existence theorem in sample surveys. Sankhya A 24:327–330
Heo S (2002) Chi-squared tests for homogeneity based on complex survey data subject to misclassification errors. Korean Commun Stat 9(3):853–864
Heo S, Eltinge JL (2003) The analysis of categorical data from a complex sample survey: chi-squared tests for homogeneity subject to misclassification error. Manuscript seen by curtsey of the authors
Herzel A (1986) Sampling without replacement with unequal probabilities sample designs with pre-assigned joint inclusion-probabilities of any order. Metron XLIV 1:49–68
Hidiroglou M, Rao JNK (1987) χ^2-tests with categorical data from complex surveys, I. II. J Off Stat 3(117–131):133–140
Holt D, Ewings PD (1989) Logistic models for contingency tables. In: Skinner CJ, Holt D, Smith TMF (eds) Analysis of complex surveys. Wiley, New York
Holt D, Scott AJ, Ewings PG (1980) Chi-squared tests with survey data. J R Stat Soc A 143:302–320
Horvitz DG, Thompson DJ (1952) A generalization of sampling without replacement from a finite universe. J Am Stat Assoc 47:175–195
Hosmer DW, Lemeshow S (2000) Applied logistic regression, 2nd edn. Wiley, New York
Isaki CT, Fuller WA (1982) Survey designs under the regression superpopulation model. J Am Stat Assoc 77:89–96
Johnson NL, Kotz S (1970) Continuous univariate distributions. Houghton Miffin, Boston
Judkin DR (1990) Fay's method of variance estimation. J Off Stat 6(3):223–239
Keyfitz N (1957) Estimates of sampling variances where two units are selected from each stratum. J Am Stat Assoc 52:503–510

Kim JK, Brick MJ, Fuller WA, Kalton G (2006) On the bias of the multiple-imputation variance estimator in survey sampling. J R Stat Soc B 68(3):509–521

Kish L (1965) Survey sampling. Wiley, New York

Kish L, Frankel MR (1970) Balanced repeated replication for standard errors. J Am Stat Assoc 65:1071–1094

Koch GG, Freeman DH, Freeman JL (1975) Strategies in multivariate analysis of data from complex surveys. Int Stat Rev 43:59–78

Korn EL, Graubard BI (1990) Simultaneous testing of regression coefficients with complex survey data: use of Bonferroni t statistics. J Am Stat Assoc 85:270–276

Koop JC (1972) On the deviation of expected values and variance of ratios without the use of infinite series expansions. Metrika 19:156–170

Krewski D, Rao JNK (1981) Inference from stratified samples: properties of the linearization, jackknife and balanced repeated replication methods. Ann Stat 9:1010–1019

Krieger AM, Pfeffermann D (1991) Maximum likelihood estimation from complex sample surveys. In Proceedings of the symposium in honor of Prof. V.P.Godambe, University of Waterloo, Canada

Lahiri DB (1954) Technical paper on some aspects of the development of the sampling design. Sankhya 14:332–362

Layard MWJ (1972) Large scale tests for the equality of two covariance matrices. Ann Math Stat 43:123–141

Le T Chap (1998) Applied categorical data analysis. Wiley, New York

Lee KH (1972) Partially balanced designs for the half-sample replication method of variance estimation. J Am Stat Assoc 67:324–334

Lee KH (1973) Using partially balanced designs for the half-sample replication method of variance estimation technique in the linear case. J Am Stat Assoc 68:612–614

Lehtonen R, Pahkinen EJ (1995) Practical methods for designs and analysis of complex surveys, 1st and 2nd edn. Wiley, New York (2004)

Lemeshow S, Epp R (1977) Properties of the balanced half-samples and the jack-knife variance estimation techniques in the linear case. Commun Stat Theory Methods 6(13):1259–1274

Levy PS (1971) A comparison of balanced half-sample replication and jack-knife estimators of the variance of the ratio estimates in complex sampling with two primary units per stratum (Mss): National Center for Health Statistics, Rockvliie

Little RJA (1982) Models for nonresponse in sample surveys. J Am Stat Assoc 77:237–250

Lloyd CJ (1999) Statistical analysis of categorical data. Wiley, New York

Lumley T (2010) Complex surveys: a guide to analysis using R. Wiley, New York

Madow WG (1948) On the limiting distributions of estimates based on samples from finite universe. Ann Math Stat 19:535–545

Mahalanobis PC (1946) Recent developments in statistical sampling in the Indian Statistical Institute. J R Stat Soc A 109:325–378

McCarthy PJ (1966) Replication. An analysis to the approach of data from complex surveys in Vital and Health Statistics, Sr. No. 2, No. 14, US Department of Health, Education and Welfare, Washington, US Government Printing Press

McCarthy PJ (1969a) Pseudo-replication: half samples. Int Stat Rev 37:239–264

McCarthy PJ (1969b) Pseudo-replication: further evaluation and application of balanced half-sample techniques. Vital and health statistics, Sr. 2, No. 31, national center for health statistics, public health service, Washington

McCarthy PJ, Snowdon CB (1985) The bootstrap and finite population sampling in vital and health statistics, Sr. 2, No. 95, public health service publication. US Government Printing Press, Washington, pp 85–139

McCullagh P, Nelder JA (1989) Generalized linear models, 2nd edn. Chapman and Hall, London

Miller RG Jr (1964) A trustworthy jackknife. Ann Math Stat 35:1584–1605

Miller RG Jr (1974) The jackknife—a review. Biometrika 61:1–15

Morel JG (1989) Linear regression under complex survey designs. Surv Methodol 15:203–223

Mosimann JE (1962) On the compound multinomial distribution, the multinomial β-distribution, and correlation among proportions. Biometrika 49:65–82

Mote VL, Anderson RL (1965) An investigation of the effect of misclassification on the properties of chi-square tests in the analysis of categorical data. Biometrika 52:95–109

Mukhopadhyay P (1972) A sampling scheme to realize a preassigned set of inclusion-probabilities of first two orders. Calcutta Stat Assoc Bull 21:87–122

Mukhopadhyay P (1975) An optimum sampling design to base HT-method of estimating a finite population total. Metrika 22:119–127

Mukhopadhyay P (1996) Inferential problems in survey sampling. New Age International Pvt. Ltd., New Delhi

Mukhopadhyay P (1998a) Small area estimation in survey sampling. Narosa Publishing House, New Delhi

Mukhopadhyay P (1998b) Theory and methods of survey sampling, 1st edn. Prentice Hall of India, New Delhi

Mukhopadhyay P (2000) Topics in survey sampling. Lecture notes in statistics, vol 153. Springer, New York

Mukhopadhyay P (2007) Survey sampling. Alpha Science International Ltd., Oxford

Mukhopadhyay P (2008) Multivariate statistical analysis. World Scientific, Singapore

Mukhopadhyay P (2009) Theory and methods of survey sampling, 2nd edn. PHI Learning, New Delhi

Mukhopadhyay P, Vijayan K (1996) On controlled sampling designs. J Stat Plan Inference 52:375–378

Murthy MN, Sethi VK (1959) Self-weighting design at tabulation stage. National sample survey working paper. No. 5 (also Sankhya B **27**, 201–210)

Murthy MN, Sethi VK (1961) Randomized rounded off multipliers. J Am Stat Assoc 56:328–334

Nandram B (2009) Bayesian tests of equality of stratified proportions for a multiple response categorical variable. Adv Multivar Stat Methods 4:337–355

Nandram B, Bhatta D, Sedransk J, Bhadra D (2013) A Bayesian test of independence in a two-way contingency table using surrogate sampling. J Stat Plan Inference 143:1392–1408

Nathan G (1969) Tests of independence in contingency tables from stratified samples. in Johnson NL, Smith H (eds) New developments in survey sampling. Wiley, New York, pp 578–600

Nathan G (1972) On the asymptotic power of tests for independence in contingency tables from stratified samples. J Am Stat Assoc 67:917–920

Nathan G (1975) Tests of independence in contingency tables from stratified proportional samples. Sankhya C 37:77–87

Nelder JA, Wedderburn RWM (1972) Generalized linear models. J R Stat Soc B 36:1–22

Neyman J (1949) Contribution to the theory of χ^2 tests, in Proceedings of the Berkeley symposium on mathematical statistics and probability. University of California Press, Berkeley, pp 239–273

Patil GP, Rao CR (1978) Weighted distributions and size-based sampling with application to wildlife populations and human families. Biometrics 34:179–189

Perviaz MK (1986) A comparison of tests of equality of covariance matrices with special reference to the case of cluster sampling. Unpublished Ph.D. thesis, University of Southampton

Plackett RG, Burman PJ (1946) The design of optimum factorial experiments. Biometrika 33:305–325

Plackett RL, Paul SR (1978) Dirichlet model for square contingency tables. Commun Stat Theory Methods A 7(10):939–952

Pfeffermann D (1988) The effect of sampling design and response mechanisms on multivariate regression-based predictors. J Am Stat Assoc 83:824–833

Pfeffermann D (1992) The role of sampling weights when modeling survey data, technical report, department of statistics, Hebrew University, Jerusalem

Pfeffermann D, LaVange L (1989) Regression models for stratified multistage cluster samples. In: Skinner CJ, Holt D, Smith TMF (eds) Complex survey analysis. Wiley, New York, pp 237–260

Quenouille MH (1949) Approximate tests of correlation in time series. J R Stat Soc B 11:68–84

Quenouille MH (1956) Note on bias in estimation. Biometrika 43:353–360

Rao CR (1947) Large sample tests of statistical hypotheses concerning several parameters with applications to problem of estimation. Proc Camb Philos Soc 44:50–57

Rao CR (1965a) Linear statistical inference and its applications. Wiley, New York

Rao CR (1965b) On discrete distributions arising out of methods of attainment. In: Patil GP (ed) Classical and contagious discrete distributions. Statistical Publishing Society, Calcutta, pp 320–332

Rao JNK (1965) On two simple properties of unequal probability sampling without replacement. J Indian Soc Agric Stat 3:173–180

Rao JNK (2006) Bootstrap methods for analyzing complex survey data. In: Proceedings of the statistics Canada symposium 2006: methodological issues in measuring population health

Rao JNK, Hartley HO, Cochran WG (1962) On a simple procedure of unequal probability sampling without replacement. J R Stat Soc B 24:482–491

Rao JNK, Lanke J (1984) Simplified unbiased variance estimators for multistage designs. Biometrika 77:387–395

Rao JNK, Nigam AK (1990) Optimum controlled sampling designs. Biometrika 77:807–814

Rao JNK, Scott AJ (1979) Chi-squared tests for analysis of categorical data from complex surveys. In: Proceedings of the American statistical association, section on survey research methods, pp 58–66

Rao JNK, Scott AJ (1981) The analysis of Categorical data from complex sample surveys: chi-squared tests for goodness of fit and independence in two-way tables. J Am Stat Assoc 76:221–230

Rao JNK, Scott AJ (1982) On chi-squared tests for multi-way tables with cell-proportions estimated from survey data. Paper presented at the Israeli association international meetings on analysis of sample survey data and on sequential analysis, Jerusalem, 14–18 June 1982

Rao JNK, Scott AJ (1984) On chi-squared tests for multi-way tables with cell proportions estimated from survey data. Ann Stat 12:46–60

Rao JNK, Scott AJ (1987) On simple adjustments to chi-squared tests with sample survey data. Ann Stat 15:385–397

Rao JNK, Scott AJ (1988) Analysis of cross-classified data from complex sample surveys. Sociol Methodol 18:213–269

Rao JNK, Scott AJ (1992) A simple method for the analysis of clustered binary data. Biometrics 48(2):577–585

Rao JNK, Shao J (1999) Jackknife variance estimation with survey data under hot deck imputation. Biometrika 79:811–822

Rao JNK, Tausi M (2004) Estimating the jackknife variance estimators under stratified multistage sampling. Commmun Stat Theory Methods 33:2087–2095

Rao JNK, Thomas DR (1988) The analysis of cross-classified data from complex sample surveys. Sociol Methodol 18:213–269

Rao JNK, Thomas DR (1989) Chi-squared tests for contingency tables. In: Skinner CJ, Holt D, Smith TMF (eds) Analysis of complex surveys. Wiley, Chichester, pp 89–114

Rao JNK, Thomas DR (1991) Chi-squared tests with complex survey data subject to misclassification error. In: Biemer PP, Grover RM, Lyberg LE, Mathiowetz NA, Sudman S (eds) Measurement errors in surveys. Wiley, New York

Rao JNK, Thomas DR (2003) Analysis of categorical response data from complex surveys: an appraisal and update. In: Chambers RL, Skinner CJ (eds) Analysis of survey data. Wiley, New York

Rao JNK, Webster JT (1966) On two methods of bias reduction in the estimation of ratios. Biometrika 53:571–577

Rao JNK, Wu CFJ (1985) Inference from stratified samples: second-order analysis of three methods for nonlinear statistic. J Am Stat Assoc 80:620–630

Rao JNK, Wu CFJ (1988) Resampling inference with complex survey data. J Am Stat Assoc 83:231–241

Rao PSRS, Rao JNK (1971) Small sample results for ratio estimation. Biometrika 58:625–630
Read TRC, Cressie N (1988) Goodness-of-fit statistics for discrete multivariate data. Springer, New York
Rizzo l (1992) Conditionally consistent estimators using only probabilities of selection in complex sample surveys. J Am Stat Assoc 87:1166–1173
Roberts G, Rao JNK, Kumar S (1987) Logistic regression analysis of sample survey data. Biometrika 74:1–12
Robert G, Ren Q, Rao JNK (2004) Using marginal mean model with data from a longitudinal survey having a complex design: some advances in methods (Chapter 20). Methodology of longitudinal surveys, Wiley, Chichester
Royall RM (1970) On finite population sampling theory under certain linear regression models. Biometrika 57:377–387
Rubin DB (1976) Inference and missing data. Biometrika 53:581–592
Rust K (1984) Techniques for estimating variances for sample surveys. Ph.D. thesis, University of Michgan
Rust K (1986) Efficient replicated variance estimation. In: Proceedings of the section on survey research method, American statistical association, Washington
Sarndal CE, Swensson B, Wretman J (1992) Model assisted survey sampling. Springer, New York
Satterthwaite FE (1946) An approximate distribution of estimates of variance components. Biometrics 2:110–114
Scott AJ (1989) Logistic regression with survey data. In: Proceedings of the section on survey research method, American statistical association, pp 25–30
Scott AJ, Rao JNK (1981) Chi-squared tests for contingency tables with proportions estimated from survey data. In: Krewski D, Platek R, Rao JNK (eds) Current topics in survey sampling. Academic Press, New York, pp 247–266
Scott AJ, Rao JNK, Thomas DR (1990) Weighted least squares and quasi-likelihood estimation for categorical data under singular models. Linear Algebr Appl 127:427–447
Scott A, Smith TMF (1969) Estimation in multistage surveys. J Am Stat Assoc 64:830–840
Scott AJ, Wu CFJ (1981) On the asymptotic distribution of ratio and regression estimators. J Am Stat Assoc 76:98–102
Shao J (1989) The efficiency and consistency of approximations to the jackknife variance estimators. J Am Stat Assoc 84:114–119
Shao J (1996) Resampling methods in sample survey (with discussion). Statistics 27:203–254
Shao J, Wu CFJ (1989) A general theory of jackknife variance estimation. Ann Stat 3:1176–1197
Shuster JJ, Downing DJ (1976) Two-way contingency tables for complex sampling schemes. Biometrika 63:271–276
Sinha BK (1973) Un sampling schemes to realize preassigned sets of inclusion-probabilities of first two orders. Calcutta Stat Assoc Bull 22:69–110
Sitter RR (1992) A resampling procedure from complex survey data. J Am Stat Assoc 87:755–765
Skinner CJ (1989) Introduction to Part A. In Skinner CJ, Holt D, Smith TMF (eds) Analysis of complex surveys. Wiley, New York
Skinner CJ, Holmes DJ, Smith TMF (1986) The effect of sample design on principal component analysis. J Am Stat Assoc 81:789–798
Skinner CJ, Holt D, Smith TMF (eds) (1989) Analysis of complex surveys. Wiley, New York
Solomon H, Stephens MA (1977) Distribution of a sum of weighted chi-square variables. J Am Stat Assoc 72:881–885
Srivastava JN, Saleh F (1985) Need of t design in sampling theory. Util Math 25:5–7
Stata Corporation (1999) Stata statistical software. Stata corporation, College Station
Sunter AB (1977) List sequential sampling with equal or unequal probabilities without replacement. Appl Stat 26:261–268
Tenenbein A (1972) A double-sampling scheme for estimating from misclassified multinomial data with applications to sampling inspection. Technometrics 14(1):187–202

Thomas DR, Rao JNK (1984) A monte carlo study of exact levels of goodness-of-fit statistics under cluster sampling. In: Proceedings of American statistical association, Section on survey research methods, pp 207–211

Thomas DR, Rao JNK (1987) Small sample comparison of level and power for simple goodness of fit statistics under cluster sampling. J Am Stat Assoc 82:630–638

Thomas DR, Singh AC, Roberts GR (1996) Tests of independence on two-way tables under cluster sampling: an evaluation. Int Stat Rev 64(3):295–311

Thompson SK (2012) Sampling, 3rd edn. Wiley, New York

Tukey JW (1958) Bias and confidence in not-quite large samples. Abstract Ann Math Stat 29:614

Valliant R (1987) Some prediction properties of balanced half samples variance estimators in single stage samples. J R Stat Soc B 49:68–81

Von Bahr B (1972) On sampling from a finite set of independent random variables. Zeitdschrift fur Wahrscheinlichkeitstheorie und Verwandte Gebiete 24:279–286

von Eye A, Mun E-Y (2013) Log-linear modeling. Wiley, New York

Wald A (1941) Asymptotically most powerful tests of statistical hypotheses. Ann Math Stat 12:1–9

Wald A (1943) Tests of stratified hypotheses concerning several parameters when the number of observations is large. Trans Am Math Soc 54:429–482

Wolter KM (1985) Introduction to variance estimation. Springer, New York

Woodruff RS, Causey BD (1976) Computerized method approximating the variance of a complicated estimate. J Am Stat Assoc 71:315–321

Wu Changbao, Rao JNK (2010) Bootstrap procedures for the pseudo empirical likelihood methods in sample surveys. Stat Probab Lett 80(19–20):1472–1478

Yates F, Grundy PM (1953) Selection without replacement within strata with probability proportional to sizes. J R Stat Soc B 15:253–261

Printed in the United States
By Bookmasters